T0296552

AUTOTROPHIC
MICRO-ORGANISMS

AUTOTROPHIC
MICRO-ORGANISMS

FOURTH SYMPOSIUM OF THE
SOCIETY FOR GENERAL MICROBIOLOGY
HELD AT THE INSTITUTION OF
ELECTRICAL ENGINEERS, LONDON
APRIL 1954

CAMBRIDGE

Published for the Society for General Microbiology

AT THE UNIVERSITY PRESS

1954

CAMBRIDGE
UNIVERSITY PRESS

University Printing House, Cambridge CB2 8BS, United Kingdom

Cambridge University Press is part of the University of Cambridge.

It furthers the University's mission by disseminating knowledge in the pursuit of education, learning and research at the highest international levels of excellence.

www.cambridge.org
Information on this title: www.cambridge.org/9781316509852

First published 1954
First paperback edition 2015

A catalogue record for this publication is available from the British Library

ISBN 978-1-316-50985-2 Paperback

CONTRIBUTORS

BAALSRUD, K., Biological Laboratory, University of Oslo.

BISSET, K. A., Department of Bacteriology, University of Birmingham.

BLINKS, L. R., Hopkins Marine Station of Stanford University, Pacific Grove, California.

BUTLIN, K. R., Microbiology Section, Chemical Research Laboratory, Teddington, Middlesex.

ELSDEN, S. R., Department of Microbiology, University of Sheffield.

FOGG, G. E., Department of Botany, University College, London.

GAFFRON, H., Institute of Radiobiology and Biophysics, University of Chicago, Chicago, Illinois.

GRACE, J. B., Department of Bacteriology, University of Birmingham.

LARSEN, H., Department of Chemistry, Norges Tekniske Högskole, Trondheim, Norway.

LASCELLES, J., Microbiology Unit, Department of Biochemistry, University of Oxford.

LEES, H., Department of Biological Chemistry, University of Aberdeen.

MEIKLEJOHN, J., Rothamsted Experimental Station, Harpenden, Hertfordshire.

POSTGATE, J. R., Microbiology Section, Chemical Research Laboratory, Teddington, Middlesex.

SYRETT, P. J., Department of Botany, University College, London.

WASSINK, E. C., Laboratory of Plant Physiological Research, Agricultural University, Wageningen, Holland.

WOLFE, M., Department of Botany, University College, London.

WOODS, D. D., Microbiology Unit, Department of Biochemistry, University of Oxford.

CONTENTS

EDITORS' PREFACE

This Symposium on 'Autotrophic Micro-organisms' has been organized along lines somewhat similar to the three previous symposia promoted and published by the Society for General Microbiology. Contributors were invited to write authoritative articles on specific subjects within the field of study of autotrophic micro-organisms. Their articles, which are collected together in this book, form the basis of short lectures by the contributors at the London meeting (April 1954). This volume is thus primarily intended to provide the background for the verbal communications and subsequent discussions, but the Editors believe that it will also prove useful as an account of the present position and trends in this expanding field of knowledge.

The period available for the preparation and editing of the individual contributions has been limited, and it has not been possible to allow authors to make major alterations or additions at the proof stage. The Editors apologize for any shortcomings or inaccuracies which may have arisen as a result.

Some contributors to the Symposium describe the nutritional type of a micro-organism in terms of a nomenclature, based on energy sources, which was put forward at a symposium held at Cold Spring Harbor in 1946. Since this terminology may be unfamiliar to many readers, it is reproduced on p. xi.

B. A. FRY
J. L. PEEL

Department of Microbiology and Agricultural Research
Council Unit for Microbiology
University of Sheffield
9 *November* 1953

NOMENCLATURE OF NUTRITIONAL TYPES OF MICRO-ORGANISMS BASED ON ENERGY SOURCES*

A. PHOTOTROPHY—energy provided chiefly by photochemical reaction.

1. *Photolithotrophy*—growth dependent on exogenous inorganic hydrogen-donors.

2. *Photo-organotrophy*—growth dependent on exogenous organic hydrogen-donors.

B. CHEMOTROPHY—energy provided entirely by dark chemical reaction.

1. *Chemolithotrophy*—growth dependent on oxidation of exogenous inorganic substances.

2. *Chemo-organotrophy*—growth dependent on oxidation or fermentation of exogenous organic substances.

C. PARATROPHY—energy provided by host cell.

* (From *Heredity and Variation in Micro-organisms, Cold Spr. Harb. Symp. quant. Biol.* **11**, 302 (1946).)

THE NO MAN'S LAND BETWEEN THE AUTOTROPHIC AND HETEROTROPHIC WAYS OF LIFE

D. D. WOODS AND JUNE LASCELLES

Microbiology Unit, Department of Biochemistry,
University of Oxford

It is our task to introduce the subject of this Symposium simply because those with expert knowledge (and who are able to be with us) are naturally making their own contributions in the more specialized papers which follow. It was felt, however, that the Symposium should begin with some more general statement of the particular problems of the autotrophic way of life, and of its relationship to the way of life of the heterotrophic organisms with which the majority of microbiologists are more familiar. For it is certainly true, purely from the more statistical point of view, that autotrophic life among micro-organisms has received far less attention than its intrinsic interest merits; to balance this it has attracted some of the pioneers—Beijerinck, Winogradsky, Kluyver, van Niel amongst others—in the development of general microbiology. The comparative lack of intensive research on this aspect of microbiology is not easy to understand. It was clear almost from the time of the discovery of the microbes concerned that their economic importance, especially in soil fertility and the cycle of nitrogen, sulphur and carbon in nature, was probably great. They were obviously organisms of the widest biochemical versatility likely to provide novel and interesting information. Their possible significance from the evolutionary point of view was often debated, yet such debate was handicapped by the rather scanty knowledge available concerning the basic life processes of many of the organisms. Happily, as the remainder of this Symposium will show, there is now developing an intense interest; the relatively few pioneers of the past may hope to find many more of their brilliant researches, and the fascinating organisms they discovered, subjected to detailed study at all the levels of investigation now available to the microbiologist.

It is presumably incumbent upon the first contributors to this discussion to attempt to explain or define at least the title of the Symposium. This is no easy matter; workers in this field have long been agreed that there is no sharp line of demarcation between, for example, autotrophic bacteria and the rest of the bacterial kingdom. For reasons that will become clear later, we would ourselves prefer to speak of the auto-trophic mode of life among micro-organisms rather than of autotrophic micro-organisms. The *ability* to live autotrophically is, as far as is known at present, more common than the *necessity* to live autotrophically. The word autotroph taken literally would seem to mean 'self-nourishing' or 'self-sustaining'; this is not very helpful without an implied qualification that the independence is in respect of more elaborate forms of carbon compounds. Only in recent years, with the discovery that certain photosynthetic bacteria are able to live with carbon dioxide as the only source of carbon and gaseous nitrogen as the only source of nitrogen, do we seem to have organisms with at least no visible means of support!

The autotrophic way of life could be defined as the ability to live and multiply on indefinite subculture in an environment containing carbon dioxide as the sole source of carbon; it necessarily follows that inorganic sources of nitrogen, sulphur and other elements are also sufficient. Such a definition implies a very high synthetic ability on the part of the organism since all new organic cell material must be fabricated ultimately from carbon dioxide and inorganic material. The definition may indeed be considered too exclusive. Is an organism to be said not to be living autotrophically simply because it cannot synthesize for itself from carbon dioxide minute traces of one or more substances of the vitamin type, which must therefore be supplied preformed? Such organisms do exist (e.g. certain strains of *Hydrogenomonas* and of Athiorhodaceae when living autotrophically); by all other criteria they are living a typically autotrophic existence, and it would seem reasonable to amend the phrase 'sole source of carbon' in the above definition to 'main or bulk source of carbon'. This is only one example of the sort of trouble that is met if any rigid definition is attempted; others will become obvious as the discussion proceeds. Werkman (1951) has indeed suggested that it is better to sense the differentiation between autotrophy and heterotrophy than to define it. Our task unfortunately is to convey this sense to others who, like ourselves, are not active workers in the field; in the absence of facilities for telepathic communication the only method would seem to be to put up a temporary definition of convenience as a general basis for discussion.

One matter should perhaps be made clear at this stage. Green plants, which photosynthesize higher carbon compounds from carbon dioxide, normally live a typically autotrophic life (at any rate as concerns bulk carbon nutrition). Photosynthesis is not, however, in any way tied to the autotrophic life; there are many bacteria which live heterotrophically with the aid of photosynthetic processes and which certainly do not use carbon dioxide as the bulk ultimate source of carbon.

There are a number of excellent and brief accounts of the history, general physiology and biochemistry of autotrophic bacteria which it would be foolish to try to duplicate here (van Niel, 1943, 1944, 1949; Umbreit, 1947, 1951; Foster, 1951; Stephenson, 1949; Koffler & Wilson, 1951; Gest, 1951). The more general implications of these organisms in the biochemical approach to the problem of evolution amongst micro-organisms has been fascinatingly discussed by Lwoff (1944), Knight (1945) and van Niel (1949) among others. Lwoff (1944) has also at-tempted to classify and subdivide further the types of life process found in various micro-organisms, though he strongly emphasizes the absence of clear-cut boundaries. It seemed, therefore, that it would be best for us to point out, as simply as possible, what appear to be the main pro-blems of the autotrophic life and how these may differ (if at all) from those of the heterotrophic life which most of us know much better. For a more detailed characterization of these problems we shall concern ourselves mainly with the properties of bacteria which have either the ability to live both heterotrophically and autotrophically, or which, in one way or another, seem to occupy the no man's land between the two ways of life. Some of these properties, now existing perhaps singly in isolated species, may, when combined in a single organism at some earlier phase of evolution, have dictated the way of life of the organism in a particular environment at that time.

The main problems to be solved seem to us to be biochemical in nature. For reasons of space we must use bacteria as main examples of the points we wish to make. In the course of a wide survey it is inevitable that we shall have to make general statements which may not be in accord with every fact gleaned from every bacterium. We can only hope to provide a very general picture which later contributors will expand and modify from their own special knowledge. In other words, we shall attempt mainly to provide a row of pegs on which later contributors may hang their hats or umbrellas. One difficulty we have is that pegs should be a rather permanent installation, whereas people may frequently purchase new hats and umbrellas, or even accidentally acquire those of other people! In order to keep the list of references within reasonable bounds we shall, whenever it is appropriate, cite one or other of the several excellent reviews as a source also of the original literature.

SOME PROBLEMS OF AUTOTROPHIC LIFE

The ability to use carbon dioxide as the main (if not the only) source of carbon for cell multiplication requires that the organism must have certain special biochemical abilities. This does not mean that all these abilities are peculiar to autotrophic organisms; some are certainly possessed by typical heterotrophs. Others whose nature is not yet known may well be possessed by other organisms. It is presumably only the combination of all such abilities in a single species which permits the autotrophic life, and it is for this reason that each such biochemical activity should be examined in detail wherever it may be found.

Firstly, the organism must have the enzyme systems necessary for bringing about an initial reaction resulting in the formation from carbon dioxide of a compound with two or more carbon atoms. This may need the actual enzymic activation of carbon dioxide itself, or only the formation or activation of some organic compound which then reacts spontaneously with carbon dioxide. At first sight it might seem that the bulk utilization of carbon dioxide requires the power to create carbon to carbon bonds between two C_1 residues. This may indeed be the case, but, on the other hand, it is necessary to remember that any culture must start from at least one organism which in turn contains many preformed higher carbon compounds. A cyclical process could therefore be envisaged (Lipmann, 1946; Calvin, 1949; Gaffron, Fager & Rosenberg, 1951; Buchanan *et al.* 1952) in which, for example, C_1 units are successively added to a pre-existing carbon chain and the original compound regenerated after splitting off the new carbon chains so built up. Similarly, even if there is *de novo* formation of C_2 units from C_1 units, such a cyclic mechanism might then proceed with the primary C_2 unit as acceptor. There is therefore no theoretical necessity for the bulk formation of carbon to carbon bonds between single carbon atoms, though it must be emphasized that this may happen.

Secondly, carbon dioxide presents the carbon atom to the cell in a highly oxidized form. On the average the carbon atoms of cell material are much less oxidized than in carbon dioxide. Their state in this respect is often taken to be represented by the empirical formula $[CH_2O]$ (cf. $[CHO_3]$ for the bicarbonate ion); van Niel (1936), however, has found evidence that in at least some organisms in the general group under discussion the average state of the carbon atom is even more reduced, and is more accurately represented by $[C_2H_3O]$. It is clear, therefore, that the overall utilization of carbon dioxide for growth requires linkage with active reducing systems, i.e. some oxidation process must occur which is coupled to the overall reduction of carbon dioxide to new cell constituents. This does not mean of course that the primary reaction undergone by carbon dioxide is necessarily a reductive one, although this may seem probable. Detailed examples of such coupled oxidative reactions will be considered later in connexion with energy requirements.

The third special problem of autotrophic life concerns the overall energy requirement for growth. Presumably more energy will be required for each new cell produced than with an organism living heterotrophically. The required reduction of the carbon atoms of carbon dioxide is an endergonic process (except when molecular hydrogen is the reducing agent); furthermore, a far greater number of new carbon to carbon bonds must be created than in heterotrophic life where more complex carbon compounds must be provided preformed. Both heterotrophs and autotrophs will of course require energy-yielding reactions for coupling to the systems which synthesize other cell constituents from the C_2, C_3 and C_4 compounds which are usually considered to be the key 'building stones'.

In biological systems the energy-yielding reactions are in the main oxidations. Oxidations of appropriate substrates could theoretically also serve as hydrogen donor systems for the reduction of carbon dioxide. Conversely, any substrate that the cell can use for the latter purpose must itself undergo oxidation and is therefore a potential source of energy to the cell. Whether or not it is an effective source of energy for autotrophic life presumably depends on the cell being able to bring about energy transfer from one system to the other, a process which, by analogy with similar events in yeast and animal tissues, would require a definite chemical link and possibly the intermediate formation of high-energy phosphate esters.

There are two main types of energy source for autotrophic life. In chemosynthetic autotrophy the oxidation of an inorganic substrate (usually rather specific for a given organism) provides both the necessary reducing system and the energy for growth on a medium containing carbon dioxide or bicarbonate as the sole source of carbon. Typical examples are given in Table 1; these organisms are usually known as the chemosynthetic- or chemo-autotrophs, but not all of them (see later) are confined to this way of life. The amount of substrate oxidized is much greater than would be required for the reduction of the amount of carbon dioxide actually incorporated into cell material. Some other bulk hydrogen acceptor is therefore required; in most cases this is molecular oxygen (that is, the organisms are strict aerobes when growing in this way), but there are anaerobic organisms known in which nitrate (*Thiobacillus denitrificans*) and sulphate (*Desulphovibrio desulphuricans*) serve this purpose. Inorganic substances oxidized by various organisms include ammonia, nitrite, sulphur compounds (e.g. hydrogen sulphide, sulphur itself, sodium thiosulphate), hydrogen and ferrous iron. There has always been a little doubt about the autotrophic status of some of the iron-oxidizing microbes (Pringsheim, 1949), partly due to the difficulty of discriminating between biological oxidation of ferrous hydroxide and its spontaneous oxidation at normal pH values. A clear-cut case is, however, provided by the recent isolation (Temple & Colmer, 1951) of *Thiobacillus ferro-oxidans*, which is able to grow in acid media where spontaneous oxidation is very slow. There appears to be no case

Table 1. *Examples of bacteria able to grow chemo-autotrophically*

Organism	Substrates oxidized	Oxidant	End-product of oxidation	Reference
Thiobacillus thio-oxidans	Sulphur, thiosulphate and other inorganic S compounds	Oxygen	Sulphate	(1), (2)
Thiobacillus denitrificans	As above	Nitrate*	Sulphate	(1), (2)
Nitrosomonas spp.	Ammonia	Oxygen	Nitrite	(1)
Nitrobacter spp.	Nitrite	Oxygen	Nitrate	(1)
Thiobacillus ferro-oxidans	Ferrous salts or thiosulphate	Oxygen	Ferric salts or sulphate	(3)
Hydrogenomonas spp.	Hydrogen	Oxygen	Water	(1)
Desulphovibrio desulphuricans	Hydrogen	Sulphate and other inorganic S compounds	Sulphide	(4)

* Nitrate is reduced to molecular nitrogen during oxidation of sulphur compounds.

References: (1) Reviews of the literature in van Niel (1943), Stephenson (1949) and Foster (1951); (2) Bunker (1936); (3) Temple & Colmer (1951); (4) Butlin, Adams & Thomas (1949).

so far discovered of an organism using the oxidation of a higher carbon compound solely as reducing system and energy source for chemo-autotrophic growth, that is, with no liberation of carbon dioxide from substrate oxidation and no incorporation of carbon other than exogenous carbon dioxide.

A second group of micro-organisms is able to live autotrophically when supplied with radiant energy; all such organisms contain a chlorophyll pigment which is essential for the photosynthetic life. From the point of view of the present discussion we shall concern ourselves mainly with bacteria which can live photo-autotrophically. Such bacteria, although deriving their energy from light, still require the presence of an inorganic substrate (sulphur compounds or hydrogen) whose oxidation provides the necessary reducing system for the overall carbon dioxide-fixation process. The organisms can live in this way anaerobically, and the amount of inorganic substrate oxidized is relatively small and in accord with the amount of carbon dioxide reduced and fixed by the cell. It might be thought that energy derived from this anaerobic oxidation might meet at least part of the total energy requirements, providing of course that the organism has the appropriate energy-transfer mechanisms. van Niel (1941) has pointed out, however, that such a contribution of energy, even if utilizable, is likely to be relatively small. The amount of substrate (e.g. hydrogen sulphide) oxidized could only account for fixation of one-eightieth of the total carbon dioxide actually fixed, assuming that the overall efficiency of the chemosynthesis were similar to that found with chemosynthetic organisms oxidizing the same substrate. With the photosynthetic autotroph *Chlorobium thiosulphatophilum* (which can utilize several substrates as reducing agent), Larsen, Yocum & van Niel (1953) have found that the amount of carbon dioxide fixed is largely independent of the nature of the oxidizable substrate, though the energy potentially available differs with the substrate. It is clear, therefore, that in photosynthetic autotrophy the bulk of the energy requirement for growth comes from radiant energy.

In photosynthesis by the higher plants, algae and other micro-organisms containing chlorophyll (e.g. diatoms) no extraneous oxidizable substrate is required to form the reducing system. The necessary overall reduction is made possible by the fact that the organisms are able to dispose of oxygen atoms through the evolution of molecular oxygen. The process is analogous in a way to the ability of a number of heterotrophic bacteria to split off molecular hydrogen and thus to carry out certain oxidations in the absence of any final hydrogen-acceptor. The evolution of oxygen is characteristic of plant photosynthesis, but does not occur with bacteria; hence the need for an oxidizable substance. Possible explanations for the differences between the photosynthetic process in bacteria and plants could not be made clear without detailed discussion at a biochemical level of current knowledge and thought concerning the mechanism of the primary photo-induced reaction; these matters will no doubt be dealt with by other contributors to the Symposium. There is, however, general agreement with van Niel's concept (1941) that the processes are fundamentally similar, and that photolysis of water provides the ultimate reducing system in plants and possibly the immediate reducing system in bacteria, though here an ultimate reducing system is required. Bacterial chlorophyll, except possibly that of green sulphur bacteria, differs slightly from plant chlorophyll in chemical structure and in its absorption spectrum (see van Niel, 1944; Rabinowitch, 1951); in consequence photosynthesis is carried out by bacteria with light of a longer wavelength, and the absorption of each quantum would raise the energy level of the receptor molecule to a less extent than in plants. This may ultimately account for the observed differences in detail in the two processes.

In the succeeding sections we shall examine the ability of various bacteria which can only live heterotrophically to utilize metabolites (especially carbon dioxide) which are particularly implicated in autotrophic life. We shall then consider points which arise from the study of organisms which can live both autotrophically and heterotrophically, drawing our main examples from the photosynthetic bacteria and the so-called hydrogen bacteria.

HETEROTROPHIC METABOLISM OF SUBSTANCES USED IN AUTOTROPHIC LIFE

Utilization of carbon dioxide in heterotrophic life

It was mentioned earlier that the ability to utilize exogenous carbon dioxide, and even to incorporate its carbon atom into cell material, was not confined to organisms living autotrophically. Heterotrophic carbon dioxide fixation has been intensively studied and the results have been exhaustively and critically reviewed by Utter & Wood (1951) and by Ochoa (1951). It is proposed here to select points which seem germane to the present discussion.

Primary fixation reactions

There are now known a number of overall reactions by which carbon dioxide is thought to be incorporated into carbon compounds containing two or more carbon atoms by bacteria living heterotrophically; the principal types of reaction are shown in Table 2. The relative simplicity of the overall chemical changes and the fact that most of them have been demonstrated with cell-free enzyme preparations, if not purified enzymes, suggests that these may be primary carbon dioxide-fixation reactions, though there is certainly not complete agreement about this. Most of them have so far been shown to occur only by isotope exchange experiments; these demonstrate the potentiality of the cell to incorporate carbon dioxide by the reaction in question, but not of course that the intact cell does so to a significant extent.

Several points call for comment from the present standpoint. There is so far no definite enzymic reaction known in which two molecules of carbon dioxide (or indeed any two C_1 compounds) are condensed to give a C_2 compound; all the known reactions require a preformed carbon chain of at least two atoms as acceptor.

In each of the reactions of Table 2 the carbon atom of carbon dioxide appears in the carboxyl (COOH) group of the product. The carbon atom has therefore undergone one stage of reduction, but is still considerably more oxidized than in the cell as a whole. This does not of course militate in any way against a carboxylation reaction as a primary fixation reaction in autotrophic life.

Table 2. *Some proposed overall reactions for heterotrophic assimilation of carbon dioxide by bacteria*

Type of reaction	Overall reaction	Experimental system	Reference
(1) $C_2 + C_1$	(a) $H_2 + CO_2 \rightleftharpoons$ formate + acetate $\rightleftharpoons$ pyruvate	Suspensions of *Bact. coli*	(1)
	(b) $H_2 + CO_2 +$ acetate $\rightleftharpoons$ pyruvate	Extracts of *Cl. butylicum*	(1)
(2) $C_3 + C_1$	(a) Pyruvate $+ CO_2 \rightleftharpoons$ oxaloacetate	Extracts of *M. lysodeikticus*	(1)
		Extracts of *Bact. coli* and *P. morganii*	(2)
	(b) Pyruvate $+ CO_2 + DPN_{red.}$* $\rightleftharpoons$ malate $+ DPN_{ox.}$	Extracts of *Lb. arabinosus*	(1)
	(c) Propionate $+ CO_2 \rightleftharpoons$ succinate	Suspensions of *Pr. pentosaceum*	(3)
(3) $C_4 + C_1$	Succinate $+ CO_2 \rightleftharpoons \alpha$-ketoglutarate	Extracts of *Bact. coli*	(1)
(4) $C_5 + C_1$	α-Ketoglutarate $+ CO_2 + TPN_{red.}$* $\rightleftharpoons$ *iso*citrate $+ TPN_{ox.}$	Extracts of *Esch. freundii*	(4)

* DPN and TPN = di- and tri-phosphopyridine nucleotide coenzymes respectively; ox. = oxidized form, red. = reduced form.

References: (1) Utter & Wood (1951); full references and a critical evaluation of the evidence will be found in this review. (2) Kaltenbach & Kalnitsky (1951*a, b*); *de novo* formation of oxaloacetate demonstrated in these systems. (3) Barban & Ajl (1951); for critical discussion of CO_2 fixation by *Propionibacterium* see also (1) and Leaver & Wood (1953). (4) Barban & Ajl (1953).

None of these reactions, as far as is known, results in the overall utilization by the microbes concerned of significant quantities of exogenous carbon dioxide as a source of carbon for new cell material. There is, however, no doubt that such reactions are of critical importance to the growth of the culture. Carbon dioxide (in traces) has long been known as an essential factor for the initiation of growth of certain bacteria living heterotrophically (Gladstone, Fildes & Richardson, 1935; see also review by Knight, 1945). There is evidence from nutritional experiments that the function of carbon dioxide may be in the initial synthesis of dicarboxylic acids such as glutamic acid, aspartic acid and their probable precursors, α-ketoglutaric and oxaloacetic acid (Lwoff &

Monod, 1947; Ajl & Werkman, 1948; Lyman, Moseley, Wood, Butler & Hale, 1947); this may well result from carboxylation reactions such as (2*a*), (2*b*) and (3) of Table 2. Similar experiments implicate carbon dioxide, perhaps less directly, at some stage in the synthesis by certain lactic acid bacteria of other amino-acids such as arginine, tyrosine and serine (Lyman *et al.* 1947; Lascelles, Cross & Woods, 1954; Cross, 1953), and in the synthesis of purine derivatives by *Streptococcus haemolyticus* (Pappenheimer & Hottle, 1940).

What evidence is there that the primary reaction undergone by carbon dioxide in autotrophic metabolism may be a carboxylation of the type listed in Table 2? With autotrophic bacteria there is almost no evidence on this point, but with algae, workers certainly seem in recent years to lean towards this idea. Later contributors will no doubt analyse the matter in detail, but briefly there is strong evidence for condensation of carbon dioxide with a C_2 compound as the primary reaction (Calvin, 1949; Gaffron *et al.* 1951). The carbon atom of the carbon dioxide appears primarily in the carboxylic group of phosphoglyceric acid, which is the product of the condensation. The nature of the C_2 acceptor is unknown, though it is not acetate or a derivative as in reaction (1) of Table 2. Cell suspensions of a photosynthetic bacterium (*Rhodospirillum rubrum*) have been found to incorporate isotope-labelled carbon dioxide into phosphoglyceric acid when incubated in hydrogen (Glover, Kamen & van Genderen, 1952), but the significance of the reaction in growth is not clear. Calvin & Massini (1952) also have evidence for the condensation of carbon dioxide with a C_3 compound to yield malate (possibly as in reaction (2)) as an early, and perhaps primary, reaction in photosynthesis by algae; they propose that such a C_3–C_1 condensation follows the C_2–C_1 condensation in the cycle of reactions by which the original C_2 compound is thought to be regenerated. Vishniac & Ochoa (1952*a*, *b*) have shown that illuminated chloroplast fragments can bring about the reduction of the phosphopyridine nucleotide coenzymes, and that this system can be coupled *in vitro* to a reductive carboxylation reaction (such as (2*b*) of Table 2) catalysed by enzymes from other sources; overall therefore, in this artificial system, the carbon of carbon dioxide is reduced and condensed with an acceptor molecule, the energy being ultimately obtained from light. A malic enzyme system from plant sources has also been found to be active in such a reconstruction (Tolmach, 1951; Arnon, 1951). It is not suggested that this is the primary carbon dioxide-fixation mechanism in plant photosynthesis (the reaction as so far demonstrated is slow), but only that it indicates interesting possibilities.

Other reactions in which carbon dioxide is a substrate

Some heterotrophic bacteria can utilize carbon dioxide as the main hydrogen-acceptor (oxidant) for anaerobic growth. The carbon atom is reduced further than in the formation of the carboxylic acid group, and in some cases there is evidence that, concurrently with reduction, there is formation of a C_2 compound (acetate) from carbon dioxide. Although the detailed mechanism of most of these reactions is still obscure, and although they do not result in the bulk fixation of carbon dioxide into cell substance, they are important from the present point of view in indicating potential methods of carbon dioxide metabolism by autotrophs.

Reduction of carbon dioxide to formic acid. The simplest case in which carbon dioxide acts as a hydrogen-acceptor, though specifically to molecular hydrogen, is the formic hydrogenlyase enzyme system of *Bacterium coli* and other organisms (see Stephenson, 1937):

$$HCOOH \rightleftharpoons H_2 + CO_2.$$

Cell suspensions not only degrade formate, but catalyse the recombination of hydrogen and carbon dioxide (Woods, 1936). This is the simplest case of a carboxylation reaction, though the detailed mechanism may be more complex than appears from the simple overall equation (Gest, 1952a). Since the energy change is small (a measurable equilibrium is reached) the reaction has no possibilities as an anaerobic growth mechanism. Its significance for the cell is rather that oxidation of formate to carbon dioxide can take place without a hydrogen-accepting system (Stephenson, 1947).

Reduction of carbon dioxide to methane. Barker (1940, 1941) has made a detailed study of the anaerobe *Methanobacterium omelianskii* which uses carbon dioxide as the bulk final hydrogen-acceptor for its energy-yielding reaction, which is the oxidation of alcohols to the corresponding fatty acids or ketones. The carbon dioxide is reduced to methane, that is, the carbon atom is brought all the way from its most oxidized to its most reduced state; furthermore, carbon dioxide is specifically essential for growth. Isotope experiments showed that the methane is entirely derived from carbon dioxide and that none comes from the carbon atoms of the alcohol (Barker, Ruben & Kamen, 1940). The generalized equation for the growth reaction is thus:

$$4H_2A + CO_2 \rightarrow 4A + CH_4 + 2H_2O.$$

Some carbon from the carbon dioxide also appears in the new cell material, but the amount is small compared with that derived from the organic substrate. Cell suspensions of the organism can bring about this reaction and also the oxidation of hydrogen with carbon dioxide as hydrogen-acceptor (Barker, 1943). There is no evidence, however, that the organism can *grow* under the latter conditions with carbon dioxide as the main source of carbon. One can see in this organism a close approach to autotrophic life in the ability both to reduce and fix carbon dioxide into cell material; moreover the oxidation of hydrogen to water is an energy-yielding reaction which can be used in autotrophic life (see later). However, the organism must fail at some stage important for the effective coupling of the various processes. The small amount of carbon from carbon dioxide which appears in the cell may be derived from an intermediate in the reduction to methane, or it may arise from an independent fixation reaction.

Reduction of carbon dioxide to acetic acid. The possibility of a reductive coupling of two molecules of carbon dioxide by bacteria with the formation of acetic acid was first indicated by the work of Wieringa (1936, 1940) with cultures of a mud anaerobe. The organism, *Clostridium aceticum*, was found by Karlsson, Volcani & Barker (1948) to have complex organic growth requirements. Growing cultures show an uptake of carbon dioxide and hydrogen with production of acetate in accordance with the equation:

$$4H_2 + 2CO_2 \rightarrow CH_3COOH + 2H_2O$$

The origin of acetate from the exogenous carbon dioxide was not demonstrated with this organism. The idea that carbon dioxide may act as hydrogen-acceptor with the formation of acetate was applied, however, by Barker (1944) to his work with *Clostridium thermoaceticum*, an organism whose main energy-yielding reaction is the fermentation of glucose with production of almost three molecules of acetate. This is difficult to envisage by known fermentation mechanisms (see Elsden, 1952), and it occurred to Barker (1944) that carbon dioxide produced at one stage of the reaction might act as internal hydrogen-acceptor and itself be reduced to a third molecule of acetate, e.g.

$$C_6H_{12}O_6 + 2H_2O \rightarrow 2CH_3COOH + 2CO_2 + 8H,$$
$$2CO_2 + 8H \rightarrow CH_3COOH + 2H_2O.$$

No exogenous supply of carbon dioxide is required by this organism, but evidence for reactions of the above type was obtained with isotopes. If the fermentation were carried out in the presence of labelled carbon dioxide, the isotope was found equally in both carbon atoms of acetic acid (Barker & Kamen, 1945). The assumption is that carbon dioxide produced as an intermediate exchanges with the externally supplied carbon dioxide, part of which therefore appears in the acetate. The equal distribution of isotope in both carbon atoms of the latter strongly suggests that both arise from carbon dioxide (but see also Wood, 1952 *a*, *b*).

The importance of this reaction is the suggestion that the bulk creation of C_2 units from carbon dioxide alone is a reaction possible with at least some bacteria. The detailed mechanism may of course involve more complex carbon compounds (later regenerated) as primary reactants with carbon dioxide.

In several other bacterial fermentations (Table 3) there is evidence (of a similar type) that endogenously produced carbon dioxide acts as hydrogen-acceptor and is converted to acetate.

Table 3. *Bacteria which may form acetic acid*
from endogenous carbon dioxide

The reviews by Elsden (1952) and Utter & Wood (1951) should be consulted for full references and discussion.

Organism	Compound fermented	End-products
Cl. acidi-urici	Uric acid	Acetic acid, NH_3, CO_2
Cl. cylindrosporum	Uric acid	Acetic acid, glycine, NH_3, CO_2
Cl. thermoaceticum	Glucose	Acetic acid
Butyribact. rettgeri	Lactate	Acetic and butyric acids, CO_2
Diplococcus glycinophilus	Glycine	Acetic acid, NH_3, CO_2

Metabolism of inorganic compounds

We may next examine briefly the ability of bacteria living heterotrophically to utilize, in one way or another, the inorganic substrates whose metabolism is characteristic of autotrophic life. The two main classes to be considered are inorganic nitrogen compounds and inorganic sulphur compounds. Problems concerning the utilization of molecular hydrogen will be dealt with in another section.

Inorganic nitrogen compounds. The autotrophic nitrifying bacteria bring about the oxidation of ammonia to nitrite and of nitrite to nitrate. Such oxidations are so far unknown in heterotrophic life. On the other hand, the reverse series of reactions (the reduction of nitrate through nitrite to ammonia) is well known (see, for example, Woods, 1938; Lascelles & Still, 1946*b*), and nitrate can be used as final hydrogen-acceptor for the anaerobic growth of *Bact. coli* (Quastel, Stephenson & Whetham, 1925). Cases are also known, both in heterotrophic and autotrophic life, in which nitrate is reduced, not to ammonia, but to molecular nitrogen (e.g. *Pseudomonas stutzeri*, Allen & van Niel, 1952; van Niel, 1953; *Thiobacillus denitrificans*, see Bunker, 1936).

The ability of many heterotrophs to use ammonia as the main, if not the only, source of nitrogen for growth presupposes that they have enzymic mechanisms for bringing about reactions between ammonia and organic molecules. It is probable that glutamic and aspartic acids are the first amino-acids to be formed by direct reaction between ammonia and α-ketoglutaric and fumaric or oxaloacetic acids.

Inorganic sulphur compounds. Oxidation of compounds such as sulphide, sulphur and thiosulphate provides the energy source and overall reducing system for a number of chemosynthetic autotrophs, and the reducing system for photosynthetic autotrophs. As with nitrogen compounds, the metabolism of inorganic sulphur compounds in heterotrophic life results mainly in their reduction rather than oxidation. The one recorded instance of oxidation by heterotrophs is the limited oxidation of thiosulphate to tetrathionate by *Pseudomonas* spp. Starkey (1935) found such an organism; recently Baalsrud & Baalsrud (unpublished observations, quoted by van Niel, 1953) isolated another, which can, however, also bring about the converse reaction and use tetrathionate as hydrogen-acceptor.

Sulphate and other sulphur compounds are reduced to sulphide by the anaerobe *Desulphovibrio desulphuricans* and related organisms (see review, Bunker, 1936); this reaction indeed forms the bulk hydrogen-acceptor system in both the heterotrophic and autotrophic life of the organism (see later). Tetrathionate is reduced to thiosulphate by *Bacterium paratyphosum* B and other organisms, and can serve as an alternative (though less efficient) hydrogen-acceptor to oxygen for the energy-yielding reactions. The logarithmic growth phase in stagnant broth culture is prolonged by the addition of this compound; a similar, though more dramatic, effect is produced by aeration (Knox, 1945).

The ability of many heterotrophs to use inorganic sulphate as the ultimate source of sulphur for the synthesis of organic sulphur-containing compounds again implies an ability to metabolize sulphate and to bring about its overall reduction. Very little is known of the mechanism by which inorganic sulphate is incorporated into organic molecules, and van Niel (1953) has suggested that the pathway of reduction may differ from that followed when sulphate acts as a bulk hydrogen-acceptor; there are mutant strains of *Bact. coli* which respond to sulphate, sulphite and sulphide respectively for growth (Lampen, Roepke & Jones, 1947), but this does not necessarily mean that they are direct intermediates.

To summarize this section of the discussion, it is clear that among obligately heterotrophic bacteria there is no intrinsic disability to metabolize the inorganic nitrogen and sulphur substrates characteristic of autotrophic metabolism. Indeed, in some cases such metabolism can be a major reaction of the cell. But in the main, heterotrophic metabolism is of a different type, reduction rather than oxidation.

'BRIDGE ORGANISMS'

We pass now to a consideration of those organisms which are able to live both heterotrophically and autotrophically, and whose properties seem likely to provide us eventually with knowledge of the detailed metabolic events which differentiate the two ways of life. The photosynthetic bacteria as a group provide an excellent illustration of how the two types of life seem to merge almost indistinguishably into one another, whilst the hydrogen bacteria (among the chemo-autotrophs) will be taken as an example of organisms, which, at the present state of knowledge, seem to live a more 'Jekyll and Hyde' existence.

Photosynthetic bacteria

There are admirable and comprehensive reviews of the physiology and biochemistry of this group of organisms (van Niel, 1941, 1944, 1949; Gest, 1951), and various aspects of their metabolism will be considered by other contributors to this Symposium. We shall follow van Niel's division of the group into three main classes based on the general mode of life and other properties of the organisms (Table 4). We also follow him in emphasizing that this is a classification of convenience and that it is impossible to define clearly any one group. Some of the biochemical properties with which we shall be concerned may only have been found in a single species of one organism, or perhaps only in a particular strain. This is unimportant from the present point of view; the discovery of a certain biochemical potentiality in even a single cell (if this were possible) might be of critical value, just as in a more biological sphere the capture of a single coelocanth causes considerable excitement among those interested in evolution.

The green sulphur bacteria have been relatively little studied compared with the other groups, especially in the light of modern views on the photosynthetic reaction. This is being remedied by detailed work on *Chlorobium thiosulphatophilum* (Larsen, 1952; Larsen *et al.* 1953). As far as is known the group lead an obligately photo-autotrophic life with inorganic sulphur compounds as hydrogen-donors for the overall reduction of carbon dioxide; some strains can also use molecular hydrogen.

With the Thiorhodaceae we again have a strictly anaerobic group of organisms which are confined to a photosynthetic life but not necessarily autotrophically. They can live autotrophically with the same type of growth reactions as *Chlorobium*; on the other hand, they can also use organic compounds as hydrogen-donors. In the latter case the carbon atoms of the organic compounds are mainly used for the synthesis of new cell material and free carbon dioxide is not an intermediate in this process. Whether or not free carbon dioxide is also assimilated depends on the nature (general state of oxidation) of the organic substrate

Table 4. *General properties of photosynthetic bacteria*

Group	Natural habitat	Ultimate reducing agent for photosynthesis	Growth in dark	Requirement for trace growth factors
(Green sulphur bacteria), e.g. *Chlorobium* spp.	Mainly sulphide containing	Sulphide [*More oxidized forms of inorganic* S]	None	None
Thiorhodaceae (purple sulphur bacteria), e.g. *Chromatium* spp.	Mainly sulphide containing	Inorganic S compounds, hydrogen, organic acids	None	None
Athiorhodaceae (purple and brown non-sulphur bacteria), e.g. *Rhodospirillum* spp., *Rhodopseudomonas* spp.	Mainly containing organic compounds	Organic compounds [*Inorganic* S *compounds, hydrogen*]	[*Aerobic only, on substrates used in light*]	One or more vitamin B-group substances

Properties listed [*italics*] are shown only by some organisms of the group.

(Muller, 1933); in some cases, e.g. malate, there is even an evolution of carbon dioxide. With an appropriate substrate (e.g. butyrate or higher fatty acids), where there is considerable concurrent fixation of exogenous carbon dioxide, the organism seems in a sense to be living a mixed heterotrophic and autotrophic life.

Among the Athiorhodaceae we find the most interesting organisms of all from the present point of view. The main growth substrates for photosynthetic life are organic compounds; whether or not exogenous carbon dioxide is also assimilated depends on the same factors as with the Thiorhodaceae when living in this way. We also have in this group, however, a case in which the organic substrate is used solely as final reducing agent for the photo-assimilation of exogenous carbon dioxide; Foster (1944) found that strains of Athiorhodaceae, which had been 'trained' to metabolize secondary alcohols as organic substrates, fixed carbon dioxide in the light according to the general equation below. This occurred both during growth and with cell suspensions; no carbon from the organic substrate appeared in the new cell material which was entirely derived from carbon dioxide:

$$2 \underset{R_2}{\overset{R_1}{\diagdown}} CHOH + CO_2 \rightarrow 2 \underset{R_2}{\overset{R_1}{\diagdown}} CO + [CH_2O] + H_2O.$$

We have here an organic compound with precisely the same limited function as the inorganic sulphur compounds which act as hydrogen-donors for the purely autotrophic life of the Thiorhodaceae; the present life process is clearly almost impossible to differentiate from autotrophy even though organic carbon is required. In recent work designed to extend knowledge of this interesting phenomenon, Siegel & Kamen (1950) isolated new strains (of *Rhodopseudomonas gelatinosa*) which could oxidize *iso*propanol (Foster's organisms were no longer available). The results were not, however, so clear-cut; *iso*propanol was again used for the fixation of carbon dioxide, but carbon from acetone (the product of *iso*propanol oxidation) was also photo-assimilated by the organism.

Some members of the Athiorhodaceae can also live photo-auto-trophically using the oxidation of inorganic sulphur compounds or hydrogen as overall reducing system. The mode of life of such organisms can scarcely be distinguished, if at all, from that of the Thiorhodaceae. In general, however, Athiorhodaceae always require traces of B-group vitamins as growth factors (*Rhodomicrobium vannielii* (Duchow & Douglas, 1949) is an exception), whereas Thiorhodaceae do not; in addition, the Thiorhodaceae are strict anaerobes, whereas many members of the Athiorhodaceae are not.

The last point brings us to a third interesting property of some members of the Athiorhodaceae. While certain of the organisms (e.g. *Rhodomicrobium vannielii, Rhodospirillum fulvum*) are strict anaerobes and confined to the photosynthetic life, be it autotrophic or heterotrophic, many of them are also able to grow aerobically in the dark. Energy is obtained not from light but by the aerobic oxidation of the same substrates used as hydrogen-donors in the anaerobic photosynthetic life. When therefore the substrate is an organic compound the organisms are living, as far as is known, a typically heterotrophic life. Strains which can live photo-autotrophically with sulphur compounds or molecular hydrogen as hydrogen-donor can also live aerobically in the dark and assimilate carbon dioxide with energy derived from the oxidation of these substances. These strains are therefore living a typical chemosynthetic life akin to the autotrophic sulphur bacteria or the hydrogen bacteria according to the substrate oxidized.

There may be present therefore in a single species the ability to live by all four general processes, i.e. photo-heterotrophically, photo-autotrophically, chemo-autotrophically and chemo-heterotrophically. There are general statements in the literature on the properties of the Athiorhodaceae (e.g. van Niel, 1944) which, taken together, imply that such an organism must exist; we have not, however, been able to track down a particular species or strain to quote by name.

When an organism can use more than one of the general growth processes, and when the conditions in the environment are also compatible with more than one mode of life, it may not be easy to decide which process the organism in fact uses, or, indeed, whether both are used concurrently. The nature of the normal environment in which organisms are found in nature should be taken into account (van Niel, 1931, 1941). Thus Thiorhodaceae (which can use both organic compounds and inorganic sulphur compounds for photosynthetic life) are found primarily in environments where sulphur compounds are present. This may point to the photo-autotrophic life being ecologically the more important; it could also be argued that the concomitant reducing conditions found in such environments are necessary for the survival of these strictly anaerobic organisms. The Athiorhodaceae, some of which have the same two possibilities for photosynthetic life, are primarily in environments where organic carbon is present. It has been seen that certain members of this group which are not strict anaerobes are able to live aerobically in the dark with the same organic substrate used for anaerobic growth in the light; van Niel (1941) has given evidence that when both light and oxygen are present the photosynthetic life predominates.

Bacteria which oxidize hydrogen

There are a number of organisms which are able to live heterotrophically using the oxidation of molecular hydrogen (to water) as energy source and overall reducing system for the fixation of carbon dioxide. With one group, the hydrogen bacteria (Knallgasbakterien), oxygen is the obligatory final hydrogen-acceptor for autotrophic life. The *Desulphovibrio* spp., on the other hand, are strict anaerobes and the only hydrogen-acceptors used for the oxidation of molecular hydrogen in autotrophic growth are inorganic sulphur compounds. Neither group is confined to the autotrophic life, for every species so far examined can also live heterotrophically.

The ability to oxidize hydrogen in the presence of a suitable acceptor (equation (i)) is widespread among obligate heterotrophs; oxygen can almost always be used as final acceptor. The initial enzymic activation of hydrogen is brought about by the hydrogenase enzyme system (equation (ii)) first described in *Bact. coli* by Stephenson & Stickland (1931):

$$H_2 + A \rightarrow H_2A, \tag{i}$$

$$H_2 \rightleftharpoons 2H^+ + 2e. \tag{ii}$$

It is natural therefore to inquire whether activation of hydrogen in autotrophic life is brought about by a similar enzyme. This appears to be so; the typical hydrogen bacterium *Hydrogenomonas facilis* (grown autotrophically) has a hydrogenase enzyme whose properties (Schatz & Bovell, 1952) are similar to those of the hydrogenase enzyme of *Bact. coli*, *Aerobacter indologenes*, *Proteus vulgaris* and *Azotobacter vinelandii*; in each case, for example, there is evidence that the enzyme has an iron-containing prosthetic group (Lascelles & Still, 1946a; Waring & Werkman, 1944; Hoberman & Rittenberg, 1943; Green & Wilson, 1953; Hyndman, Burris & Wilson, 1953). The hydrogenase of a photosynthetic heterotroph (*Rhodospirillum rubrum*) is also in this group (Gest, 1952b).

Hydrogen bacteria. Recent work, especially with the new species *Hydrogenomonas facilis* (Schatz & Bovell, 1952), has provided some most interesting results from the point of view of the relationship between autotrophic and heterotrophic life. For a time it appeared from the work of Kluyver & Manten (1942) with *Hydrogenomonas flava* that cells grown aerobically under heterotrophic conditions (on lactate) did not contain hydrogenase; the enzyme seemed to be associated specifically with autotrophic life. Wilson, Stout, Powelson & Koffler (1953) found the same thing with *H. facilis* if the cells were grown in air; if, however, the oxygen concentration was reduced to 5 % the enzyme was present. Coupled with the observation by many workers that hydrogenase is rapidly inactivated by oxygen, these results make it probable that hydrogenase is present in cells grown either autotrophically or heterotrophically, but is inactivated if aeration is too great.

Oxidation of both molecular hydrogen and organic carbon (lactate) can occur simultaneously and additively with cell suspensions of both the above species (Kluyver & Manten, 1942; Wilson *et al.* 1953). The oxidation systems used in autotrophic and heterotrophic life can therefore co-exist in the same cell and do not impede each other. Cell suspensions of *H. facilis* can, like *Bact. coli*, oxidize hydrogen with nitrate or methylene blue as well as oxygen as final hydrogen-acceptor; the organism can, however, only fix carbon dioxide or grow autotrophically when oxygen is acceptor. The energy from the oxidation of hydrogen by nitrate cannot apparently be linked to the carbon dioxide-fixation reaction (Schatz, 1952). A most significant point has recently emerged from experiments on the heterotrophic metabolism of *H. facilis*. Schatz, Isenberg & Trelawney (1953) have briefly reported that during lactate oxidation the amount of carbon dioxide assimilated is 'far more than one would expect in the ordinary heterotrophic fixation of carbon dioxide'. This suggests that oxidation of lactate, as well as hydrogen, may be coupled to the carbon dioxide-fixation process. If this is so the situation is analogous in a way to that with photosynthetic Thiorhodaceae when utilizing organic hydrogen-donors. The heterotrophic life of *Hydrogenomonas* may be a mixture of normal heterotrophic mechanisms and what are effectively (in spite of the organic hydrogen-donor) autotrophic mechanisms.

It has been pointed out by Koffler & Wilson (1951) that the two typical heterotrophs *Bact. coli* and *Azotobacter vinelandii* are able both to oxidize hydrogen with oxygen and to fix carbon dioxide during heterotrophic growth. There is even some evidence (Hyndman *et al.* 1953) that oxidation of hydrogen by the latter organism can be coupled to the synthesis of the high-energy phosphate esters which are well known as intermediates in energy-transfer reactions. Yet these organisms cannot live autotrophically as do the hydrogen bacteria; the latter must therefore have a different kind of method of creating carbon to carbon bonds, or an energy-transfer mechanism which differs at least in detail from those so far known in heterotrophs.

Sulphate-reducing bacteria. Until recently *Desulphovibrio desulphuricans* and related species were known only as heterotrophs obtaining energy for anaerobic growth by oxidizing substrates (e.g. lactate) with sulphate and related compounds as final hydrogen-acceptor. Cell suspensions were known to oxidize hydrogen by a similar mechanism, and Butlin *et al.* (1949) showed that strictly autotrophic growth could be supported by this reaction. At present the main interest of this group as an intermediate type is that both heterotrophic and autotrophic life use, and in fact demand, the same final hydrogen-acceptor system.

Other evidence for basically similar biochemical mechanisms
in the autotrophic and heterotrophic life of bacteria

Detailed knowledge of biochemical events in autotrophic bacteria is mainly restricted to the overall energy-yielding reactions. This is not surprising, since it is only here that there are clearly defined reaction products to analyse. The only other major metabolite, carbon dioxide, is assimilated quantitatively into new cell material, a process which is closely associated with overall growth of the culture. The technical difficulties are in a way analogous, though much more severe, than those encountered with certain heterotrophic bacteria which show, even in non-growing cell suspension, marked oxidative assimilation (Clifton, 1951); in such organisms no proper balance sheet relating substrate and products can be constructed because a variable, and sometimes rather high, proportion of the substrate-carbon is assimilated into cell substance. It is not without interest that such assimilation is often particularly marked in *Pseudomonas* spp., which seem to be closely related morphologically both to hydrogen bacteria and to some of the photosynthetic bacteria.

There are a few rather isolated observations, apart from those discussed in earlier sections, which are in accord with the idea of closely similar biochemical mechanisms in autotrophic and heterotrophic life.

(1) The amino-acid composition of the autotrophic nitrifying organism *Nitrosomonas* is closely similar to that of *Corynebacterium diphtheriae* (Hofman, 1953).

(2) Many substances of the vitamin B-group type are known to have ultimate function in cell metabolism as part of the structure of coenzymes or prosthetic groups of enzymes concerned in the metabolism of key intermediates in heterotrophic metabolism. Members of the vitamin B group are known to be synthesized by the obligate autotroph *Thiobacillus thio-oxidans* (O'Kane, 1942) and to be required as trace growth factors by certain Athiorhodaceae which can live a photo-autotrophic life (Hutner, 1946, 1950).

(3) Several strict autotrophs have the ability to metabolize (though not to grow upon) higher carbon compounds. *T. thio-oxidans*, for example, can break down endogenous polysaccharide cell constituents (Vogler, 1942; LePage, 1943), while members of the Thiorhodaceae can, in the dark, oxidize or ferment (albeit slowly) carbon compounds that they photosynthesize in the light (Gaffron, 1935; Nakamura, 1937, 1939). Larsen (1951) has found that the strict autotroph *Chlorobium thiosulphatophilum* photosynthesizes succinate from propionate and carbon dioxide; propionate cannot replace sulphide for growth.

(4) Umbreit and his colleagues (see Umbreit, 1947, 1951) found evidence that the oxidation of sulphur by *T. thio-oxidans* was linked to the carbon dioxide-fixation reaction through the intermediate formation of the high-energy phosphate esters which are traditionally associated with energy-transfer mechanisms in heterotrophic life.

Baalsrud & Baalsrud (1952) have recently criticized this evidence mainly on quantitative grounds; they have also been unable to find adequate evidence for such a view with the autotrophic *Thiobacillus* spp. which they studied themselves.

(5) There is some evidence, mainly from experiments on cyanide sensitivity, that the oxidation of sulphur by *T. thio-oxidans* may require a terminal hydrogen-transfer system of the traditional cytochrome type found in the aerobic oxidation of organic substrates by heterotrophs (Vogler, LePage & Umbreit, 1942). The detailed study of the oxidation of sulphur compounds by the strict autotroph *Thiobacillus thioparus* recently begun by Vishniac (1952) may provide further evidence on this matter. Heavy metals, including iron, are also probably involved in the oxidation of ammonia by *Nitrosomonas* spp. (Lees, 1952; Meiklejohn, 1953).

EPILOGUE

We cannot attempt to draw any logical and general conclusions from the work that has been surveyed. It is very clear that a much deeper biochemical and physiological analysis of the metabolism of the fascinating biological material already available (and perhaps yet to be discovered) will be necessary before the secret of autotrophic life is discovered. It is the purpose of this Symposium to set before you some of the recent advances at various levels of microbiological investigation. From a very general examination of present knowledge one gains the impression (also expressed by Koffler & Wilson, 1951) that the key to autotrophic life is most probably to be found, not in the details of energy-yielding or synthetic reactions, but in the mechanism by which they are coupled and by which chemical energy is transferred from one to another.

We have rather avoided one aspect of the passage between autotrophic and heterotrophic life. A good case can be made (see, for example, Lwoff, 1944; Knight, 1945) that evolution among microorganisms may have occurred by loss of function, the line passing from simple to more complex nutritional requirements. In this sense the autotrophs would be among the most primitive types, and the borderline cases that have been discussed might have great evolutionary significance. Yet, as has been pointed out by van Niel (1943) and Umbreit (1947) among others, the autotrophs, though simple in the nutritional sense, have a greater range of biochemical accomplishment than any other living creatures. To account in turn for their origin something must be imagined analogous to the birth of the Goddess Athené who, you may remember, sprang forth fully armed (in war-gear golden and bright) from the head of Zeus.* This lady is described as being, amongst other things, bright-eyed and inventive. If, on this evidence, we may consider her metaphorically as an autotroph, it is certain that her bright eyes are proving an ever-increasing attraction to biochemists and micro-biologists; her guile, however, has so far defeated them.

In preparing this general survey we have drawn largely upon the thought and ideas, as well as the experiments, of leading workers in the field. Their several stimulating and thoughtful reviews are invaluable, and it is certain that we are indebted to them collectively to a far greater extent than can be acknowledged by mere citation of references to their publications.

REFERENCES

AJL, S. J. & WERKMAN, C. H. (1948). Replacement of CO_2 in heterotrophic metabolism. *Arch. Biochem.* **19**, 483.

ALLEN, M. B. & VAN NIEL, C. B. (1952). Experiments on bacterial denitrification. *J. Bact.* **64**, 397.

ARNON, D. I. (1951). Extracellular photosynthetic reactions. *Nature, Lond.* **167**, 1008.

BAALSRUD, K. & BAALSRUD, K. S. (1952). The role of phosphate in CO_2 assimilation of *Thiobacilli*. In *Phosphorus Metabolism*, **2**, 544. Baltimore: Johns Hopkins Press.

BARBAN, S. & AJL, S. J. (1951). Interconversion of propionate and succinate by *Propionibacterium pentosaceum*. *J. biol. Chem.* **192**, 63.

BARBAN, S. & AJL, S. J. (1953). Interconversion of *iso*citrate and alpha-ketoglutarate by *Escherichia freundii*. *J. Bact.* **66**, 68.

BARKER, H. A. (1940). Studies upon the methane fermentation. IV. The isolation and culture of *Methanobacterium omelianskii*. *Leeuwenhoek ned. Tijdschr.* **6**, 201.

BARKER, H. A. (1941). Studies on the methane fermentation. V. Biochemical activities of *Methanobacterium omelianskii*. *J. biol. Chem.* **137**, 153.

BARKER, H. A. (1943). Studies on the methane fermentation. VI. The influence of carbon dioxide concentration on the rate of carbon dioxide reduction by molecular hydrogen. *Proc. nat. Acad. Sci., Wash.* **29**, 184.

BARKER, H. A. (1944). On the role of carbon dioxide in the metabolism of *Clostridium thermoaceticum*. *Proc. nat. Acad. Sci., Wash.* **30**, 88.

BARKER, H. A. & KAMEN, M. D. (1945). Carbon dioxide utilization in the synthesis of acetic acid by *Clostridium thermoaceticum*. *Proc. nat. Acad. Sci., Wash.* **31**, 219.

BARKER, H. A., RUBEN, S. & KAMEN, M. D. (1940). The reduction of radioactive carbon dioxide by methane-producing bacteria. *Proc. nat. Acad. Sci., Wash.* **26**, 426.

BUCHANAN, J. G., BASSHAM, J. A., BENSON, A. A., BRADLEY, D. F., CALVIN, M., DAUS, L. L., GOODMAN, M., HAYES, P. M., LYNCH, V. H., NORRIS, L. T. & WILSON, A. T. (1952). The path of carbon in photosynthesis. XVII. Phosphorus compounds as intermediates in photosynthesis. In *Phosphorus Metabolism*, **2**, 440. Baltimore: Johns Hopkins Press.

* *The Homeric Hymns*, XXVIII, 5. We have relied on the translation by J. Edgar (1891). Edinburgh: James Thin. We are unable to trace the line of thought which led us to this analogy which may have been used before. It was also used, independently, by Dr K. A. Bisset in a broadcast talk while this paper was in publication.

BUNKER, H. J. (1936). *A review of the physiology and biochemistry of the sulphur bacteria. Spec. Rep. Chem. Res., Lond.,* no. 3. London: H.M.S.O.

BUTLIN, K. R., ADAMS, M. E. & THOMAS, M. (1949). The isolation and cultivation of sulphate-reducing bacteria. *J. gen. Microbiol.* **3,** 46.

CALVIN, M. (1949). The path of carbon in photosynthesis. In *Reilly Lectures,* **2.** Notre Dame: University of Notre Dame.

CALVIN, M. & MASSINI, P. (1952). The path of carbon in photosynthesis. XX. The steady state. *Experientia,* **8,** 445.

CLIFTON, C. E. (1951). Assimilation by bacteria. In *Bacterial Physiology* (ed. C. H. Werkman and P. W. Wilson), ch. 17. New York: Academic Press.

CROSS, M. J. (1953). *Leuconostoc citrovorum* factor and the synthesis of serine by micro-organisms. *VIth Int. Congr. Microbiol. Abstr.* p. 121.

DUCHOW, E. & DOUGLAS, H. C. (1949). *Rhodomicrobium vannielii,* a new photo-heterotrophic bacterium. *J. Bact.* **58,** 409.

ELSDEN, S. R. (1952). Bacterial fermentation. In *The Enzymes* (ed. J. B. Sumner and K. Myrbäck), **2,** pt. 2, ch. 65. New York: Academic Press.

FOSTER, J. W. (1944). Oxidation of alcohols by non-sulphur photosynthetic bacteria. *J. Bact.* **47,** 355.

FOSTER, J. W. (1951). Autotrophic assimilation of carbon dioxide. In *Bacterial Physiology* (ed. C. H. Werkman and P. W. Wilson), ch. 11. New York: Academic Press.

GAFFRON, H. (1935). Über die Kohlensäureassimilation der roten Schwefelbakterien. II. *Biochem. Z.* **279,** 1.

GAFFRON, H., FAGER, E. W. & ROSENBERG, J. L. (1951). Intermediates in photosynthesis: formation and transformation of phosphoglyceric acid. In *Carbon Dioxide Fixation and Photosynthesis. Symp. Soc. exp. Biol.* **5,** 262.

GEST, H. (1951). Metabolic patterns in photosynthetic bacteria. *Bact. Rev.* **15,** 183.

GEST, H. (1952*a*). Molecular hydrogen: oxidation and formation in cell-free systems. In *Phosphorus Metabolism,* **2,** 522. Baltimore: Johns Hopkins Press.

GEST, H. (1952*b*). Properties of cell-free hydrogenases from *Escherichia coli* and *Rhodospirillum rubrum. J. Bact.* **63,** 111.

GLADSTONE, G. P., FILDES, P. & RICHARDSON, G. M. (1935). Carbon dioxide as an essential factor in the growth of bacteria. *Brit. J. exp. Path.* **16,** 335.

GLOVER, J., KAMEN, M. D. & VAN GENDEREN, H. (1952). Studies on the metabolism of photosynthetic bacteria. XII. Comparative light and dark metabolism of acetate and carbonate by *Rhodospirillum rubrum. Arch. Biochem. Biophys.* **35,** 384.

GREEN, M. & WILSON, P. W. (1953). Hydrogenase and nitrogenase in *Azotobacter. J. Bact.* **65,** 511.

HOBERMAN, H. D. & RITTENBERG, D. (1943). Biological catalysis of the exchange reaction between water and hydrogen. *J. biol. Chem.* **147,** 211.

HOFMAN, T. (1953). The biochemistry of the nitrifying organisms. 3. Composition of *Nitrosomonas. Biochem. J.* **54,** 293.

HUTNER, S. H. (1946). Organic growth essentials of the aerobic non-sulphur photosynthetic bacteria. *J. Bact.* **52,** 213.

HUTNER, S. H. (1950). Anaerobic and aerobic growth of purple bacteria (Athiorhodaceae) in chemically defined media. *J. gen. Microbiol.* **4,** 286.

HYNDMAN, L. A., BURRIS, R. H. & WILSON, P. W. (1953). Properties of hydrogenase from *Azotobacter vinelandii. J. Bact.* **65,** 522.

KALTENBACH, J. P. & KALNITSKY, G. (1951*a*). The enzymatic formation of oxalacetate from pyruvate and carbon dioxide. I. *J. biol. Chem.* **192,** 629.

KALTENBACH, J. P. & KALNITSKY, G. (1951*b*). The enzymatic formation of oxalacetate from pyruvate and carbon dioxide. II. *J. biol. Chem.* **192,** 641.

KARLSSON, J. L., VOLCANI, B. E. & BARKER, H. A. (1948). The nutritional requirements of *Clostridium aceticum. J. Bact.* **56**, 781.

KLUYVER, A. J. & MANTEN, A. (1942). Some observations on the metabolism of bacteria oxidizing molecular hydrogen. *Leeuwenhoek ned. Tijdschr.* **8**, 71.

KNIGHT, B. C. J. G. (1945). Growth factors in microbiology. Some wider aspects of nutritional studies with micro-organisms. *Vitamins and Hormones,* **3**, 105.

KNOX, R. (1945). The effect of tetrathionate on bacterial growth. *Brit. J. exp. Path.* **26**, 146.

KOFFLER, H. & WILSON, P. W. (1951). The comparative biochemistry of molecular hydrogen. In *Bacterial Physiology* (ed. C. H. Werkman and P. W. Wilson), ch. 16. New York: Academic Press.

LAMPEN, J. O., ROEPKE, R. R. & JONES, M. J. (1947). Studies on the sulphur metabolism of *Escherichia' coli*. III. Mutant strains of *Escherichia coli* unable to utilise sulphate for their complete sulphur requirements. *Arch. Biochem.* **13**, 55.

LARSEN, H. (1951). Photosynthesis of succinic acid by *Chlorobium thiosulphatophilum. J. biol. Chem.* **193**, 167.

LARSEN, H. (1952). On the culture and general physiology of the green sulphur bacteria. *J. Bact.* **64**, 187.

LARSEN, H., YOCUM, C. S. & VAN NIEL, C. B. (1953). On the energetics of the photosyntheses in green sulphur bacteria. *J. gen. Physiol.* **36**, 161.

LASCELLES, J., CROSS, M. J. & WOODS, D. D. (1954). The folic acid and serine nutrition of *Leuconostoc mesenteroides* P 60 (*Streptococcus equinus* P 60). *J. gen. Microbiol.* **10**, 267.

LASCELLES, J. & STILL, J. L. (1946a). Utilization of molecular hydrogen by bacteria. *Aust. J. exp. Biol. med. Sci.* **24**, 37.

LASCELLES, J. & STILL, J. L. (1946b). The reduction of nitrate, nitrite and hydroxylamine by *E. coli. Aust. J. exp. Biol. med. Sci.* **24**, 159.

LEAVER, F. W. & WOOD, H. G. (1953). Evidence from fermentation of labelled substrates which is inconsistent with present concepts of propionic acid fermentation. *J. cell. comp. Physiol.* **41**, Suppl. p. 225.

LEES, H. (1952). The biochemistry of the nitrifying organisms. I. The ammonia oxidizing systems of *Nitrosomonas. Biochem. J.* **52**, 134.

LEPAGE, G. A. (1943). The biochemistry of autotrophic bacteria. The metabolism of *Thiobacillus thio-oxidans* in the absence of oxidizable sulphur. *Arch. Biochem.* **1**, 255.

LIPMANN, F. (1946). Acetyl phosphate. *Advanc. Enzymol.* **6**, 231.

LWOFF, A. (1944). *L'Évolution Physiologique. Étude des Pertes de Fonctions chez les Microorganismes.* Paris: Hermann.

LWOFF, A. & MONOD, J. (1947). Essai d'analyse du rôle de l'anhydride carbonique dans la croissance microbienne. *Ann. Inst. Pasteur,* **73**, 323.

LYMAN, C. M., MOSELEY, O., WOOD, S., BUTLER, B. & HALE, F. (1947). Some chemical factors which influence the amino acid requirements of the lactic acid bacteria. *J. biol. Chem.* **167**, 177.

MEIKLEJOHN, J. (1953). Iron and the nitrifying bacteria. *J. gen. Microbiol.* **8**, 58.

MULLER, F. M. (1933). On the metabolism of the purple sulphur bacteria in organic media. *Arch. Mikrobiol.* **4**, 131.

NAKAMURA, H. (1937) Über die Kohlensäureassimilation von *Rhodospirillum giganteum.* Beiträge zur Stoffwechselphysiologie der Purpurbakterien. II. *Acta Phytochim., Tokyo,* **9**, 231.

NAKAMURA, H. (1939). Weitere Untersuchungen über den Wasserstoffumsatz bei den Purpurbakterien, nebst einer Bemerkung über die gegenseitige Beziehung zwischen *Thio-* und *Athiorhodaceen.* Beiträge zur Stoffwechselphysiologie der Purpurbakterien. V. *Acta Phytochim., Tokyo,* **11**, 109.

VAN NIEL, C. B. (1931). On the morphology and physiology of the purple and green sulphur bacteria. *Arch. Mikrobiol.* **3**, 1.

VAN NIEL, C. B. (1936). On the metabolism of the Thiorhodaceae. *Arch. Mikrobiol.* **7**, 323.

VAN NIEL, C. B. (1941). The bacterial photosyntheses and their importance for the general problem of photosynthesis. *Advanc. Enzymol.* **1**, 263.

VAN NIEL, C. B. (1943). Biochemical problems of the chemoautotrophic bacteria. *Physiol. Rev.* **23**, 338.

VAN NIEL, C. B. (1944). The culture, general physiology, morphology, and classification of the non-sulphur purple and brown bacteria. *Bact. Rev.* **8**, 1.

VAN NIEL, C. B. (1949). The comparative biochemistry of photosynthesis. In *Photosynthesis in Plants* (ed. J. Franck and W. E. Loomis), ch. 22. Ames: Iowa State College Press.

VAN NIEL, C. B. (1953). Introductory remarks on the comparative biochemistry of micro-organisms. *J. cell. comp. Physiol.* **41**, Suppl. p. 9.

OCHOA, S. (1951). Biological mechanisms of carboxylation and decarboxylation. *Physiol. Rev.* **31**, 56.

O'KANE, D. J. (1942). The presence of growth factors in the cells of the autotrophic sulphur bacteria. *J. Bact.* **43**, 7.

PAPPENHEIMER, A. M. & HOTTLE, G. A. (1940). Effect of certain purines and CO_2 on growth of strain of group A haemolytic streptococcus. *Proc. Soc. exp. Biol., N.Y.* **44**, 645.

PRINGSHEIM, E. G. (1949). Iron bacteria. *Biol. Rev.* **24**, 200.

QUASTEL, J. H., STEPHENSON, M. & WHETHAM, M. D. (1925). Some reactions of resting bacteria in relation to anaerobic growth. *Biochem. J.* **19**, 304.

RABINOWITCH, E. I. (1951). *Photosynthesis and Related Processes.* **2**, pt. 1. New York: Interscience Publishers.

SCHATZ, A. (1952). Uptake of carbon dioxide, hydrogen and oxygen by *Hydrogenomonas facilis.* *J. gen. Microbiol.* **6**, 329.

SCHATZ, A. & BOVELL, C. (1952). Growth and hydrogenase activity of a new bacterium, *Hydrogenomonas facilis.* *J. Bact.* **63**, 87.

SCHATZ, A., ISENBERG, H. D. & TRELAWNEY, G. (1953). Chemoautotrophic assimilation of carbon dioxide by *Hydrogenomonas facilis* with lactate as hydrogen-donor. *VIth Int. Congr. Microbiol. Abstr.* p. 161.

SIEGEL, J. M. & KAMEN, M. D. (1950). Studies on the metabolism of photosynthetic bacteria. VI. Metabolism of *iso*propanol by a new strain of *Rhodopseudomonas gelatinosa.* *J. Bact.* **59**, 693.

STARKEY, R. L. (1935). Products of the oxidation of thiosulphate by bacteria in mineral media. *J. gen. Physiol.* **18**, 325.

STEPHENSON, M. (1937). Formic hydrogenlyase. *Ergebn. Enzymforsch.* **6**, 139.

STEPHENSON, M. (1947). Some aspects of hydrogen transfer. *Leeuwenhoek ned. Tijdschr.* **12**, 33.

STEPHENSON, M. (1949). *Bacterial Metabolism*, 3rd ed. London: Longmans, Green.

STEPHENSON, M. & STICKLAND, L. H. (1931). Hydrogenase: a bacterial enzyme activating molecular hydrogen. I. The properties of the enzyme. *Biochem. J.* **25**, 205.

TEMPLE, K. L. & COLMER, A. R. (1951). The autotrophic oxidation of iron by a new bacterium: *Thiobacillus ferro-oxidans.* *J. Bact.* **62**, 605.

TOLMACH, L. J. (1951). The influence of triphosphopyridine nucleotide and other physiological substances on oxygen evolution from illuminated chloroplasts. *Arch. Biochem. Biophys.* **33**, 120.

UMBREIT, W. W. (1947). Problems of autotrophy. *Bact. Rev.* **11**, 157.

UMBREIT, W. W. (1951). Significance of autotrophy in comparative physiology. In *Bacterial Physiology* (ed. C. H. Werkman and P. W. Wilson), ch. 19. New York: Academic Press.

UTTER, M. F. & WOOD, H. G. (1951). Mechanisms of fixation of carbon dioxide by heterotrophs and autotrophs. *Advanc. Enzymol.* **12**, 41.

VISHNIAC, W. (1952). The metabolism of *Thiobacillus thioparus*. I. The oxidation of thiosulphate. *J. Bact.* **64**, 363.

VISHNIAC, W. & OCHOA, S. (1952a). Fixation of carbon dioxide coupled to photochemical reduction of pyridine nucleotides by chloroplast preparations. *J. biol. Chem.* **195**, 75.

VISHNIAC, W. & OCHOA, S. (1952b). Reduction of pyridine nucleotides in photosynthesis. In *Phosphorus Metabolism*, **2**, 467. Baltimore: Johns Hopkins Press.

VOGLER, K. G. (1942). The presence of an endogenous respiration in the autotrophic bacteria. *J. gen. Physiol.* **25**, 617.

VOGLER, K. G., LEPAGE, G. A. & UMBREIT, W. W. (1942). Studies on the metabolism of autotrophic bacteria. I. The respiration of *Thiobacillus thio-oxidans* on sulphur. *J. gen. Physiol.* **26**, 89.

WARING, W. S. & WERKMAN, C. H. (1944). Iron deficiency in bacterial metabolism. *Arch. Biochem.* **4**, 75.

WERKMAN, C. H. (1951). Assimilation of carbon dioxide by heterotrophic bacteria. In *Bacterial Physiology* (ed. C. H. Werkman and P. W. Wilson), ch. 12. New York: Academic Press.

WIERINGA, K. T. (1936). Over het verwijnen van waterstof en koolzuur onder anaerobe voorwaarden. *Leeuwenhoek ned. Tijdschr.* **3**, 263.

WIERINGA, K. T. (1940). The formation of acetic acid from carbon dioxide and hydrogen by anaerobic spore-forming bacteria. *Leeuwenhoek ned. Tijdschr.* **6**, 251.

WILSON, E., STOUT, H. A., POWELSON, D. & KOFFLER, H. (1953). Comparative biochemistry of the hydrogen bacteria. I. The simultaneous oxidation of hydrogen and lactate. *J. Bact.* **65**, 283.

WOOD, H. G. (1952a). A study of carbon dioxide fixation by mass determination of the types of ^{13}C-acetate. *J. biol. Chem.* **194**, 905.

WOOD, H. G. (1952b). Fermentation of 3,4-^{14}C- and 1-^{14}C-labelled glucose by *Clostridium thermoaceticum*. *J. biol. Chem.* **199**, 579.

WOODS, D. D. (1936). Hydrogenlyases. IV. The synthesis of formic acid by bacteria. *Biochem. J.* **30**, 515.

WOODS, D. D. (1938). The reduction of nitrate to ammonia by *Clostridium welchii*. *Biochem. J.* **32**, 2000.

THE NATURE AND RELATIONSHIPS OF AUTOTROPHIC BACTERIA

K. A. BISSET AND JOYCE B. GRACE

Department of Bacteriology, University of Birmingham

I. GENERAL INTRODUCTION

No generally accepted definition of an autotrophic bacterium has yet been propounded. According to Umbreit (1947) a true autotroph can obtain energy only by the oxidation of a specific inorganic substrate and relies exclusively on carbon dioxide as a source of carbon. Such a definition unquestionably excludes a considerable proportion of those bacteria commonly regarded as autotrophic, including evidently the photosynthetic groups, and indeed only *Thiobacillus* and the true nitrifying bacteria really appear to qualify. In this survey, however, we have examined all those groups to which the term autotrophic is normally applied at the present time, including the photosynthetic bacteria and those credited with obtaining some part of their energy by the oxidation of iron, sulphur, hydrogen, nitrogen and carbon monoxide. In many cases, recent work has shown that their autotrophy may be incomplete or facultative, and doubt has been cast on the ability of some, notably the filamentous iron bacteria (Pringsheim, 1949 b), to utilize the energy provided by the oxidation of inorganic compounds. As at present accepted, the autotrophic bacteria also include a variety of other organisms, some of which almost certainly have no autotrophic faculties, whereas others are not bacteria, but belong more properly to other groups of autotrophic micro-organisms.

The morphology of many of the autotrophic bacteria was originally described some fifty years ago. At that time, because of their apparently unique metabolism, they were regarded as distinct from all other bacteria and were classified primarily on a physiological basis, according to the substrate from which they obtain their energy. The original descriptions were often clear and detailed, and subsequent workers, most of whom have been more interested in the physiology than in the morphology of these organisms, have seldom attempted to revise the early classification.

The physiological groups have been subdivided into genera and species by such morphological characters as motility and the size and shape of the individual cells or cell aggregates, but their variations of form, under different environmental conditions, have not been taken fully into account. Such classifications, therefore, are needlessly complex, and many of the described genera are almost certainly cultural variants. This has been conclusively demonstrated in the case of the filamentous iron bacteria (Pringsheim, 1949 b), and the purple and brown, non-sulphur bacteria (van Niel, 1944). In addition, the physiological groups are so defined as to include a heterogeneous collection of micro-organisms other than true or typical bacteria, including both genuine autotrophs and heterotrophic or predatory contaminants of the crude cultures in which autotrophs have commonly been studied.

Much of the literature which deals with the morphology of the autotrophic bacteria is old and difficult to elucidate, and some of the interpretations suggested here may prove to be ill-founded. But it is the purpose of this study to show that most autotrophic bacteria are typical bacteria of primitive morphology according to the evolutionary criteria of Bisset (1950 b, 1952), and that the contemporary rather confused situation with regard to their nature and relationships has arisen from unjustifiable over-classification on morphological grounds, and the inclusion, by purely physiological criteria, of algae and even protozoa, with true bacteria. This confusion has, indeed, been responsible to some extent for the relatively slow development of an acceptable systematology for bacteria in general. The literature upon the subject is inexhaustible and we have not attempted to quote, or to elucidate the status of every genus or species which has been described, but rather to discover, where possible, the nature of the foundation on which this enormous structure has been imposed.

II. AUTOTROPHIC NITRIFYING BACTERIA

(1) *Ammonia-oxidizing bacteria*

In 1862 Pasteur suggested that micro-organisms were responsible for the conversion of ammonia to nitrate in the soil, and his theory was proved to be correct by the experiments of Schloesing & Muntz (1877 a, b). Schloesing & Muntz (1879) and Warington (1884) observed that nitrification often took place in two stages: the conversion of ammonia to nitrite, and of nitrite to nitrate.

S. Winogradsky (1890, 1891, 1892), S. & H. Winogradsky (1933) and H. Winogradsky (1937), described organisms of five genera, *Nitrosomonas*, *Nitrosococcus*, *Nitrosospira*, *Nitrosocystis* and *Nitrosogloea*, all of which, they claimed, oxidized ammonia to nitrite. Of these, *Nitrosomonas* alone has been repeatedly isolated and described by later workers, and it appears to be the commonest, if not the only, bacterium capable of carrying out this oxidation. Many workers who have cultured and described it (Boullanger & Massol, 1903; Nelson, 1931; Kingma Boltjes, 1935; Hanks & Weintraub, 1936; Meiklejohn, 1950) agree with S. Winogradsky (1890) that the cells are small, Gram-negative, oval or rod-shaped and are motile by a long, polar flagellum; thus, morphologically, *Nitrosomonas* is a typical small eubacterium (Bisset, 1950a, 1952).

Two cyst-forming genera, *Nitrosocystis* and *Nitrosogloea*, were also described, the former including one species, *Nitrosocystis javanensis* (S. Winogradsky, 1931), which was originally designated as *Nitrosomonas* (S. Winogradsky, 1892). These are of considerable interest, since, with the exception of *Nitrosocystis javanensis*, nearly all the species originally described appear, from the figures of S. & H. Winogradsky (1933), to have been myxobacteria of the genus *Sorangium* (Grace, 1951). The cells are tightly packed together in cysts, which form large aggregates. In some species the cysts appear to be surrounded by a small sheath, but in every case they are composed of numerous, tightly packed spherical or oval cells. This appearance is identical with that of fruiting bodies of *Sorangium* species, and the 'nitrifying' genera, as illustrated by H. Winogradsky (1937, figs. 2–5), also resemble *Sorangium* fruiting bodies in the various appearances observable in preparations stained by classical methods and photographed *in situ*, or stained by cytological methods. In the former the 'cells' (in fact, the component microcysts) appear wedge-shaped and in pairs or tetrads; in the latter they appear oval or spherical and are very tightly packed together. In our experience, not only myxobacteria, but a wide variety of other contaminants, from streptomyces to amoebae and nematode worms, occur in similar cultures on silica gel, which cannot be regarded as in any way very selective for true nitrifying bacteria.

Although they appear beyond question to be myxobacteria, the question arises whether '*Nitrosocystis*' and '*Nitrosogloea*' were simply contaminants or whether species of *Sorangium* exist which can oxidize ammonia to nitrite. S. Winogradsky (personal communication to Hanks & Weintraub, 1936) did not regard his cultures as pure, but only as suitable for his studies on nitrification and his evidence for nitrification by these organisms is quite inconclusive when it is understood that he was unaware of the possible occurrence of bacteria predatory on the true nitrifying bacteria in crude culture. He does not appear to have been acquainted with *Sorangium*, since these distinctive myxobacteria are not mentioned in his collected studies (S. Winogradsky, 1945). The myxobacterial swarm is often difficult to detect, although the coloured fruiting bodies are clearly visible to the naked eye and can readily be mistaken for colonies. The techniques employed by S. Winogradsky for the detection of contaminating bacteria would fail to reveal their existence. S. Winogradsky (1945) himself pointed out that stained pre-parations could only be used for confirmation of the purity of the cultures, and to demonstrate whether contaminating micro-organisms were present, he inoculated a loopful of his culture into nutrient broth and incubated at 30° for 10 days. Most later workers have used this method for determining the purity of their cultures, although some (Gibbs, 1919; Gowda, 1924) have pointed out that it is quite unreliable. Myxobacteria are difficult to cultivate on standard laboratory nutrient media, and their presence might well remain unsuspected.

The view that *Nitrocystis* and *Nitrosogloea* are myxobacteria is sup-ported by Imsenecki (1946), who demonstrated that *Sorangium* species are very common contaminants on plate cultures of nitrifying bacteria, and that they are unable to oxidize ammonia to nitrite. He suggested that there is a symbiotic relationship between the myxobacteria and *Nitrosomonas*, the former retaining moisture for the nitrifying bacteria in the mucous material of the cysts. This seems extremely unlikely, however, as there is very little mucoid material in these fruiting bodies. It is probable, therefore, that in the case of *Nitrosocystis* and *Nitroso-gloea*, S. Winogradsky was dealing with mixed cultures of *Sorangium* and *Nitrosomonas*, the latter being responsible for the nitrifying properties of the culture.

Kingma Boltjes (1935), Hanks & Weintraub (1936) and Meiklejohn (1950), however, regard *Nitrosocystis* as 'hard-colony' variants of *Nitrosomonas* which are found under certain environmental conditions. Kingma Boltjes claimed to have isolated in pure culture these variants of *Nitrosomonas*, but some of his photomicrographs (especially Figs. 15 and 16) closely resemble those of *Nitrosocystis* by H. Winogradsky (1937). Hanks & Weintraub undoubtedly obtained *Nitrosomonas* in pure culture, but the 'hard colonies' which they describe do not resemble the complex cysts of '*Nitrosocystis coccoides*'. In *Nitrosocystis javanensis*, which S. Winogradsky originally assigned to *Nitrosomonas*, swarm cells with polar flagella are liberated from the compact cell masses. This species probably corresponds to the 'hard colonies' of Hanks & Weintraub and other workers, and may be a variant of *Nitrosomonas*.

The two remaining genera of ammonia-oxidizing bacteria listed by S. Winogradsky (1891) and S. & H. Winogradsky (1933) have been described very rarely or never by other workers, and are impossible to identify from the descriptions given. *Nitrosococcus*, originally described in 1892, possesses large non-motile spherical cells, which were stated by Omelianski (1907) to be Gram-positive, and *Nitrosospira* (S. Winogradsky, 1931), was considered to be a spirochaete, but has not been observed subsequently.

(2) *Nitrite-oxidizing bacteria*

In contrast to the *Nitrosomonas* group, the morphology of nitrite-oxidizing bacteria has been only briefly examined. S. Winogradsky (1892) and H. Winogradsky (1937) described two genera, *Nitrobacter* and *Nitrocystis*, of which only the former has been confirmed as a nitrite-oxidizer by later workers.

The cells of *Nitrobacter* were described by S. Winogradsky (1892) as small, short rods which were difficult to stain with basic dyes. A similar description was given by Gibbs (1919), who added that they were capsulated and non-motile. They are Gram-negative (Omelianski, 1907). Another species of *Nitrobacter* was described by Nelson (1931). This differed from S. Winogradsky's species principally in that the cells were actively motile by a long polar flagellum, and thus closely resembled *Nitrosomonas*.

The second genus, *Nitrocystis*, was described by H. Winogradsky (1937). From the description and photomicrographs it closely resembles *Nitrosocystis* and must also be presumed to be a contaminating myxo-bacterium of the genus *Sorangium*.

(3) *The systematic position of the nitrifying bacteria*

Bergey's Manual (1948) now lists these nitrifying genera together with *Thiobacillus* in the Nitrobacteriaceae, on the grounds that they can develop in inorganic media and are strictly or facultatively autotrophic. Only two genera of nitrifying bacteria (*Nitrosomonas* and *Nitrobacter*) have been consistently reported by different workers, and the ability to oxidize ammonia to nitrite and nitrite to nitrate has been confirmed in these cases only. They are almost identical morphologically; both are typical, small Gram-negative bacteria, and with the exception of *Nitrobacter winogradskyi*, which is non-motile, possess a long single polar flagellum. This type of morphology is also characteristic of many pseudomonads from which these nitrifying bacteria differ only in their physiology. And, indeed, the pseudomonads are strikingly versatile in their modes of life, and some are believed to be facultatively chemo-synthetic.

The remaining genera were described, in almost every case, by S. & H. Winogradsky alone, and have been isolated only rarely and dubiously by other workers. *Nitrosocystis*, *Nitrosogloea* and *Nitrocystis* are almost certainly myxobacteria predatory upon the true autotrophs, and as the only apparent alternative is that they are cultural variants of *Nitroso-monas* and *Nitrobacter*, the former genera must be considered invalid.

III. CHEMOSYNTHETIC SULPHUR BACTERIA

(1) *Introduction*

The 'sulphur bacteria' were first defined as a group by S. Winogradsky (1888) to include the chemosynthetic and photosynthetic micro-organisms which oxidized inorganic sulphur compounds and deposited elementary sulphur within the cells. Several other micro-organisms which carried out this reaction were described by later workers, and by definition the group now includes most of the bacteria and also some algae and possibly protozoa that metabolize inorganic sulphur, almost irrespective of their morphology. Most workers who have studied this group (Molisch, 1907; Waksman, 1922; van Niel, 1931, 1944) have realized that it contains a heterogeneous collection of organisms, but many bacteriologists still appear to regard them as a natural group. Many of the sulphur bacteria were described in the latter half of the nineteenth century, and their discovery coincided with the dominance of the theory of pleomorphism in bacteriology. Consequently the various morphological types were often regarded as stages in the life cycle of only one or two bacteria (Ray Lankester, 1873; Warming, 1876; Zopf, 1882); later workers, in contrast, regarded each morphological variation as a distinct species. The pure culture studies of van Niel (1931) on the red and green photosynthetic sulphur bacteria show that a variety of morphological forms may be observed in a single culture, and that the existing classifications of these groups are needlessly complicated.

Sulphur-oxidizing micro-organisms are of four physiological types which do not correspond with any morphological relationship:

(*a*) chemosynthetic micro-organisms which oxidize inorganic sulphur compounds and deposit sulphur as globules within the cell;

(*b*) chemosynthetic bacteria which oxidize inorganic sulphur compounds and deposit sulphur outside the cell;

(*c*) red photosynthetic sulphur bacteria, which can utilize hydrogen sulphide as a hydrogen donor;

(*d*) green photosynthetic sulphur 'bacteria', which can utilize only hydrogen sulphide as a hydrogen donor.

The photosynthetic types are discussed in § IV.

Numerous micro-organisms which obtain their energy by the oxidation of hydrogen sulphide to sulphur have been described. In *Bergey's Manual* (1948) the filamentous forms are included as 'alga-like bacteria' in Beggiatoaceae in the order Chlamydobacteriales, while most of the remainder are placed together in the Achromatiaceae as an appendix to this order. This, classification differs from most of the earlier systems, which included these chemosynthetic micro-organisms, together with the red photosynthetic sulphur bacteria, in the Thiobacteriales (S. Winogradsky, 1888; Buchanan, 1917; Bavendamm, 1924; Ellis, 1932; *Bergey's Manual*, 5th ed., 1939). Both are unsatisfactory, and ignore the heterogeneity of this group, some members of which are non-motile or motile by flagella, while others possess the creeping type of motility characteristic of the Myxophyceae and myxobacteria.

(2) *Beggiatoaceae*

The family Beggiatoaceae as listed in the 6th ed. (1948) of *Bergey's Manual* now includes the four genera, *Beggiatoa*, *Thioploca*, *Thiothrix* and *Thiospirillopsis*, which are classified together because they are 'alga-like unbranching filaments which may contain sulphur globules when growing in the presence of sulphides' (*Bergey's Manual*, 1948). By this definition, three distinct and apparently unrelated morphological forms are included within the same family.

Beggiatoa and *Thioploca* are filamentous, multicellular organisms which resemble one another very closely. In *Thioploca* the filaments are arranged in bundles surrounded by a gelatinous sheath (Lauterborn, 1907; Wislouch, 1912). The filaments of both genera are composed of numerous regular cells, the terminal cells being usually rounded. The arrangement of the cells closely resembles that of the *Oscillatoria* group of the Myxophyceae, but not that of the filamentous bacteria. The electron micrographs of Johnson & Baker (1947) do not reveal any morphological resemblance to bacteria (Bisset, 1951). In *Beggiatoa* and *Thioploca* motility is of the creeping type which requires a solid surface for progression (Engelman, 1879; Correns, 1897; Ullrich, 1926, 1929; Niklitschek, 1934). This type of motility is found in *Oscillatoria* and many other members of the Myxophyceae (Fritsch, 1932, 1945; Pringsheim, 1949b), but it is unknown in the bacteria, with the exception of myxobacteria, with which *Beggiatoa* has no morphological resemblance. It is generally agreed that *Beggiatoa* and *Thioploca* belong morphologically to the Myxophyceae (van Tieghem, 1880; van Niel, 1944; Pringsheim, 1949b), and several species of the former were originally classed as *Oscillatoria*, e.g. *Beggiatoa alba* (*Oscillatoria alba*

Vaucher, 1803, cited in *Bergey's Manual*, 1948), but there has been some disagreement on this point. Guilliermond (1926), for example, failed to find the differentiation of the protoplasm into the inner and outer regions, which is said to be characteristic of many of the Myxophyceae. *Beggiatoa* and *Thioploca*, however, are colourless, and the outer region of protoplasm, described in the Myxophyceae, is chiefly defined by the location of the photosynthetic pigment in this region (Pringsheim, 1949 *b*). The main objection to their inclusion among the algae is their lack of photosynthesizing pigments, but this cannot be regarded as a fundamental distinction, since colourless species are known to occur among other genera of the Myxophyceae and to regard these apochlorotic sulphur-depositing species, such as *Beggiatoa* and *Thioploca*, as bacteria on this basis alone is quite unjustifiable. In addition, it has been observed that *Oscillatoria* may contain sulphur globules when growing in high concentrations of hydrogen sulphide (Hinze, 1903; Nakamura, 1937 *b*).

Thiothrix is usually classified with *Thioploca* and *Beggiatoa*, but whereas these have obvious affinities with the Myxophyceae, *Thiothrix* more closely resembles the true bacterial chlamydobacteria. The non-motile, multicellular filaments are enclosed in a delicate sheath and are attached to solid surfaces by a gelatinous holdfast. They reproduce by detachment of the terminal cells, either singly or in short chains, which settle on the solid support and form fresh filaments. S. Winogradsky (1888) and Wille (1902) claimed that the 'conidia' exhibited a creeping motility, but more recent observations have not confirmed this (Keil, 1912; Pringsheim, 1949 *b*). Pringsheim (1949 *b*) objects to the inclusion of *Thiothrix* in the Chlamydobacteriales on the grounds that the reproductive cells are, apparently, non-motile in contrast to those of the majority of the members of this group, although he recognizes that, unlike *Beggiatoa*, it has no apparent resemblance to the Myxophyceae. However, several genera of chlamydobacteria, e.g. *Polyspora* and *Crenothrix* (Cohn, 1870), have non-motile reproductive cells and there seems little reason for the exclusion of *Thiothrix* from this bacterial group, which contains several autotrophic forms, and is fundamentally much less complex than has been supposed (Bisset, 1952).

A fourth type of filamentous sulphur micro-organism, *Thiospirillopsis*, was described by Uphof (1927). The spirally wound filaments are multicellular and exhibit a creeping motility, with a spiral motion. Morphologically, they resemble *Spirulina* in the Myxophyceae and, to some extent, *Spirochaeta* and *Achroonema* (Brunel, 1949; Skuja, 1948). *Thiospirillopsis* requires a solid surface for progressive motility and,

although it is impossible to classify it exactly from the original description, it obviously resembles the Myxophyceae more closely than the
bacteria.

(3) *Achromatiaceae*

The non-filamentous, colourless, sulphur-depositing organisms differ
considerably from the Beggiatoaceae and are usually included in a
separate family, the Achromatiaceae. This family contains a heterogeneous collection of flagellated and non-flagellated micro-organisms.
Bergey's Manual (1948) lists three genera, *Achromatium*, *Thiovulum* and
Macromonas. The non-pigmented spirilla which deposit sulphur intracellularly were also originally classified in this family, e.g. *Thiospira*
(Buchanan, 1917; Bavendamm, 1924; *Bergey's Manual*, 1939), but they
have since been grouped with the spirilla (*Bergey's Manual*, 1948) to
which they are obviously related morphologically. We have found
Thiospira to be a typical bacterial spirillum with paired nuclear bodies
and polar flagella. Numerous other genera, e.g. *Hillhousia* (West &
Griffiths, 1909, 1913), *Monas* (Warming, 1876), *Thiophysa* (Hinze,
1903), *Thiosphaerella* (Nadson, 1914) and *Microspira* (Gicklehorn, 1920),
have been described, but most if not all of these appear to be identical
with one or other of the three genera now listed.

The cells of *Achromatium* were originally described by Schewiakoff
(1893) as large spherical or ellipsoidal structures containing numerous
sulphur granules. In this organism the stainable 'central body',
characteristic of the Myxophyceae, can be clearly seen and is recorded
also by Delaporte (1939). Virieux (1913) stated that reproduction took
place by fission or by formation of a large number of endospores, but
the most characteristic property of the cells was their jerky gliding movement that required the presence of a surface on which to move. A similar
genus *Hillhousia* was recorded by West & Griffiths (1909, 1913), who
described the cells as 'peritrichously flagellated'. The small flagella were
readily detached from the organism and then exhibited independent
motility for a short time. As Pringsheim (1949*b*) pointed out, West &
Griffiths undoubtedly mistook small bacteria clinging to the outside of
the cells for flagella.

Three further genera, *Thiophysa* and *Thiovulum* (Hinze, 1903, 1913)
and *Thiosphaerella* (Nadson, 1914), have also been described, but appear
to be synonymous with *Achromatium*. Because of their size and gliding
type of motility, which can readily be observed, these organisms cannot
be classified as bacteria. They closely resemble *Synechococcus* in the
Myxophyceae (Pringsheim, 1949*b*), and are probably apochlorotic

chemosynthetic members of that genus as *Beggiatoa* is of *Oscillatoria* (Stanier & Van Niel, 1941).

Monas Warming, which is considered by *Bergey's Manual* (1948) to be synonymous with *Thiovulum*, is nevertheless quite distinct from the other 'sulphur bacteria'. The cells are relatively large (5–20μ in diameter), spherical, oval or pear-shaped and contain an eccentric vesicular nucleus which is stainable by iron haematoxylin. The sulphur granules are confined to one pole. Hinze (1903) described *Monas* as belonging morphologically to the flagellate Protozoa, and our own observations are in accordance with this. It is entirely unlike either the Myxophyceae according to the criteria of Fritsch (1945) or the bacteria (Bisset, 1952).

The third genus *Macromonas* (Utermohl & Koppe, 1925) was created to include actively motile cells which contained calcium carbonate granules as well as sulphur. The cells are slightly curved, ellipsoidal or cylindrical and are motile by a single polar flagellum. *Microspira vacillans* and *Pseudomonas bipunctata* (Gicklehorn, 1920) are probably synonymous, and the morphology of all these genera closely resembles that of the photosynthetic sulphur organism *Chromatium* (see § IV (1)), of which group they are probably apochlorotic chemosynthetic non-pigmented representatives.

(4) *Thiobacillus*

After the description of the 'sulphur bacteria' by S. Winogradsky (1888), several typical chemosynthetic bacteria were described which oxidized inorganic sulphur compounds and deposited sulphur outside the cell. Beijerinck (1904) gave the generic name *Thiobacillus* to these bacteria, and numerous species have been described (Nathanson, 1902; Beijerinck, 1904; Waksman & Joffe, 1922; Emoto, 1928, 1929, cited in *Bergey's Manual*, 1948; Starkey, 1934, 1935; Lipman & McLees, 1940; Colmer, Temple & Hinkle, 1950). The distinction between these species is almost entirely physiological and many are probably synonymous. The best known and well-established species are *Thiobacillus thioparus*, Beijerinck (1904), *Thiobacillus thio-oxidans*, Waksman & Joffe (1922) and *Thiobacillus denitrificans*, Beijerinck (1904).

Morphologically, they are almost identical and, like the nitrifying bacteria and some of the photosynthetic 'non-sulphur' bacteria, closely resemble *Pseudomonas* species. *T. thio-oxidans* was originally stated to be Gram-positive, but later workers (Starkey, 1935; Knaysi, 1943) did not confirm this. Most of these bacteria are motile by a single polar flagellum, although non-motile species have been recorded (*Thiobacillus novellus*, Starkey, 1934). *T. thio-oxidans* was originally reported as non-motile, but Umbreit & Anderson (1942) observed flagellated cells in electron micrographs and Knaysi (1943) also noticed occasional rapidly moving cells. Thus the genus *Thiobacillus* comprises a group of typical small Gram-negative eubacteria which is included in the Nitrobacteriaceae on the basis of their autotrophic metabolism (*Bergey's Manual*, 1948), but which also resembles many of the genera of this family in its pseudomonad-like morphology. The generic name *Thiobacillus* is misleading, as none of the species is Gram-positive or forms endospores; *Thiobacterium* (Lehmann & Neumann, 1927) and *Sulfomonas* (Orla-Jensen, 1909) would appear to be more suitable.

IV. PHOTOSYNTHETIC BACTERIA

(1) *Red photosynthetic sulphur bacteria—Thiorhodaceae*

The family Thiorhodaceae was originally defined by Molisch (1907) to include the conspicuous red photosynthetic micro-organisms which under natural conditions contain sulphur globules. S. Winogradsky (1888) was the first to make a detailed study of these organisms. He examined their morphology in slide cultures and classified them into twelve genera, mainly by the character and mode of development of the cell aggregates. In complete contrast to earlier workers, he regarded each variant as a new genus or species. Some subsequent workers attempted to alter and condense this classification (Molisch, 1907; Bavendamm, 1924; Ellis, 1932), and their lack of success was due largely to the fact that they did not obtain pure cultures. Consequently, they based their classifications on characteristics such as cell size, the number of planes in which the cells divided, and pigmentation, all of which have since been shown to be dependent upon environmental conditions (van Niel, 1931; Manten, 1942).

In most classifications, the Thiorhodaceae are divided into the aggregate-forming and non-aggregate-forming genera. The simple forms, which we will discuss first, include *Thiospirillum* and *Chromatium*. The former, the purple sulphur spirilla, have been observed by many workers, but they have never been obtained in pure culture, and the descriptions of the various species are often incomplete. They appear to be large typical spirilla and, where this character is recorded, possess tufts of flagella at each pole.

The nature and relationships of the second genus, *Chromatium*, are more difficult to determine. The actively motile cells are ovoid or curved and usually possess a tuft of polar flagella. Bütschli (1890) and Dangeard (1909) claimed to have observed a central region in the cells, which

Dangeard considered to resemble both the nucleus of flagellates and the central body of the Myxophyceae. We have found that the nucleus of *Chromatium* is difficult to demonstrate, but appears as a small compact structure, in contrast with the nucleus of the flagellates, which is usually large, vesicular and readily stained with basic dyes. The cells of *Chromatium* are often comma-shaped, and they then closely resemble the plump single cells of some species of *Spirillum*, both in morphology and in movement; the longer cells show a flexibility similar to that of *Spirillum*. Because of its flagella *Chromatium* cannot be classified with the Myxophyceae (Fritsch, 1945). Some workers have considered it to be more closely related to the Protozoa (Cohn, 1875; S. Winogradsky, 1888), but our observations suggest that it is a bacterium, typical if rather large, and probably related to *Spirillum* which it also resembles in its type of flagellation. This is also the opinion of Kluyver & van Niel (1936). Differentiation between species of *Chromatium* has in the past been based almost entirely on the size of the cells, which van Niel (1931) and Manten (1942) have shown quite clearly to be a variable factor. The small motile cells of *Chromatium* are almost indistinguishable from those of *Thiocystis*, which is more fully discussed below.

The genus *Rhabdomonas* was created by Cohn (1875) to include the irregular, swollen cells and filaments which he often observed in his cultures, some of which were motile by means of a tuft of polar flagella. S. Winogradsky (1888), Nadson (1913) and Ellis (1932) considered that it was a morphological variant of *Chromatium*, while Lauterborn (1915), Buder (1919) and Bavendamm (1924) regarded it as a distinct genus. The observations and drawings of van Niel (1931) show that cells resembling *Rhabdomonas* may appear under certain environmental conditions in cultures of *Thiocystis* as well as of *Chromatium*. Our own observations are in complete agreement with this, and it is interesting to note that many of the 'involution' forms closely resemble the swollen cells which appear in many Gram-negative bacteria when these undergo the L-cycle mode of reproduction (Dienes, 1946; Klieneberger-Nobel, 1949).

The formation of cell aggregates is the characteristic feature of the second group of purple sulphur bacteria, and appearance and mode of development of these has been used as a basis for their classification. The aggregates may vary in shape from an irregular tightly packed mass of cells, either capsulated as in *Thiocapsa* or non-capsulated as in *Thiopolycoccus*, to the beautiful net-like structure present in *Thiodictyon* (S. Winogradsky, 1888). An interesting feature of many of these structures is their strong resemblance to certain of the cell aggregates found in the Myxophyceae. Thus the tabular colonies of *Thiopedia* which are composed of numerous tetrads of cells closely resemble those of *Merismopedia*, and *Thiothece*, in which the cells are separated from one another by a relatively large capsule, resembles *Aphanothece*. Unfortunately, very little information is available about the morphology either of these purple sulphur bacteria or of the corresponding Myxophyceae. The drawings by S. Winogradsky (1945) suggest that the individual cells of the purple genera quite closely resemble those of *Thiocystis* (van Niel, 1931) which is a well-established bacterial genus, and the resemblances between these aggregates and those of the algae may be due simply to evolutionary convergence in two groups which have adopted a similar mode of life. The arrangement of the cells in the aggregates is obviously dependent on the number of planes in which the cells divide and the presence or absence of capsules. S. Winogradsky (1888) and some subsequent workers believed that these were constant features and they therefore regarded each variant as a separate genus. However, van Niel (1931) isolated in pure culture several strains of a sulphur bacterium, which he identified as *Thiocystis*, and noted that the appearance of the aggregates varied from day to day. Thus on one occasion they resembled the descriptions of *Thiocystis violacea* and on the following day those of *Amoebobacter roseus*. He concluded that the appearance of the aggregates was without significance, and that most of the genera are synonymous. This is undoubtedly true in the majority of cases, and these aggregate-forming purple sulphur bacteria all appear to be very similar in those of their characters which are constant. They may, however, be divided into three groups according to their reported mode of development and distribution.

Representatives of the first group, including *Thiocystis* and a few similar genera, produce swarm cells with polar flagella, from which new aggregates are reproduced. Some of these swarmers resemble those of *Chromatium*. The second group is very similar, with the exception that the aggregates are reported to reproduce by passive disintegration. The descriptions of genera of this type, e.g. *Thiocapsa* and *Thiopolycoccus*, so closely resemble *Thiocystis* that it is difficult to believe that they are anything other than closely related forms, the motile cells of which have not yet been observed. The entire life cycle of members of the first, and probably of the second group also, is characteristic of the true chlamydobacteria, as described by Bisset (1952), The possession of typically bacterial swarm cells is particularly significant.

The two genera in the third group, *Thiodictyon* and *Amoebobacter*, are unusual and superficially quite distinct. The compact aggregates are indistinguishable from those of the two preceding groups in shape and appearance, but their method of development is reported to differ. In *Thiodictyon* the cell mass opens out into a net-like structure, simpler but reminiscent of that of *Hydrodictyon* in the Chlorophyceae. Occasionally, groups of cells separate themselves from the colony by active movements (S. Winogradsky, 1888). The cycle in *Amoebobacter* is similar, but the aggregates are enclosed in a capsule which bursts and liberates the cells. The resulting colony continually changes shape and moves over the supporting surface. This phenomenon has never been fully described, and it must be assumed that the individual cells are capable of some form of creeping motility. There are certain superficial resemblances between this life cycle and that of the myxobacteria, but the morphology of the *Amoebobacter* is entirely distinct. Some species are described as possessing large, ovoid cells, whereas others have broad rods. If their crawling motility is confirmed, then the former must be regarded as Myxophyceae, although the latter, which are apparently indistinguishable from *Thiodictyon*, may well be identical with this organism, which our own observations suggest is a variant of *Chromatium*. Several aggregates found in pond mud and closely resembling those of *Thiodictyon* were observed in slide culture. In all instances, the individual cells gradually became separated from one another and in two instances actively motile polar-flagellated cells, similar to those of *Chromatium*, were liberated. No creeping movement of the cells or cell aggregates was observed. However, in the present state of our information, the taxonomic position of *Amoebobacter* itself remains obscure.

Thus the Thiorhodaceae includes three types of true bacteria, *Spirillum*, *Chromatium* and *Thiocystis*. The last two groups include most of the aggregate-forming genera which, according to the criteria of Bisset, can be classed as chlamydobacteria.

(2) *Red and brown photosynthetic non-sulphur bacteria: 'Athiorhodaceae'*

The Athiorhodaceae comprises a small group of photosynthetic bacteria originally described and classified by Molisch (1907). These bacteria differ from the Thiorhodaceae in that they can utilize fatty acids as hydrogen-donors in preference to sulphide; most of them require growth factors (van Niel, 1944), and cannot develop in a completely inorganic medium, although many strains can use sulphide or thiosulphate (Gaffron, 1933; Nakamura, 1937a; van Niel, 1944), in which cases sulphur is deposited outside the cell. There must be considerable doubt concerning the validity of any classification which separates some of these from morphologically similar members of the Thiorhodaceae, especially as there exist organisms with morphological and physiological characters intermediate between the two (van Niel, 1931).

Molisch (1907) grouped these bacteria in seven genera on the basis of the shape and motility of the individual cells and on the formation of cell aggregates. This classification was adopted by Orla-Jensen (1909), Buchanan (1917) and *Bergey's Manual* (1939) without any major alteration.

The researches of van Niel (1944) have shown clearly that the Athiorhodaceae are another example of an over-classified group. During his investigations of 150 strains, van Niel observed that they were all Gram-negative and, contrary to the findings of Molisch (1907), motile by polar flagella. He distinguished only two morphological types, small rod-shaped cells placed in the genus *Rhodopseudomonas*, and spiral cells placed in the genus *Rhodospirillum*. The various types described by Molisch were found to be species or cultural variants of these two genera. It is interesting to note that in *Rhodopseudomonas gelatinosa* (*Rhodocystis*, Molisch, 1907) the cells are arranged regularly in a gelatinous sheath and only an occasional cell is motile. These motile cells may be regarded as swarmers and the behaviour of this species is characteristic of the chlamydobacteria.

A third very interesting morphological type, *Rhodomicrobium vannielii*, has been recently described (Duchow & Douglas, 1949; Murray & Douglas, 1950). This is an unusual stalked organism consisting of ovoid cells which remain attached to one another by slender threads. These may branch and give rise to a ramifying mycelial mass. The cells contain a central chromatinic body which divides prior to the budding of the daughter cell. Morphologically, *Rhodomicrobium* closely resembles *Hyphomicrobium* (Stutzer & Hartleb, 1899; Ruhlman, 1897; Kingma Boltjes, 1936), which has been described in cultures of *Nitrobacter*, but the exact taxonomic position of these genera is uncertain. The photomicrographs by Duchow & Douglas (1949) show that this photosynthesizing micro-organism closely resembles the bacteria cytologically but is not a typical caulobacterium.

(3) *Green photosynthetic sulphur bacteria—Chlorobacteriaceae*

The term 'green bacteria' was originally a descriptive one applied to small green micro-organisms which were found in a variety of habitats and which could not be readily assigned to the Myxophyceae. The term is now reserved almost exclusively for the green photosynthetic micro-organisms which utilize hydrogen sulphide as a hydrogen-donor and deposit sulphur outside the cell. They are often found in association with red sulphur bacteria (Szafter, 1910; van Niel, 1931).

Several early observations exist concerning green bacteria, e.g. *Bacterium viride, Bacillus virens* (van Tieghem, 1880) and *Bacterium chlorinum* (Engelmann, 1882), but it is now impossible to identify them from these descriptions. S. Winogradsky (1888) observed green bacteria in his cultures of sulphur micro-organisms, but did not realize that they metabolized hydrogen sulphide.

As is so often the case, descriptions of green micro-organisms have for the most part been based on observations of mixed cultures, and only one species, *Chlorobium limicola*, seems to have been studied in any detail. It was described by Ewart (1897 *a*, *b*) as *Streptococcus varians* and later by Nadson (1912). Four strains of this organism were obtained in pure culture by van Niel (1931). It was found to be Gram-negative, non-motile and to exist in a variety of morphological forms ranging from small coccal cells to spiral, large, vacuolated structures. Lauterborn (1915) and Geitler & Pascher (1925) described several micro-organisms of similar morphology to *Chlorobium* and which formed aggregates of cells. Aggregates have been described ranging from irregular masses in *Sorochloris* (Geitler & Pascher, 1925) to net-like structures in *Pelodictyon* (Lauterborn, 1915) and *Clathrochloris* (Geitler & Pascher, 1925); these may also form a layer around the cells of other micro-organisms such as amoebae and flagellates (Lauterborn, 1915; Buder, 1913). Rather surprisingly, these associations between 'green bacteria' and other micro-organisms have frequently been accorded generic status, although their taxonomic significance is quite unproven.

Most of these green organisms are classified with the bacteria, but their natural relationships are very doubtful. Many workers consider that they are related to the algae, especially the Myxophyceae (Dangeard, 1894; Schmidle, 1901). Van Niel (1931), on the other hand, on the basis of observations dealing with four strains, regarded almost all the green micro-organisms described by Lauterborn & Pascher and other workers as synonymous with *Chlorobium limicola*, which he considered to be a true bacterium.

Certain characteristics of the 'green bacteria' suggest a relationship with the Myxophyceae. The cells are usually non-motile or may possibly possess a creeping motility; for example, in *Pelodictyon* the aggregates open out into net-like structures (cf. *Thiodictyon*, § IV (1)). In many there is a differentiation of the protoplasm into a central region, and an outer region containing the pigment (Lauterborn, 1915). This is characteristic of many Myxophyceae, but also, under certain conditions, of Thiorhodaceae.

The golden brown pigmentation of some of these organisms is distinct from that of many Myxophyceae which are usually blue-green, but it is also known to occur under certain environmental conditions in the latter. On the other hand, some 'green bacteria', e.g. *Tetrachloris*, are distinctly blue-green.

The family Chlorobacteriaceae seems thus to be another physiological group which contains a heterogeneous collection of apparently unrelated micro-organisms. The green pseudomonad-type cells with polar flagella described by Czurda & Maresch (1937) are almost certainly bacteria. Green spirilla with polar flagella have also been claimed (Ewart, 1897*b*; Benecke, 1912). The term Cyanochloridinae (Pringsheim, 1949*b*) is reserved for the small green micro-organisms such as *Tetrachloris*, which have a blue-green rather than a yellow-green pigmentation and may be very small Myxophyceae. Such genera as *Chroostipes* are usually regarded as true Myxophyceae and assigned to genera of the Chroococcaceae. A more recent paper by Pringsheim (1953) is also in accordance with these general views.

V. THE IRON- AND MANGANESE-OXIDIZING BACTERIA

Numerous bacteria which can oxidize ferrous and manganous compounds have been described; most deposit the iron and manganese oxides in the form of a colloidal sheath.

The literature dealing with the biology of these organisms has been recently reviewed by Pringsheim (1949*a, c*). He has shown that, as in the case of most of the autotrophs, there has been a complete lack of co-ordination among the research workers, and every slight morphological variation has been regarded as representing a new genus or species. Thus a complex classification, especially of the filamentous iron bacteria, has been gradually built up. These bacteria, which are propagated by the liberation of polar flagellated swarm cells are included, together with many unrelated micro-organisms, in the Chlamydobacteriales (*Bergey's Manual*, 1948) (see § III (2)). Numerous genera, e.g. *Sphaerotilus*, *Leptothrix* and *Cladothrix*, have been described, but it has been shown that almost all of them are cultural variants of two species, *Sphaerotilus discophorus* and *Sph. natans* (Pringsheim, 1949*a*). Both are typical filamentous chlamydobacteria in contrast to many members classified in the order Chlamydobacteriales (*Bergey's Manual*, 1948) (see § III (2), (3)).

A second interesting genus is *Gallionella*, in which the ferric hydroxide is deposited in the form of a stalk. The cells are comma- or bean-shaped, like those of *Vibrio*, and the stalk is secreted along the inner curve of the bacterium. Because the bacterial cell is slightly spiral, the resulting stalk is twisted, although this is usually attributed to a rotary movement of the cells. Motility has never been observed in this genus. *Gallionella* is usually assigned to the caulobacteria or stalked bacteria. The stalk in *Gallionella* is a mucous secretion, whereas in most other caulobacteria it is part of the cell itself. Some of the environmental variants of *Sphaerotilus* closely resemble *Gallionella*, and it is suggested that the genus is more closely related to the chlamydobacteria than to the true caulobacteria (Bisset, 1952).

Several unicellular iron bacteria, e.g. *Siderocapsa treubii* (Molisch, 1909), have been reported, but they have never been cultivated in the laboratory, and it is impossible to make any comment on their taxonomic position.

VI. THE HYDROGEN-OXIDIZING BACTERIA— *HYDROGENOMONAS* AND *DESULPHOVIBRIO*

Several strains of bacteria which can oxidize hydrogen have been reported (Kaserer, 1906; Niklewski, 1910, 1914), and as in *Thiobacillus* (see § III (4)), differentiation is based almost entirely on physiological properties. *Bergey's Manual* (1948) now includes them in the Nitrobacteriaceae together with *Thiobacillus* and the nitrifying bacteria. Many of the descriptions of the species of *Hydrogenomonas* are incomplete, but the majority of these organisms are small Gram-negative rods, which are either non-motile, or motile by a single polar flagellum. This morphology, which is typical of *Pseudomonas* (Bisset, 1952) is also characteristic of most species of *Thiobacillus*, *Nitrosomonas* and *Nitrobacter*.

It has been shown that pure cultures of sulphate-reducing bacteria can be grown in a completely mineral medium with hydrogen as the sole oxidizable substance, and they must, therefore, be included here (Butlin & Adams, 1947; Butlin, Adams & Thomas, 1949). Most of these bacteria are small slightly curved Gram-negative rods with a single polar flagellum. Typical spiral cells are only occasionally observed. They have been designated as *Spirillum* (Beijerinck, 1895), *Bacillus* (Saltet, 1900), *Microspira* (Migula, 1900; van Delden, 1903) and *Vibrio* (Holland, 1920). Kluyver & van Niel (1936) included these sulphate-reducing bacteria in a separate genus *Desulphovibrio*, and this has been adopted by *Bergey's Manual* (1948). These bacteria are typical small vibrios, and the allocation of a separate genus to them is questionable, as many of the species described appear to be only variants of the type species *D. desulphuricans*. An interesting morphological feature is that many strains form heat-resistant endospores when they are cultivated at 55°. At this temperature the cells are less spiral and show a type of peritrichous flagellation. Starkey (1938) placed these strains in a separate genus, *Sporovibrio*.

VII. THE CARBON MONOXIDE-OXIDIZING BACTERIA

A Gram-positive bacterium which oxidizes carbon monoxide has been reported; its discovery is attributed to Beijerinck & van Delden (1903), who isolated a small non-motile rod, which they named *Bacillus oligocarbophilus*, and which grew in an inorganic medium with carbon dioxide as the sole source of carbon. They were unable, however, to discover the source from which it obtained its energy. Later Kaserer (1906) isolated a bacterium which he regarded as morphologically identical with Beijerinck's organism despite the paucity of the original description. He claimed that it oxidized carbon monoxide as a source of energy although he gave no experimental proof. A more complete morphological description was given by Lantzsch (1922), who also believed that carbon monoxide could be utilized in this manner. He showed that the bacterium, which was acid-fast, existed in two morphological forms, as long filaments and as small, almost coccoid cells, and he renamed it '*Actinomyces oligocarbophilus*'. His description, however, closely resembles *Nocardia*, which is Gram-positive, slightly acid-fast and has a transitory filamentous phase (Morris, 1951), and which is often found to exist saprophytically upon very refractory food sources. If, as we consider rather unlikely, further experimental work shows that such bacteria can obtain their energy autotrophically by the oxidation of carbon monoxide, they will provide a unique example of autotrophy in the higher bacteria.

VIII. DISCUSSION

It appears, beyond any reasonable doubt, that the 'autotrophic bacteria', as at present listed in *Bergey's Manual*, 6th ed. (1948), include a heterogeneous collection of bacteria, algae (especially Myxophyceae) and possibly protozoa. Even among the bacteria, examples can be found of what are probably heterotrophic or micropredatory contaminants, credited with autotrophic powers, and granted generic rank on this supposition. The number of genera of autotrophs which must be regarded as invalid, because they are, in fact, either redescriptions of known organisms in other biological groups, or non-autotrophic bacteria, is only exceeded by those whose validity is questionable upon the ground that they are minor, environmental variants of single species.

It cannot be too strongly emphasized that the overwhelming proportion of the morphologically peculiar and unusual 'genera' listed as autotrophic bacteria appear to have no real existence. They are minor variants, or are not autotrophic, or are not bacteria.

Of those genera which are bacterial in phylogeny and also of autotrophic physiology, nearly all are simple and typically bacterial in morphology, and belong to one of the groups which, according to the recent classification and evolutionary scheme of Bisset (1950 *b*, 1952) are relatively primitive and adapted to an aquatic environment, i.e. spirilla and pseudomonads; on the other hand, the more complex forms are almost all assignable to the chlamydobacteria, which, according to the same classification, are only slightly less primitive in character, and are also adapted to an aquatic existence, although, in this case, sessile, except for the flagellated swarm cells. The basic morphology of chlamydobacteria is not significantly different from that of eubacteria, and there is no justification for placing them in a separate order (Bisset, 1950 *a*). Similarly, the simple morphology of the autotrophic spirilla and pseudomonads makes apparent the falsity of the complex classifications which separate them from the heterotrophic forms, to which they are so obviously related, and groups them (in company with algae and possibly protozoa) in artificial orders, defined almost exclusively by physiological criteria. Such classifications are not acceptable in any other science, and should not be employed in bacteriology alone.

The unsatisfactory nature of the present classification of autotrophic bacteria has not remained entirely unnoticed in the past. Molisch (1907), Pribram (1929) and Buchanan (1917) realized that the physiological groups are morphologically heterogeneous, while the more recent classifications of Kluyver & van Niel (1936) and Pringsheim (1949 *a*) deal adequately with the allocation of bacterial forms to appropriate families among the Eubacteriales, and rationalize the artificially complex definition of genera in the iron bacteria.

Thus, in summary, it may be said that among the nitrifying autotrophs, *Nitrosomonas* and *Nitrobacter* are pseudomonads, and that the morphologically complex genera of nitrifiers are probably misdescriptions of micropredatory contaminants. The chemosynthetic *Thiobacillus* and *Hydrogenomonas* are also pseudomonads. The iron-oxidizing bacteria are simply chlamydobacteria, with a mucous sheath (*Sphaerotilus*) or a stalk of mucus impregnated with iron (*Gallionella*). The photosynthetic sulphur and non-sulphur bacteria are pseudomonads or spirilla, with some chlamydobacterial forms, consisting of sessile aggregates with flagellated swarm cells.

A few autotrophic bacteria, such as *Macromonas* and *Chromatium*, are intermediate in morphology between spirilla and pseudomonads.

Thus autotrophy appears to be quite commonly found among representatives of a variety of groups of simple and mainly aquatic bacteria. Although autotrophic properties have also been attributed to Gram-positive bacteria of a more advanced type, the evidence upon which this is based is often extremely unconvincing.

Many autotrophic bacteria are morphologically indistinguishable from related heterotrophic genera, although certain sulphur bacteria can be distinguished very readily by their size, and by the presence of intracellular sulphur globules and photosynthetic pigments. When the sulphur is deposited intracellularly the cells are often larger than those of similar heterotrophic bacteria, and there is some evidence that sulphur deposition may depend upon cell size (van Niel, 1931). Although the photosynthetic pigments differ slightly from those of green plants (Metzner, 1922), and although bacterial photosynthesis is an unusual and almost unique process in that sulphides or organic acids replace water as hydrogen-donor, these characters are also possessed by certain algae, and are not, therefore, so peculiar as to permit the establishment of major taxonomic groups on this basis alone. The minor biochemical and physiological characters which have also been used for purposes of classification are even less admissible, and the inclusion of obvious algae in bacterial families, purely upon physiological grounds, is indefensible. There remain several groups of apparently autotrophic micro-organisms the affiliations of which cannot be easily determined. Some of these, e.g. the green sulphur bacteria, are almost certainly not of a single origin but contain both algae and typical bacteria; others, such as *Amoebobacter*, are exceedingly difficult to classify, in the present state of our information. But there is little doubt that further research will elucidate the nature and relationships of these also.

IX. SUMMARY

The majority of autotrophic bacteria are simple, probably primitive eubacteria, either spirilla, pseudomonads or similar forms with mucoid sheaths which may be regarded as chlamydobacteria.

The complex and morphologically peculiar forms which have been described as autotrophic bacteria are for the most part members of the Myxophyceae or are possibly protozoa, or are erroneous descriptions of heterotrophic and micropredatory contaminants in mixed cultures.

REFERENCES

BAVENDAMM, W. (1924). Die farblosen und roten Schwefelbakterien des Süss und Salz-wassers. *Pfanzenforschung*, **2**.

BEIJERINCK, M. W. (1895). Ueber Spirillum desulfuricans als Ursache von Sulfatreduktion. *Zbl. Bakt.* (2. Abt.), **1**. 1.

BEIJERINCK, M. W. (1904). Über die Bakterien welche sich im Dunkeln mit Kohlensäure als Kohlenstoffquelle ernähren können. *Zbl. Bakt.* (2. Abt.), **11**, 593.

BEIJERINCK, M. W. & VAN DELDEN, A. (1903). Über eine farblose Bakterie deren Kohlenstoffnahrung aus der atmosphärischen Luft herrührt. *Zbl. Bakt.* (2. Abt.), **10**, 33.

BENECKE, W. (1912). *Bau und Leben der Bakterien.* Leipzig und Berlin: Teubner.

Bergey's Manual of Determinative Bacteriology, 5th ed. (1939). London: Baillière, Tindall and Cox.

Bergey's Manual of Determinative Bacteriology, 6th ed. (1948). London: Baillière, Tindall and Cox.

BISSET, K. A. (1950*a*). *The Cytology and Life-History of Bacteria.* Edinburgh: Livingstone.

BISSET, K. A. (1950*b*). Evolution in bacteria and the significance of the bacterial spore. *Nature, Lond.* **166**, 431.

BISSET, K. A. (1951). Morphology and cytology of bacteria. *Ann. Rev. Microbiol.* **5**, 1.

BISSET, K. A. (1952). *Bacteria.* Edinburgh: Livingstone.

BOULLANGER, E. & MASSOL, L. (1903). Études sur les microbes nitrificateurs. I. *Ann Inst. Pasteur*, **17**, 492.

BRUNEL, J. (1949). *Achroonema spiroideum* Skuja of the Trichobacteriales discovered simultaneously in Sweden and Canada. *Contr. Inst. bot. Univ. Montréal*, **64**, 21.

BUCHANAN, R. E. (1917). Studies in the nomenclature and classification of the bacteria. 2. The primary subdivisions of the Schizomycetes. *J. Bact.* **2**, 155.

BUDER, J. (1913). Chloronium mirabile. *Ber. dtsch. bot. Ges.* **31**, 80.

BUDER, J. (1919). Zur Biologie des Bacteriopurpurins und der Purpurbakterien. *Jb. wiss. Bot.* **58**, 526.

BUTLIN, K. R. & ADAMS, M. E. (1947). Autotrophic growth of sulphate-reducing bacteria. *Nature, Lond.* **160**, 154.

BUTLIN, K. R., ADAMS, M. E. & THOMAS, M. (1949). The isolation and cultivation of sulphate-reducing bacteria. *J. gen. Microbiol.* **3**, 46.

BÜTSCHLI, O. (1890). *Sur la Structure des Bactéries et Organismes Voisins.* Leipzig: Engelmann.

COHN, F. (1870). Über den Brunnenfaden (*Crenothrix polyspora*) mit Bermerkungen über die mikroskopische Analyse des Brunnenwassers. *Beitr. Biol. Pfl.* **1**, 108.

COHN, F. (1875). Untersuchungen über Bakterien. II. *Beitr. Biol. Pfl.* **1**, 3, 141.

COLMER, A. R., TEMPLE, K. L. & HINKLE, M. E. (1950). An iron-oxidising bacterium from the acid-drainage of bituminous coal mines. *J. Bact.* **59**, 317.

CORRENS, C. (1897). Über die Membran und die Bewegung der Oscillarien. *Ber. dtsch. bot. Ges.* **15**, 139.

CZURDA, V. & MARESCH, E. (1937). Beitrag zur Kenntnis der Athiorhodobakterien-Gesellschaften. *Arch. Mikrobiol.* **8**, 99.

DANGEARD, P. A. (1894). Observations sur le groupe des bactéries vertes. *Botaniste*, **4**, 1.

DANGEARD, P. A. (1909). Note sur la structure d'une bactériacée, le *Chromatium okenii*. *Bull. Soc. bot. Fr.* **56**, 291.

50   K. A. BISSET AND JOYCE B. GRACE

DELAPORTE, B. (1939). Recherches cytologiques sur les bactéries et les Cyanophycées. *Rev. gén. Bot.* **51**, 615.

VAN DELDEN (1903). Beitrag zur Kenntnis der Sulfatereduktion durch Bakterien. *Zbl. Bakt.* (2. Abt.), **11**, 81.

DIENES, L. (1946). Complex reproductive processes in bacteria. *Cold Spr. Harb. Symp. quant. Biol.* **11**, 51.

DUCHOW, E. & DOUGLAS, H. C. (1949). *Rhodomicrobium vannielii*, a new photoheterotrophic bacterium. *J. Bact.* **58**, 409.

ELLIS, D. (1932). *Sulphur Bacteria*. London: Longmans Green.

ENGELMANN, T. W. (1879). Über die Bewegungen der Oscillarien und Diatomeen. *Bot. Ztg*, **37**, 49.

ENGELMANN, T. W. (1882). Zur Biologie der Schizomyceten. *Bot. Ztg*, **40**, 320.

EWART, A. J. (1897*a*). On the evolution of oxygen from coloured bacteria. *J. linn. Soc. Lond.* **33**, 123.

EWART, A. J. (1897*b*). Bacteria with assimilatory pigments in the tropics. *Ann. Bot., Lond.* **11**, 486.

FRITSCH, F. E. (1932). *A Treatise on British Freshwater Algae*. Cambridge University Press.

FRITSCH, F. E. (1945). *The Structure and Reproduction of the Algae*, 1 and 2. Cambridge University Press.

GAFFRON, H. (1933). Über den Stoffwechsel der schwefelfreien Purpurbakterien. *Biochem. Ztg*, **260**, 1.

GEITLER, L. & PASCHER, A. (1925). Cyanochloridinae = Chlorobacteriaceae. In Pascher's *Süsswasser-Flora Deutschlands, Osterreichs und der Schweiz*, Heft **12**, 451.

GIBBS, W. M. (1919). Isolation and study of nitrifying bacteria. *Soil Sci.* **8**, 427.

GICKLEHORN, J. (1920). Über neue farblose Schwefelbakterien. *Zbl. Bakt.* (2. Abt.), **50**, 415.

GOWDA, R. N. (1924). Nitrification and the nitrifying organisms. *J. Bact.* **9**, 251.

GRACE, J. B. (1951). Myxobacteria mistaken for nitrifying bacteria. *Nature, Lond.* **168**, 117.

GUILLIERMOND, A. (1926). Sur la structure des *Beggiatoa* et leurs relations avec les Cyanophycées. *C.R. Soc. Biol., Paris*, **194**, 579.

HANKS, R. H. & WEINTRAUB, R. L. (1936). Pure culture isolation of ammonia-oxidising bacteria. *J. Bact.* **32**, 653.

HINZE, G. (1903). *Thiophysa volutans*, ein neues Schwefelbakterium. *Ber. dtsch. bot. Ges.* **21**, 309.

HINZE, G. (1913). Beiträge zur Kenntnis der farblosen Schwefelbakterien. *Ber. dtsch. bot. Ges.* **31**, 189.

HOLLAND, D. F. (1920). The families and genera of the bacteria. V. Generic index of the commoner forms of bacteria. *J. Bact.* **5**, 225.

IMSENECKI, A. (1946). Symbiosis between myxobacteria and nitrifying bacteria. *Nature, Lond.* **157**, 877.

JOHNSON, F. H. & BAKER, R. F. (1947). Electron and light microscopy of *Beggiatoa*. *J. cell. comp. Physiol.* **30**, 141.

KASERER, H. (1906). Die Oxydation des Wasserstoffes durch Mikro-organismen. *Zbl. Bakt.* (2. Abt.), **16**, 681.

KEIL, F. (1912). Beiträge zur Physiologie der farblosen Schwefelbakterien. *Beitr. Biol. Pfl.* **11**, 335.

KINGMA BOLTJES, T. Y. (1935). Untersuchungen über die nitrifizierenden Bakterien. *Arch. Mikrobiol.* **6**, 79.

KINGMA BOLTJES, T. Y. (1936). Über *Hyphomicrobium vulgare* Stutzer & Hartleb. *Arch. Mikrobiol.* **7**, 188.

KLIENEBERGER-NOBEL, E. (1949). Origin, development and significance of L-forms in bacterial cultures. *J. gen. Microbiol.* **3**, 434.

KLUYVER, A. J. & VAN NIEL, C. B. (1936). Prospects for a natural system of classification of Bacteria. *Zbl. Bakt.* (2. Abt.), **94**, 369.

KNAYSI, G. (1943). A cytological and microchemical study of *Thiobacillus thiooxidans*. *J. Bact.* **46**, 451.

LANTZSCH, K. (1922). *Actinomyces oligocarbophilus* (*Bacillus oligocarbophilus* Beij.). Seine Formwechsel und seine Physiologie. *Zbl. Bakt.* (2. Abt.), **57**, 309.

LAUTERBORN, R. (1907). Eine neue Gattung der Schwefelbakterien (*Thioploca Schmidlei* nov.gen., nov.spec.). *Ber. dtsch. bot. Ges.* **25**, 238.

LAUTERBORN, R. (1915). Die sapropelische Lebewelt. Ein Beitrag zur Biologie des Faulschlammes natürlicher Gewasser. *Verh. naturh.-med. Ver. Heidelb.*, N.F., **13**, 395.

LEHMANN, K. B. & NEUMANN, R. O. (1927). *Bakteriologische Diagnostik*, 7. Aufl. Bd. 2. München: Lehmanns Verlag.

LIPMAN, C. B. & MCLEES, E. (1940). A new species of sulphur oxidising bacteria from a coprolite. *Soil Sci.* **50**, 429.

MANTEN, A. (1942). The isolation of *Chromatium okenii* and its behaviour in different media. *Leeuwenhoek ned. Tijdschr.* **8**, 164.

MEIKLEJOHN, J. (1950). The isolation of *Nitrosomonas europoea* in pure culture. *J. gen. Microbiol.* **4**, 185.

METZNER, W. (1922). Über die Farbstoffe der grünen Bakterien. *Ber. dtsch. bot. Ges.* **40**, 125.

MIGULA, W. (1900). *System der Bakterien*, **1** and **2**. Jena: Fischer.

MOLISCH, H. (1907). *Die Purpurbakterien nach neuen Untersuchungen*. Jena: Fischer.

MOLISCH, H. (1909). *Die Eisenbakterien*. Jena: Fischer.

MORRIS, E. O. (1951). Observations on the life-cycle of the *Nocardia*. *J. Hyg.*, Camb. **49**, 175.

MURRAY, R. G. E. & DOUGLAS, H. C. (1950). The reproductive mechanism of *Rhodomicrobium vannielii* and the accompanying nuclear changes. *J. Bact.* **59**, 157.

NADSON, G. A. (1912). Mikrobiologische Studien. I. *Chlorobium limicola* Nads., ein grüner Mikroorganismus mit inaktivem Chlorophyll. *Bull. Jard. bot. St-Pétersb.* **12**, 83.

NADSON, G. A. (1913). Über Schwefelorganismen des Hapsaler Meerbusens. *Bull. Jard. bot. St-Pétersb.* **13**, 106.

NADSON, G. A. (1914). Über die Schwefelbakterien: *Thiophysa* und *Thiosphaerella*. *J. Microbiol.*, St-Pétersb. **1**, 72.

NAKAMURA, H. (1937a). Über die Photosynthese bei der schwefelfreien Purpurbakterie *Rhodobazillus palustris*. Beiträge zur Stoffwechselphysiologie der Purpurbakterien. I. *Acta phytochim.*, Tokyo, **9**, 189.

NAKAMURA, H. (1937b). Über das Auftreten des Schwefelkugelchens im Zellinern von einigen niederen Algen. *Bot. Mag.*, Tokyo, **51**, 529.

NATHANSON, A. (1902). Über eine neue Gruppe von Schwefelbakterien und ihren Stoffwechsel. *Mitt. zool. Sta. Neapel*, **15**, 665.

NELSON, D. H. (1931). Isolation and chacterization of *Nitrosomonas* and *Nitrobacter*. *Zbl. Bakt.* (2. Abt.), **83**, 280.

VAN NIEL, C. B. (1931). On the morphology and physiology of the purple and green sulphur bacteria. *Arch. Mikrobiol.* **3**, 1.

VAN NIEL, C. B. (1944). Culture, general physiology, morphology and classification of the non-sulphur, purple and brown bacteria. *Bact. Rev.* **8**, 1.

NIKLEWSKI, B. (1910). Über die Wasserstoffoxydation durch Mikroorganismen. *Jb. wiss. Bot.* **48**, 113.

NIKLEWSKI, B. (1914). Über die Wasserstoffaktivierung durch Bakterien unter besonderer Berücksichtigung der neuen Gattung *Hydrogenomonas agilis. Zbl. Bkt.* (2. Abt.), **40**, 430.

NIKLITSCHEK, A. (1934). Das problem der Oscillatorien-Bewegung. I. Die Bewegungserscheinungen der Oscillatorien. *Beih. Bot. Zbl.* (2. Abt.), **52**, 205.

OMELIANSKI, W. (1907). Kleinere Mitteilungen über Nitrifikationsmikroben. *Zbl. Bakt.* (2. Abt.), **19**, 263.

ORLA-JENSEN (1909). Die Hauplinien des natürlichen Bakteriensystems. *Zbl. Bakt.* (2. Abt.), **22**, 305.

PRIBRAM, E. (1929). A contribution to the classification of microorganisms. *J. Bact.* **18**, 361.

PRINGSHEIM, E. G. (1949a). The filamentous iron bacteria *Sphaerotilus, Leptothrix, Cladothrix* and their relation to iron and manganese. *Phil. Trans.* B, **233**, 453.

PRINGSHEIM, E. G. (1949b). The relationship between bacteria and Myxophyceae. *Bact. Rev.* **13**, 47.

PRINGSHEIM, E. G. (1949c). Iron bacteria. *Biol. Rev.* **24**, 200.

PRINGSHEIM, E. G. (1953). Taxonomy of green bacteria. *Nature, Lond.* **172**, 167.

RAY LANKESTER, E. (1873). On a peach coloured bacterium, *Bacterium rubescens. Quart. J. micro. Sci.* **13**, 408.

RUHLMAN, W. (1897). Ueber ein Nitrosobakterium mit neuen Wuchsformen. *Zbl. Bakt.* **3**, 228.

SALTET, R. H. (1900). Ueber Reduktion von Sulfaten in Brackwasser durch Bakterien. *Zbl. Bakt.* (2. Abt.), **6**, 648.

SCHEWIAKOFF, W. (1893). Über einen neuen Bakterienähnlichen Organismus des Süsswassers. *Verh. naturh.-med. Ver. Heidelb.* **5**.

SCHLOESING, T. & MUNTZ, A. (1877a). Sur la nitrification par les ferments organisés. *C.R. Acad. Sci., Paris*, **84**, 301.

SCHLOESING, T. & MUNTZ, A. (1877b). Sur la nitrification par les ferments organisés. *C.R. Acad. Sci., Paris*, **85**, 1018.

SCHLOESING, T. & MUNTZ, A. (1879). Recherches sur la nitrification. *C.R. Acad. Sci., Paris*, **89**, 891.

SCHMIDLE, W. (1901). Über drei Algengenera. *Ber. dtsch. bot. Ges.* **19**, 10.

SKUJA, H. (1948). Taxonomie des Phytoplanktons eininger Seen in Uppland. Uppsala, Schweden. *Symb. bot. Upsaliens.* **9**, 3.

STANIER, R. Y. & VAN NIEL, C. B. (1941). The main outlines of bacterial classification. *J. Bact.* **42**, 437.

STARKEY, R. L. (1934). Cultivation of organisms concerned in the oxidation of thiosulphate. *J. Bact.* **28**, 365.

STARKEY, R. L. (1935). Isolation of some bacteria which oxidise thiosulphate. *Soil Sci.* **39**, 207.

STARKEY, R. L. (1938). A study of spore formation and other morphological characteristics of *Vibrio desulfuricans. Arch. Mikrobiol.* **9**, 268.

STUTZER, A. & HARTLEB, R. (1899). Untersuchungen über die bei der Bildung von Saltpeter beobachteten Mikroorganismen. *Zbl. Bakt.* (2. Abt.), **4**, 678.

SZAFTER, W. (1910). Zur kenntnis der Schwefelflora in der Umgebung von Lemberg. *Bull. Int. Acad. Cracovie (Classe des Sci. Math. Nat. Sci.),* Ser. B, **3**, 161.

VAN TIEGHEM, P. (1880). Observations sur des bacteriacées vertes, sur des Phycochromacées blanches et sur les affinités de ces deux familles. *Bull. Soc. bot. Fr.* **27**, 174.

ULLRICH, H. (1926). Über die Bewegungen von *Beggiatoa mirabilis* und *Oscillatoria jenensis.* I. *Planta*, **2**, 295.

ULLRICH, H. (1929). Über die Bewegungen der Beggiatoaceen und Oscillatoriaceen. II. *Planta*, **9**, 144.

UMBREIT, W. W. (1947). Problems of autotrophy. *Bact. Rev.* **11**, 157.

UMBREIT, W. W. & ANDERSON, T. F. (1942). A study of *Thiobacillus thio-oxidans* with the electron microscope. *J. Bact.* **44**, 317.

UPHOF, J. C. TH. (1927). Zur Oekologie der Schwefelbakterien in den Schwefelquellen Mittelfloridas. *Arch. Hydrobiol.* **18**, 71.

UTERMOHL, H. & KOPPE, V. (1925). Limnologische Phytoplanktonstudien: Die Besiedelung ostholsteinischer Seen mit Schwebpflanzen. *Arch. Hydrobiol.* Suppl. vol. **5**, 233.

VIRIEUX, J. (1913). Recherches sur l'*Achromatium oxaliferum. Ann. Sci. nat. Bot.*, Sér. 9, **17**, 264.

WAKSMAN, S. A. (1922). Micro-organisms concerned with the oxidation of sulphur in the soil. I. Introductory. *J. Bact.* **7**, 231.

WAKSMAN, S. A. & JOFFE, J. S. (1922). Micro-organisms concerned with the oxidation of sulphur in the soil. II. *Thiobacillus thio-oxidans*, a new sulphur oxidising organism isolated from the soil. *J. Bact.* **7**, 239.

WARMING, E. (1876). Om nogle ved Danmarks Kyster levende Bakterier. *Vidensk. Medd. naturh. Foren. Kbh.* **20** (French Abstract), 1.

WARINGTON, R. (1884). Nitrification. III. *J. chem. Soc.* **45**, 637.

WEST, G. S. & GRIFFITHS, B. M. (1909). *Hillhousia mirabilis*, a giant sulphur bacterium. *Proc. roy. Soc.* B, **81**, 398.

WEST, G. S. & GRIFFITHS, B. M. (1913). The lime-sulphur bacteria of the genus *Hillhousia. Ann. Bot., Lond.*, **27**, 83.

WILLE, N. (1902). Über Gasvacuolen bei einer Bakterie. *Biol. Zbl.* **22**, 257.

WINOGRADSKY, H. (1937). Contribution à l'étude de la microflora nitrificatrice des boues activées. *Ann. Inst. Pasteur*, **58**, 325.

WINOGRADSKY, S. (1888). *Beitrage zur Morphologie und Physiologie der Bakterien.* I. *Schwefelbakterien.* Leipzig: Felix.

WINOGRADSKY, S. (1890). Recherches sur les organismes de la nitrification. 1, 2, 3. *Ann. Inst. Pasteur*, **4**, 213.

WINOGRADSKY, S. (1891). Recherches sur les organismes de la nitrification. 4, 5. *Ann. Inst. Pasteur*, **5**, 92.

WINOGRADSKY, S. (1892). Contribution à la morphologie des organismes de la nitrification. *Arch. Biol. St-Pétersb.* **1**, 127.

WINOGRADSKY, S. (1931). Nouvelles recherches sur les microbes de la nitrification. *C.R. Acad. Sci., Paris*, **192**, 1004.

WINOGRADSKY, S. (1945). *Microbiologie du Sol.* Paris: Masson.

WINOGRADSKY, S. & WINOGRADSKY, H. (1933). Études sur la microbiologie du sol. VII. La nitrification. *Ann. Inst. Pasteur*, **50**, 350.

WISLOUCH, S. M. (1912). *Thioploca ingrica* nov.spec. *Ber. dtsch. bot. Ges.* **30**, 470.

ZOPF, W. (1882). *Zur Morphologie der Spaltpflanzen.* Leipzig: Engelmann.

SOME ASPECTS OF THE PHYSIOLOGY
OF THIOBACILLI

KJELL BAALSRUD

Biological Laboratory, University of Oslo

INTRODUCTION

The concept of chemo-autotrophy was formulated by Winogradsky as early as 1887 on the basis of his studies of the filamentous sulphur bacteria (Winogradsky, 1887). In this century more attention has been paid to another group of organisms oxidizing sulphur, namely, the tiny, rod-shaped, colourless thiobacilli. Since Nathansohn (1902) reported on the first pure culture of a thiobacillus, these bacteria have been repeatedly used in studies of problems of autotrophy. Their common occurrence, their willingness to grow in pure cultures, and their physiological characteristics have attracted workers in the field of pure as well as of applied microbiology, and though much has been added to our knowledge of the thiobacilli during the last decades, we still consider them an important subject in comparative biochemical studies.

The present article will be restricted to the 'true thiobacilli', a term which we apply to the obligate autotrophic members of the genus. Although a number of such species has been reported (*Bergey's Manual*, 1948), they may all be regarded as different strains of three main species, *Thiobacillus thioparus*, *Thiobacillus thio-oxidans*, and *Thiobacillus denitrificans*.

These organisms have the following morphological and physiological characteristics in common. The cells are small, rod-shaped, Gram-negative, non-sporeforming, and generally very motile. They are able to live and grow in the dark with carbon dioxide as the sole carbon source, obtaining their energy by the oxidation of reduced sulphur compounds. Normally the substrate is oxidized completely to sulphate. No organic carbon compound has been found to replace carbon dioxide as carbon source.

T. thioparus and *T. thio-oxidans* are generally distinguished by their different tolerance to acidity, the former preferring a slightly alkaline or neutral environment, the latter growing best in the region pH 1–5. These two bacteria should, however, be regarded as two extremes among a group of similar organisms rather than as two distinct species. From a series of elective cultures (Baalsrud & Baalsrud, 1952) six different aerobic thiobacilli were isolated; if incubated in a neutral thiosulphate medium, these were inactivated when, due to acid production, the pH of the medium had fallen to 5·2, 5·0, 5·0, 4·6, 3·6 and 1·5, respectively. The list may be extended by a well-known *T. thio-oxidans* strain (Waksman & Joffe, 1922) which is still active at a pH below zero.

Whereas *T. thioparus* and *T. thio-oxidans* are dependent on the presence of molecular oxygen for the oxidation of the inorganic substrate, *T. denitrificans* can live anaerobically, using nitrate as the ultimate oxidizing agent. It is not an anaerobe in the strict sense of the word; like other nitrate reducers it can be grown aerobically, but the enzyme system reducing nitrate is thereby inactivated (Baalsrud & Baalsrud, 1954).

NUTRIENT REQUIREMENTS

Carbon

As would be expected from the characterization of the thiobacilli as obligate chemo-autotrophic organisms, all their cell constituents are synthesized from carbon dioxide as the sole carbon source; the carbon dioxide requirement is absolute. All attempts to adapt these bacteria to a heterotrophic mode of life have been negative, and at present there exists no conclusive evidence that carbon dioxide can be replaced by organic compounds.

Cases have been reported in which the presence of organic compounds appeared to have positive effects on growth in an inorganic medium. Thus Vishniac (1949) observed a considerable increase in the yield of cell nitrogen of *T. thioparus* when succinate was added to the inorganic medium; the increase was, however, independent of the succinate concentration. Similarly, Starkey (1925), working with *T. thio-oxidans*, discovered that in a sulphur medium to which some glucose was added growth was accompanied by the disappearance of glucose. In this case the general correlation between the glucose that disappeared and the sulphuric acid produced from sulphur indicated that glucose entered the cell's metabolism. Yet, in neither of these cases can it be concluded that the organic compound present served as a carbon source.

Nitrogen

The general experience is that the presence of ammonium salts is necessary for the development of the thiobacilli in a synthetic medium. Two exceptions to this observation have been reported. The first pure culture of a thiobacillus, *T. thioparus*, was obtained in a mineral medium provided with nitrate as the only nitrogenous compound (Nathansohn, 1902). Likewise, Lieske's medium for *T. denitrificans* contained only nitrate (Lieske, 1912). The ability of the aerobic species to develop with nitrate as the sole nitrogen source has never been confirmed, and in recent experiments with *T. denitrificans*, none of four denitrifying isolates would grow in the absence of ammonium salts (Baalsrud & Baalsrud, 1954).

The lack of a reduced nitrogen source may also have been the reason why several investigators did not succeed in growing *T. denitrificans* in pure cultures (Beijerinck, 1920; Gehring, 1915; Sijderius, 1946). The existence of strains that can use nitrate for growth should of course not be denied, but it appears more probable that the positive results of Nathansohn and Lieske were due to the presence of small amounts of ammonium salts, added as impurity with the other ingredients of the medium.

Inorganic requirements

The exact mineral requirements have not been worked out for any member of the genus *Thiobacillus*.

A positive effect of traces of manganese and iron on the growth of aerobic thiobacilli was demonstrated by Starkey (1934*a*); this observation was recently confirmed by Vishniac (1952) in experiments with *T. thioparus*. In earlier experiments these two elements, as well as other necessary trace minerals, were probably provided in the form of impurities in other chemicals.

Excellent growth has been obtained in simple media prepared with tap water; in this case the aerobic species required ammonia, phosphate, potassium and magnesium in addition to the oxidizable substrate, whereas development of *T. denitrificans* also depended on the addition of iron salts (Baalsrud & Baalsrud, 1954). The last finding is in accord with the view that cytochromes are associated with the denitrification process. Under reducing conditions and in the absence of nitrate the presence of cytochromes in cell suspensions of *T. denitrificans* could be demonstrated by the strong absorption bands at 522 and 548 mμ. No such absorption bands were exhibited by suspensions of the aerobic species.

A common property of the thiobacilli is their ability to tolerate relatively high concentrations of phosphate. This makes it easy to prepare media with such buffering capacity that during substrate oxidation the pH is kept within a favourable range.

SUBSTRATE OXIDATION

A considerable part of the reported investigations on thiobacilli deals with the oxidation of the reduced sulphur compounds.

In nature sulphides and elementary sulphur are the available energy sources, and one is certain to observe abundant growth of thiobacilli in habitats rich in these substrates. For laboratory cultures, thiosulphate is usually preferable as substrate, except in cases where the desired pH is low enough to cause a spontaneous decomposition of thiosulphate. Other polythionates have also been shown to serve as oxidizable substrates; in experiments with *T. thioparus* Vishniac (1952) demonstrated that tetrathionate, trithionate and dithionate could all be oxidized.

The reduced sulphur compounds which serve as substrates are completely oxidized to sulphate by all members of the genus *Thiobacillus*; consequently the production of acid is characteristic of these bacteria. It is generally assumed that the oxidation proceeds through a number of steps, and Vishniac's experiments showed conclusively that tetrathionate and dithionate were both formed during thiosulphate oxidation by *T. thioparus*. However, the lack of satisfactory analytical methods makes it extremely difficult to elucidate the intermediate steps in the oxidation. The problem is further complicated by the fact that, during oxidation, various sulphur compounds may occur as a result of spontaneous chemical reactions between the intermediary oxidation products. Such a phenomenon is the precipitation of sulphur which is typical of *Thiobacillus* cultures in thiosulphate media; it was suggested by Nathansohn (1902) and shown conclusively by Vishniac (1952) in experiments with *T. thioparus* that the sulphur does not arise by a purely biological mechanism.

Though no longer regarded as belonging to the thiobacilli (Starkey, 1934b; Sijderius, 1946), the type of thiosulphate oxidizing bacteria first reported by Trautwein (1921) and designated by him as facultatively autotrophic, should be mentioned here. These bacteria, of which a number of strains have been isolated, are heterotrophic organisms containing a tetrathionase system which catalyses the reaction

$$2S_2O_3^{2-} \rightarrow S_4O_6^{2-} + 2e.$$

With a bacterium of this type, isolated from elective cultures for aerobic thiobacilli, it was possible to show the reversibility of the above reaction (see van Niel, 1953). The tetrathionase system has been investigated by Pollock & Knox (1943) during their studies of tetrathionate reduction by anaerobic coliform bacteria. There is reason to believe that this enzyme system is not uncommon among heterotrophic organisms. Experiments with bacteria of the Trautwein type have provided no evidence that the energy from the oxidation of thiosulphate to tetrathionate is utilized for growth, and it also appears doubtful whether autotrophic thiosulphate oxidizing bacteria can utilize the energy from this oxidation step (Baalsrud & Baalsrud, 1954).

Recently, Temple & Colmer (1951) have described a new thiobacillus which is physiologically very similar to *T. thio-oxidans* except that given an acid environment, it can obtain energy for growth by the oxidation of ferrous to ferric iron. This bacterium, properly named *Thiobacillus ferro-oxidans*, is to date the only organism unambiguously shown to maintain itself by the oxidation of iron.

ASSIMILATION OF CARBON DIOXIDE

In order to demonstrate the chemo-autotrophic nature of the thiobacilli, quantitative determinations of carbon dioxide assimilation were carried out at an early stage. By comparing the extent of carbon dioxide assimilation with that of substrate oxidation, a measure of the metabolic efficiency is obtained which is particularly valuable for comparisons with other autotrophic processes. The earlier experiments were carried out over periods of several days, the carbon dioxide assimilation being determined by combustion analysis of the organic carbon content of the culture at the beginning and at the end of the experiment, whereas the more recent data were obtained in short-term experiments performed by conventional manometric methods.

Representative data are shown in Table 1, the 'efficiency' being expressed as the ratio of the amount of oxygen required for complete oxidation of the substrate to the amount of carbon dioxide reduced to organic matter. The variations in the data of Table 1 are strikingly large. It appears that generally a lower O_2/CO_2 ratio is encountered in short manometric experiments than in growth experiments extending over longer periods of time; this is to be expected, as various secondary, energy-consuming metabolic reactions may assert themselves to a large extent in the long-term experiments. Besides, under the well-controlled conditions of a manometric experiment, environmental factors such as pH and carbon dioxide tension (which is obviously a limiting factor under growth conditions) may be kept at more favourable levels.

A special comment must, however, be made on the data of Vogler & Umbreit, pertaining to *T. thio-oxidans*, which indicate a higher 'efficiency' than has been reported for any other autotrophic organism. At first sight it may seem probable that the carbon dioxide taken up in the experiments of Vogler & Umbreit was not assimilated, but merely fixed as carboxyl groups. This explanation is apparently ruled out by one of the experiments, referred to in Table 1, which was especially designed to demonstrate that during substrate oxidation carbon dioxide could replace oxygen as an oxidant and thus was actually reduced. Though it is not possible to account for the discrepancies of the data listed in Table 1 on the basis of methodological considerations, the extremely low O_2/CO_2 ratios reported by Vogler & Umbreit should certainly not be accepted without reservations.

Table 1. *Molecular ratio of amount of oxygen used for substrate oxidation to amount of carbon dioxide assimilated by* Thiobacillus *species*

Bacterium	Substrate	Oxidant	O_2/CO_2*	Reference
Prolonged growth experiments:				
T. denitrificans	$S_2O_3^{2-}$	NO_3^-	9	Lieske, 1912
T. thio-oxidans	S	O_2	18	Waksman & Starkey, 1923
T. thioparus	$S_2O_3^{2-}$	O_2	19	Starkey, 1935
Manometric experiments:				
T. thio-oxidans	S	O_2	2·9†	Vogler, 1942, p. 113
T. thio-oxidans	S	O_2	1·5‡	Vogler & Umbreit, 1942
T. thio-oxidans	$S_2O_3^{2-}$	O_2	9–26	Baalsrud & Baalsrud,
T. thioparus	$S_2O_3^{2-}$	O_2	average: 15	1952
T. denitrificans	$S_2O_3^{2-}$	NO_3^-	4·6–11	Baalsrud & Baalsrud, 1952

* The quantities of nitrate used by *T. denitrificans* are calculated as oxygen according to the equation

$$2HNO_3 \rightarrow H_2O + N_2 + 2\tfrac{1}{2}O_2.$$

† The lowest figure that can be computed from a series of experiments in which carbon dioxide was demonstrated to act as oxidizing agent in sulphur oxidation.

‡ Obtained by an experiment in which substrate oxidation and carbon dioxide uptake were separated in time.

As already indicated, a manometric determination of the carbon dioxide uptake does not necessarily provide any information as to whether the gas absorbed is actually assimilated. In manometric assimilation experiments (Baalsrud & Baalsrud, 1952, 1954) from which data are given in Table 1, a true carbon dioxide assimilation was conclusively demonstrated in the following manner.

Cell suspensions of the aerobic species *T. thioparus* and *T. thio-oxidans* would oxidize a known quantity of thiosulphate at a high and constant rate. In the absence of carbon dioxide, the ultimate oxygen uptake always corresponded accurately to that theoretically needed for complete oxidation of the substrate, but in the presence of carbon dioxide, the amount of oxygen consumed was smaller by a factor corresponding to the carbon dioxide uptake. The latter was determined in separate experiments as the difference between the total amounts of carbon dioxide present at the beginning and at the end of the oxidation period. From this it was concluded that the carbon dioxide taken up was converted into reduced products. On the basis of the assumption that the reduction proceeds to the carbohydrate level, in which case oxygen and carbon dioxide would be equivalent as oxidants, an indirect method for a quantitative determination of the carbon dioxide assimilation was developed. In this second type of experiment the amount of oxygen used for the complete oxidation of a known quantity of thiosulphate in the presence of carbon dioxide *at constant pressure* was subtracted from that used in the absence of carbon dioxide. The observed difference, representing the amount of carbon dioxide assimilated, was always in very good agreement with the carbon dioxide uptake determined by the direct method.

With cell suspensions of *T. denitrificans* carbon dioxide assimilation could also be demonstrated manometrically in experiments of short duration; the average uptake, computed on the basis of a unit quantity of oxidant used, was about twice as high as that encountered in the experiments with the aerobic species. However, because the extent of substrate utilization by suspensions of *T. denitrificans* could not be accurately determined, these manometric experiments gave no information as to the fate of the assimilated carbon dioxide. This was investigated by employing ^{14}C-labelled carbon dioxide under conditions which were otherwise identical with those used for the determination of the carbon dioxide uptake. After a 40 min. oxidation period, about half of the radioactive carbon taken up by the suspension could be found in complex cell substances of low solubility. From this it was concluded that cell suspensions of *T. denitrificans* carry out a reduction of the absorbed carbon dioxide.

It should be stressed that the extent of carbon dioxide uptake by cell suspensions of the three thiobacilli is of the same order of magnitude as that observed in growing cultures. The high assimilation values obtained with *T. denitrificans* are in agreement with the observation that in growing cultures the cell yield of this bacterium, computed on the basis of the amount of substrate oxidized, is approximately twice as high as that of *T. thioparus* or *T. thio-oxidans* (cf. the data of Lieske (1912), and of Waksman & Starkey (1923) in Table 1).

ENERGY TRANSFER

For our understanding of the autotrophic mode of life the problem of energy transfer in thiobacilli seems most important. We know that energy is liberated by the oxidation of reduced sulphur compounds and then utilized for the assimilation of carbon dioxide. By what kind of mechanism is the energy transferred from the one process to the other?

A number of heterotrophic organisms are known to carry out an oxidation of reduced sulphur compounds, yet they do not utilize the energy released by oxidation for metabolic purposes. Thus it is not the energy-yielding process in itself that is unique, but rather the ability of the thiobacilli, as well as of other sulphur bacteria, to link together this process and their energy-requiring metabolic activities. From the point of view of comparative biochemistry it appears conceivable that an elucidation of the mechanism whereby energy is transferred by chemo-autotrophic organisms might also offer a key to another problem of autotrophy, viz. the utilization of radiant energy for the assimilation of carbon dioxide in photosynthesis.

It seemed that an important step in this direction had been taken about ten years ago when Vogler and Umbreit claimed to have obtained information concerning the mechanism of energy transfer by *T. thiooxidans*. In experiments with this organism Vogler (1942) was able to separate in time the oxidation of sulphur and the assimilation of carbon dioxide. Cell suspensions that had previously oxidized sulphur in an environment strictly free of carbon dioxide were capable, in the absence of either oxidizable substrate or oxygen (i.e. without the simultaneous occurrence of any energy-furnishing reaction), of taking up considerable amounts of carbon dioxide. It was further demonstrated by Vogler & Umbreit (1942) that, under such experimental conditions, changes in the inorganic phosphate content of the suspension medium occurred. The oxidation of sulphur in the absence of carbon dioxide was accompanied by a decrease in the amount of inorganic phosphate; when the suspension was later exposed to carbon dioxide the concentration of inorganic phosphate was restored to its initial level. Finally, a substance with the properties of an adenosine triphosphate could be isolated from the cells (LePage & Umbreit, 1943 *b*). The interpretation placed on these observations was that, during sulphur oxidation, energy was stored in the cells in such a way that it could be utilized for assimilation when carbon dioxide became available. On the basis of the changes in the inorganic phosphate content it was postulated that the energy was stored in the form of energy-rich phosphate compounds, presumably ATP. The same mechanism was assumed to be operative under normal experimental conditions, i.e. in the presence of carbon dioxide, but in this case one would not expect the energy-rich chemical mediator to accumulate as it would be used up in the simultaneous assimilatory reactions. The concept of energy transfer via high-energy phosphate bonds was believed to be of general validity for chemo-autotrophic organisms, and furthermore, it was proposed that it also occurred in photosynthesis (Umbreit, 1947, 1951).

The results of Umbreit and his co-workers were reported only a few years after the discovery of the participation of high-energy phosphate bonds in basic metabolism, and they were received as information of great value in comparative studies. The experimental approach seemed extremely promising; yet the matter was not at that time pursued further.

Recent attempts to confirm the theory put forward by Umbreit's group have been unsuccessful. These studies, which have already been reported in detail in an earlier publication (Baalsrud & Baalsrud, 1952), were initiated by a close examination of the experiments on which the theory was originally founded. It appeared that the suggested mechanism of energy coupling implied certain requirements which were not met by the results reported by Umbreit's group. Since the concept has been so widely accepted, it seems worth while to sum up the main points of criticism.

According to Vogler & Umbreit (1942), the assimilation of carbon dioxide in thiobacilli is accomplished by means of the energy liberated by the splitting of high-energy phosphate bonds. A chemosynthetic organism synthesizes all its cell material from carbon dioxide. Such a 'true' assimilation obviously involves a reduction process. Labile phosphate bonds have been invoked to provide the energy for the fixation of carbon dioxide; the conversion of carbon dioxide to cell material would, however, also require reducing substances. There is no evidence that such substances were available in the separation experiments of Vogler & Umbreit.

Even if a simple carbon dioxide fixation, rather than a true assimilation, took place, the observed phosphate changes were much too small to account for the amounts of carbon dioxide taken up. While in carboxylation reactions the molar ratio of carbon dioxide fixed to $\sim$P utilized has been found to be $1:1$, the results of Vogler & Umbreit represent a ratio of $47:1$. In the absence of oxidizable substrate, cells of *T. thio-oxidans* assimilated as much as 40 μl. carbon dioxide per 100 μg. of cell nitrogen; these data show that the 'normal' ratio of 1 mole of carbon dioxide fixed per mole of ATP could only be realized if all the cell nitrogen were present as ATP. As has been pointed out earlier the extent of carbon dioxide assimilation was much larger than that encountered by other workers; this, too, makes the data appear to be of doubtful value.

The comparative studies referred to above were carried out with the three *Thiobacillus* species *T. thioparus*, *T. thio-oxidans* and *T. denitrificans*. The results can be summarized as follows.

(1) Cell suspensions of the three species assimilated carbon dioxide during oxidation of the substrate. The assimilation was of the same order of magnitude as that encountered with growing cultures and was shown to represent an actual reduction of carbon dioxide.

(2) Cell suspensions which had oxidized thiosulphate in the absence of carbon dioxide did not fix carbon dioxide in detectable amounts after the oxidation had been completed.

(3) During thiosulphate oxidation by the three species a decrease in the inorganic phosphate concentration of the medium was observed. After the total amount of substrate had been oxidized, the inorganic phosphate concentration appeared to return to its initial level. The changes in inorganic phosphate concentration were extremely small. They occurred during actual carbon dioxide assimilation, as well as during substrate oxidation in the absence of carbon dioxide.

In contrast to the observations of Vogler & Umbreit (1942) these results indicated that carbon dioxide assimilation is dependent on a simultaneous oxidation process. The fact that it was not possible to demonstrate any storage of energy constitutes a serious objection to the postulated mechanism, as the latter was based on the separation in time of substrate oxidation and carbon dioxide assimilation.

On the other hand, the observed changes in inorganic phosphate were in good agreement with the results of Vogler & Umbreit, both qualitatively and quantitatively. However, these workers drew their conclusions from the observation that the oxidation of sulphur in the absence of carbon dioxide was accompanied by a decrease in inorganic phosphate concentration and that inorganic phosphate was again released when carbon dioxide was introduced into the system. As part of the recent studies, control experiments were carried out, demonstrating that inorganic phosphate was also released when the suspension was not subsequently exposed to carbon dioxide, and that the same reversal changes took place during substrate oxidation and concomitant carbon dioxide assimilation. These last observations evidently invalidate the theory of a direct relation between release of inorganic phosphate and carbon dioxide assimilation.

Thus it must be concluded that neither the observations reported by Umbreit's group nor our own recent studies with thiobacilli provide sufficient experimental evidence that the oxidation of the inorganic substrate is linked with the assimilation of carbon dioxide by high-energy phosphate compounds serving as chemical mediators.

It appears conceivable that energy transfer in chemo-autotrophic organisms must be accomplished by a chemically intelligible mechanism, and admittedly, the concept founded by Umbreit's group may still have its value as a working hypothesis. However, if one considers that a true assimilation of carbon dioxide, as encountered in autotrophic organisms, must include certain reduction steps, it seems equally reasonable to postulate that the energy from the oxidation of the inorganic substrate is made available by a series of oxido-reduction reactions. Such a mechanism would not exclude the participation of high-energy phosphate bonds, as the generation of such bonds by oxido-reduction reactions has been demonstrated (e.g. Lehninger, 1951). Vishniac & Ochoa (1952) have already suggested a similar approach to the problem of photo-synthesis. The observation, reported by these workers, that the biological oxidation of thiosulphate can be coupled with the reduction of triphos-phopyridine nucleotide, may be taken as evidence that the formation of reduced organic compounds is a primary event in the transfer of energy obtained by inorganic oxidations.

CARBON METABOLISM

Present information concerning the composition of *Thiobacillus* cells as well as metabolic reactions demonstrated with cell preparations suggest that the main metabolic pattern is the same as that found in other organisms. A number of the known vitamins (O'Kane, 1942), amino-acids (Frantz, Feigelman, Werner & Smythe, 1952), and phosphorylated compounds (LePage & Umbreit, 1943a) have been identified in cell preparations. By irradiation with ultraviolet light Rittenberg & Grady (1950) obtained a *T. thio-oxidans* mutant which required thiamine for growth. The presence of common enzyme systems has also been in-dicated by specific reactions. Thus R. C. Bard (personal communication) found that cell extracts of the Starkey strain of *T. thioparus* contained aldolase and glyceraldehyde phosphate dehydrogenase. With dried cell preparations of *T. thioparus* Vishniac (1949) demonstrated an oxygen consumption on the addition of glucose-1-phosphate. The amount of oxygen consumed far exceeded that theoretically necessary for the com-plete oxidation of the added glucose-1-phosphate, indicating that the latter compound caused an oxidation of some other material present in the cell preparation.

Though none of these observations appear to violate the principle of metabolic uniformity among living organisms, there remains one major problem, the solution of which will determine whether the concept of comparative biochemistry, in its broadest sense, should be extended to the chemo-autotrophic mode of life. This problem pertains to the fact that the thiobacilli are unable to synthesize their cell material from organic carbon sources.

As a working hypothesis van Niel (1943) suggested that in chemo-synthesis the transformation of carbon dioxide into cell material is accomplished via one or a very few key organic compounds, and our present knowledge of carbon dioxide assimilation by photosynthetic and heterotrophic organisms strongly supports this view. It should therefore be possible to provide the chemo-autotrophic organisms with organic media in which they would develop in the absence of oxidizable inorganic substrate. Up to the present time all attempts in this direction have, however, failed. Typical key compounds, such as a great number of the common sugars, alcohols, amino-acids, mono-, di- and tri-carboxylic acids, as well as complex substances like peptone and yeast extract, have all given completely negative results in growth experiments. This does not necessarily mean that all these compounds should no longer be considered. It is possible that the negative results have been due to unfavourable environmental conditions, or to insufficient permeability of the cell wall. However, in view of the fact that the cells are permeable to large polar ions such as $S_2O_3^{2-}$ and $S_4O_6^{2-}$, the latter explanation does not seem too likely.

A useful approach to this main problem of chemo-autotrophy might be a thorough investigation of the internal metabolism of such organisms. By a deliberate search for known enzyme systems in cell-free extracts it should be possible to reconstruct the pattern of carbon metabolism; this might, in time, be used as a basis for composing media that could be tried in growth experiments. It seems that *T. denitrificans*, which is extremely easy to cultivate on a large scale, would be well suited for experimentation along these lines.

A more direct method of elucidating key compounds would be to study the transformations of carbon dioxide into cell material by means of labelled carbon dioxide. The technique developed by Calvin *et al.* (1951) is undoubtedly applicable to chemosynthetic organisms, and, again, *T. denitrificans* should be pointed out as a most promising test organism.

REFERENCES

BAALSRUD, K. & BAALSRUD, K. S. (1952). The role of phosphate in CO_2 assimilation of thiobacilli. In *Phosphorus Metabolism*, **2**, 544. Baltimore: Johns Hopkins Press.

BAALSRUD, K. & BAALSRUD, K. S. (1954). Studies on *Thiobacillus denitrificans*. *Arch. Mikrobiol.* **20**, 34.

BEIJERINCK, M. W. (1920). Chemosynthesis and denitrification with sulphur as source of energy. *Proc. Acad. Sci. Amst.* **22**, 899.

Bergey's Manual of Determinative Bacteriology, 6th ed. (1948). London: Baillière, Tindall and Cox.

CALVIN, M., BASSHAM, J. A., BENSON, A. A., LYNCH, V. H., OUELLET, C., SCHOU, L., STEPKA, W. & TOLBERT, N. E. (1951). Carbon dioxide assimilation in plants. In *Carbon Dioxide Fixation and Photosynthesis. Symp. Soc. exp. Biol.* **5**. Cambridge University Press.

FRANTZ, I. D., FEIGELMAN, H., WERNER, A. S. & SYMTHE, M. P. (1952). Biosynthesis of seventeen amino-acids labelled with ^{14}C. *J. biol. Chem.* **195**, 423.

GEHRING, A. (1915). Beiträge zur Kenntnis der Physiologie und Verbreitung denitrifizierender Thiosulfat-Bakterien. *Zbl. Bakt.* (2. Abt.), **42**, 402.

LEHNINGER, A. L. (1951). Phosphorylation coupled to oxidation of dihydrodiphosphopyridine nucleotide. *J. biol. Chem.* **190**, 345.

LEPAGE, G. A. & UMBREIT, W. W. (1943a). Phosphorylated carbohydrate esters in autotrophic bacteria. *J. biol. Chem.* **147**, 263.

LEPAGE, G. A. & UMBREIT, W. W. (1943b). The occurrence of adenosine-3-triphosphate in autotrophic bacteria. *J. biol. Chem.* **148**, 255.

LIESKE, R. (1912). Untersuchungen über die Physiologie denitrifizierender Schwefelbakterien. *Ber. dtsch. bot. Ges.* **30**, 12.

NATHANSOHN, A. (1902). Über eine neue Gruppe von Schwefelbacterien und ihren Stoffwechsel. *Mitt. zool. Sta. Neapel*, **15**, 655.

VAN NIEL, C. B. (1943). Biochemical problems of the chemo-autotrophic bacteria. *Physiol. Rev.* **23**, 338.

VAN NIEL, C. B. (1953). Introductory remarks on the comparative biochemistry of micro-organisms. *J. cell. comp. Physiol.* **41**, Suppl. 1, 11.

O'KANE, D. J. (1942). The presence of growth factors in the cells of the autotrophic sulphur bacteria. *J. Bact.* **43**, 7.

POLLOCK, M. R. & KNOX, R. (1943). Bacterial reduction of tetrathionate. *Biochem. J.* **37**, 476.

RITTENBERG, S. C. & GRADY, R. P. (1950). Induced mutants of *Thiobacillus thiooxidans* requiring organic growth factors. *J. Bact.* **60**, 509.

SIJDERIUS, R. (1946). Heterotrophe bacterien, die thiosulfaat oxydeeren. Thesis, Amsterdam.

STARKEY, R. L. (1925). Concerning the carbon and nitrogen nutrition of *Thiobacillus thio-oxidans*, an autotrophic bacterium oxidizing sulphur under acid conditions. *J. Bact.* **10**, 165.

STARKEY, R. L. (1934a). Cultivation of organisms concerned in the oxidation of thiosulphate. *J. Bact.* **28**, 365.

STARKEY, R. L. (1934b). The production of polythionates from thiosulphate by micro-organisms. *J. Bact.* **28**, 387.

STARKEY, R. L. (1935). Products of the oxidation of thiosulphate by bacteria in mineral media. *J. gen. Physiol.* **18**, 325.

TEMPLE, K. L. & COLMER, A. R. (1951). The autotrophic oxidation of iron by a new bacterium: *Thiobacillus ferro-oxidans*. *J. Bact.* **62**, 605.

TRAUTWEIN, K. (1921). Beitrag zur Physiologie und Morphologie der Thionsäurebakterien (Omelianski). *Zbl. Bakt.* (2. Abt.), **53**, 513.

UMBREIT, W. W. (1947). Problems of autotrophy. *Bact. Rev.* **11**, 157.

UMBREIT, W. W. (1951). Significance of autotrophy for comparative physiology. In *Bacterial Physiology* (ed. C. H. Werkman and P. W. Wilson), p. 566. New York: Academic Press.

VISHNIAC, W. (1949). On the metabolism of the chemolitho-autotrophic bacterium *Thiobacillus thioparus* Beijerinck. Ph.D. Thesis, Stanford University.

VISHNIAC, W. (1952). The metabolism of *Thiobacillus thioparus*. I. The oxidation of thiosulphate. *J. Bact.* **64**, 363.

VISHNIAC, W. & OCHOA, S. (1952). Reduction of pyridine nucleotides in photosynthesis. In *Phosphorus Metabolism*, **2**, 467. Baltimore: Johns Hopkins Press.

VOGLER, K. G. (1942). Studies on the metabolism of autotrophic bacteria. II. The nature of the chemosynthetic reaction. *J. gen. Physiol.* **26**, 103.

VOGLER, K. G. & UMBREIT, W. W. (1942). Studies on the metabolism of autotrophic bacteria. III. The nature of the energy storage material active in the chemosynthetic process. *J. gen. Physiol.* **26**, 157.

WAKSMAN, S. A. & JOFFE, J. S. (1922). Micro-organisms concerned in the oxidation of sulphur in the soil. II. *Thiobacillus thio-oxidans*, a new sulphur oxidizing organism isolated from the soil. *J. Bact.* **7**, 239.

WAKSMAN, S. A. & STARKEY, R. L. (1923). On the growth and respiration of sulphur-oxidizing bacteria. *J. gen. Physiol.* **5**, 285.

WINOGRADSKY, S. (1887). Über Schwefelbakterien. *Bot. Ztg*, **45**, 489.

SOME ASPECTS OF THE PHYSIOLOGY
OF THE NITRIFYING BACTERIA

JANE MEIKLEJOHN

Rothamsted Experimental Station, Harpenden, Hertfordshire

The nitrifying bacteria, which form nitrate in soils, manure heaps, dirty stables and cowsheds, and sewage-disposal plants, are a group of organisms of the greatest economic importance, but very little is as yet known about them. Other autotrophic organisms have been more intensively studied, notably *Thiobacillus thio-oxidans*, which has been the subject of the most beautifully detailed researches by Umbreit and his colleagues (Umbreit, 1947). Unfortunately, the nitrifiers are not nearly so amenable to experiment as *T. thio-oxidans*. The latter tolerates a much higher acidity than most organisms and oxidizes sulphur, an insoluble energy source metabolized by only a limited range of organisms. On the other hand, the nitrifiers grow under conditions which would be highly suitable for the growth of many free-living heterotrophs, if only enough organic matter were present. The organic matter which is lacking at first in cultures of the nitrifiers is supplied in increasing amounts by them as they grow, so that, the better a culture of a nitrifier has grown, the more likely is a contaminant, once introduced, to establish itself. Furthermore, the energy sources of the nitrifiers, ammonia and nitrite, are not only soluble and available to other organisms, but are readily assimilated by micro-organisms. (Ammonia, and probably nitrite too, is also readily assimilated by some species of higher plants.) The nitrifiers grow very slowly in culture and this may of course be due to our failure to discover the optimum cultural conditions; personally, however, I think it is an inevitable consequence of the autotrophic mode of life.

In any case, it is most inconvenient for the experimenter, as heterotrophic contaminants overgrow the nitrifiers so easily. As a result, the nitrifiers, though easy enough to grow in mixed culture, are very difficult to isolate in pure culture, and exact studies of their physiology have therefore been few and far between.

In this paper an attempt will be made to describe what is known about the physiology of the nitrifying bacteria, and to relate it to the natural conditions in which they occur.

THE ECOLOGY OF NITRIFICATION

The normal habitat of the nitrifying bacteria is the soil, and their economic importance at the present time is due to the fact that nitrate is the form in which nitrogen is most easily assimilated by the majority of cultivated plants. Unfortunately, nitrate is also the form in which nitrogen is most easily lost, as it is very soluble and is therefore washed down by rain out of reach of the roots of growing plants. It is also easily reduced by denitrifying bacteria, which are very common in soil, to nitrogen gas, which escapes into the atmosphere.

Nitrifying bacteria have been obtained from soils from all parts of the world, including the Arctic and Antarctic (Winogradsky, 1904; Jensen, 1951), though nitrification must be very slow at high latitudes, as the cold of an English winter is enough to bring it almost to a standstill (Warington, 1884). It is known from pure culture studies that the nitrifiers are of two kinds, one oxidizing ammonia to nitrite, and the other oxidizing nitrite to nitrate. Since nitrite is seldom found in soils, it can be assumed that the two kinds of nitrifier nearly always occur together. Nitrate is formed, and nitrifying bacteria therefore presumably occur, in all fertile soils, nitrification being favoured by adequate supplies of calcium and phosphate. Some types of acid soil, not suitable for agriculture, do not contain nitrifying bacteria; acid peaty soil and certain soils carrying coniferous forest are of this type (Arnd, 1919; Hesselman, 1917). However, soil acidity does not itself always prevent nitrification. For example, a Uganda soil with a surface pH of about 5 was found to accumulate nitrate at an exceptionally rapid rate (Griffith & Manning, 1949).

Nitrate is formed in manure heaps, and this process may lead to the loss of a large part of the nitrogen in the manure, if the heaps are exposed to rain after standing for some time. Nitrification is an important part of the process of sewage disposal, being the final step in the conversion of the nitrogen compounds of the sewage into a soluble, harmless, fully oxidized form. The quantities of nitrate formed by nitrification in soils are not very large; 60 parts of nitrate-nitrogen per million is a high figure for soil. But there are natural sites where nitrate accumulates to such an extent that it crystallizes out on the surface. The existence of these sites was well known from the fourteenth century onward, as the crude nitrate (saltpetre) was used for the manufacture of gunpowder and was therefore a valuable product. The places where saltpetre was found all contained large quantities of nitrogenous organic matter; they included the sites of deserted villages, especially in hot countries like Egypt and India, burial mounds on old battlefields, the earth of disused graveyards, and the walls of cellars and of dirty stables and cowsheds 'impregnated with the excrementitious *effluvia* of the animals that inhabited them' (Boerhaave, 1753). It is important in considering the ecology of the nitrifiers to distinguish between nitrate formation and nitrate accumulation. Suitable conditions for nitrate formation, which is the process carried out by the nitrifying bacteria, must be present initially in all saltpetre sites; but once the nitrate is formed, it can be concentrated by purely physical processes, leaching and evaporation, under conditions which would prevent the formation of any more nitrate. The conditions suitable for the formation of nitrate were worked out in France in the late eighteenth century, when saltpetre was needed for gunpowder in the Napoleonic wars. Heaps of soil mixed with manure and lime were watered with waste water and urine, and after about two years the nitrate was extracted from them with hot water (Boussingault, 1844). These 'nitre-beds' contained large quantities of organic matter and were housed in wooden sheds to protect them from the sun and rain. Protection from the rain prevents the soluble nitrates from being washed away, but protection from the sun is not necessary. On the contrary, once the nitrate has been formed, exposure to sunlight enables the nitrate to be concentrated by evaporation.

THE NITRIFYING BACTERIA

(1) *Discovery of their nature*

Nitrification was looked upon by the scientific world of the eighteenth and early nineteenth centuries as a purely chemical process, the starting-point of which was a matter of dispute. The existence of living organisms which produced nitrate by oxidation was first demonstrated by the now famous experiment of Schloesing & Müntz (1877). They showed that the formation of nitrate in sewage trickling down a long tube full of sand was stopped by the vapour of chloroform, and was therefore due to a living agency. Warington, in a series of remarkable studies for which he has never been given sufficient credit, discovered the main facts about the nitrification process in soil and the causative agents. In 1878 he repeated Schloesing & Müntz's experiment with soil, obtaining the same result, and in the following year he demonstrated that no nitrate was produced in an incubated ammonium chloride solution unless it had been previously inoculated with soil. He showed that ammonia was the only starting-point for nitrification, and that about 94 % of the ammonia supplied in a liquid culture was converted into nitrate. Finally, Warington (1891) made the remarkable discovery that the process of nitrification, which apparently takes place by only one step in nature, is in reality divided into two stages, the oxidation of ammonia to nitrite, and the oxidation of nitrite to nitrate. He showed conclusively that the two stages were carried out by different organisms, and that both of them could take place in the absence of organic matter. In fact, Warington came very near to the discovery of the autotrophic nature of the nitrifiers. Unfortunately, he was denied the means of making pure cultures, though his descriptions of the bacteria seen in his cultures are such as to leave very little doubt that they were *Nitrosomonas* and *Nitrobacter*. The publication of the results of his final study was so long delayed that it was preceded by the appearance of Winogradsky's first paper on nitrification, although Warington's work had been completed some years previously.

Before he began work on nitrification, Winogradsky had studied the iron and the sulphur bacteria, and the whole conception of autotrophy in bacteria was due to his brilliant and original mind. The idea that living organisms existed which derived their energy from the oxidation of inorganic material was a very strange one, but Winogradsky showed quite conclusively that it was true of the iron and the sulphur bacteria. When he began work on the nitrifiers and found that they could grow for an indefinite number of transfers on media devoid of organic matter, he was certain that these organisms were also autotrophic (Winogradsky, 1890). He was able to overcome the difficulties of obtaining them in pure culture by the ingenious use of silica gel as an inorganic basis for solid media, and was able to show that they were in fact autotrophic, using carbon dioxide as their sole source of carbon, and the oxidation of either ammonia or nitrite as their source of energy (Winogradsky, 1891).

The reactions concerned are:

$$2NH_3 + 3O_2 \rightarrow 2HNO_2 + 2H_2O + 79 \text{ kcal.}$$

and $\qquad HNO_2 + \frac{1}{2}O_2 \rightarrow HNO_3 + 21 \cdot 6 \text{ kcal.}$

The oxygen uptake of growing pure cultures of both ammonia- and nitrite-oxidizers has been measured, and it can all be accounted for by assuming that one or other of these two reactions is the only one taking place (Meyerhof, 1916; Bömeke, 1939).

(2) *Morphology and classification*

There are three genera of bacteria which are definitely known to be autotrophic nitrifiers. Species of *Nitrosomonas* and *Nitrosococcus* oxidize ammonia to nitrite, and species of *Nitrobacter* oxidize nitrite to nitrate. (*Nitrosococcus* is a large coccus, $1 \cdot 5-2\mu$. in diameter

(Winogradsky, 1904), and as little work has been done concerning its physiology, it will not be mentioned further.)

Nitrosomonas. Three species have been described. They all consist of small oval cells, about 1·5 by 1·0μ., imbedded in a zoogloea which forms around particles of solid carbonate in the usual culture medium, leaving the supernatant clear. In some strains, 'swarmers', motile cells with a single polar flagellum appear regularly in cultures after 1–2 weeks' incubation. Spores are not formed. The colonies on silica gel or washed agar plates are very characteristic; they are very small, only about 100μ. in diameter, and have a curious glassy refringence. Some are round or oval, and some have a 'starfish' shape with projecting points at different levels (Kingma Boltjes, 1935; Meiklejohn, 1950).

Nitrobacter. Two species, one motile and the other not, have been described. *Nitrobacter winogradskyi*, the non-motile species, has oval cells about 1·0 by 0·8μ. and does not form spores. The colonies on washed agar plates are about 200μ. in diameter, and are circular, soft and transparent if on the surface of the agar, and lentil-shaped if buried (Kingma Boltjes, 1935; Meiklejohn, 1953a).

Other genera have been obtained in culture by the later technique of Winogradsky & Winogradsky (1933), which is not designed to give pure cultures. *Nitrosospira*, *Nitrosocystis* and *Nitrosogloea* were described as ammonia-oxidizers, and *Nitrocystis* as a nitrite-oxidizer. None of these organisms has, so far as I know, been isolated by any other worker. However, Imsenecki (1946) claimed that the organism from Swedish forest soils (Romell, 1927), which Winogradsky called *Nitrosocystis*, was not a nitrifier. Imsenecki tried the same method of cultivation and found that *Nitrosomonas* appeared first and was then overgrown by a myxo-bacterium, *Sorangium symbioticum*, which formed visible hard fruiting bodies, closely resembling the bodies formed by Romell's *Nitrosocystis*. Winogradsky also placed some strains which he had previously classified as *Nitrosomonas* in the genus *Nitrosocystis*, on the grounds that they formed hard colonies, not soft ones, on silica gel. However, as Kingma Boltjes (1935) obtained both hard and soft colonies in cultures of *Nitrosomonas* derived from single cells, it seems extremely probable that these strains were true *Nitrosomonas* and should not have been reclassi-fied (Meiklejohn, 1953b).

(3) *Cultural conditions*

The conditions which favour the growth of the nitrifiers in culture are, with few exceptions, those which would be expected from a study of their ecology. The nitrifiers are strictly aerobic, they adhere closely to solid particles, they are adversely affected by light, are very resistant to drying, and are fairly tolerant of changes in temperature and acidity.

It might indeed be expected that organisms whose only sources of energy are reactions involving free oxygen would be strictly aerobic, and Winogradsky (1890) found, at the very beginning of his work on the nitrifiers, that nitrification would only take place under fully aerobic conditions. Any device for increasing the air supply to liquid cultures, such as the introduction of a stream of air bubbles (Lees, 1952), or the circulation of the medium over solid particles (Boullanger & Massol, 1904), will increase nitrification. Nitrification in incubated portions of soil is depressed if the air circulated over them is diluted with nitrogen (Amer & Bartholomew, 1951). The oxygen uptake of pure cultures of *Nitrosomonas* is reversibly reduced if the partial pressure of oxygen in the Warburg apparatus is diminished to one-third of an atmosphere or below. But, curiously enough, pure oxygen, though it permits full respiration of *Nitrosomonas* for 2–3 hr., eventually causes an irreversible drop in the rate of respiration (Meyerhof, 1916).

The media usually employed for the culture of the nitrifiers contain an excess of solid calcium or magnesium carbonate, and it has been observed by Winogradsky, and by all subsequent workers, that the supernatant of a culture is usually quite clear, as the cells are all adherent to the carbonate particles at the bottom of the medium. In the absence of solid carbonate, nitrifiers attach themselves to the bottom and sides of the culture vessel, as many of the common water bacteria do. It seems very probable that *Nitrosomonas*, at any rate, is closely attached to the soil particles, as Lees & Quastel (1946) showed that ammonia was not oxidized in soil cultures unless it had first been adsorbed on to the surface of the soil colloids.

Light seems to have an adverse effect on the nitrifiers: Boussingault (1844) says, in the course of his description of the French nitre-beds, 'La pratique a enseigné que la nitrification s'opère mieux à l'ombre', and remarks that this was the reason why the beds were built inside wooden huts. Warington (1878) was the first to find that nitrification proceeded more rapidly in cultures placed in a dark cupboard than on an open bench, and the same has been found to be true by all subsequent workers; it is also true of portions of soil incubated in the laboratory (Waksman & Madhok, 1937).

Estimates of the optimum temperature for the nitrifiers in culture range from 30 to 36°, and it is probable that the optimum is not sharply defined. Exposure to 53–55° for 10 min. kills *Nitrosomonas*, and exposure for the same length of time to 56–58° kills *Nitrobacter* (Gibbs,

1919). Warington (1879) found that cultures derived from an English garden soil would not nitrify at 40°, and it is most interesting that the same was found to be true of cultures derived from an arable soil in Uganda, where air and soil temperatures are very much higher than they are in England (Meiklejohn, 1953c). Both *Nitrosomonas* and *Nitrobacter* are highly resistant to drying in soil. They have been known to survive for 7 years in bottled soil samples, and it is probable that they survive much longer than this (Gibbs, 1919). They do not, however, resist the drying up of liquid cultures; but do survive for 2 years if water is added from time to time (Bömeke, 1949).

The optimum hydrogen-ion concentration for the growth of both species in culture has generally been shown to be on the alkaline side of neutrality, but, like the optimum temperature, it is not sharply defined. Meyerhof (1916), working with cultures from Winogradsky's laboratory, measured the oxygen uptake of *Nitrosomonas* and *Nitrobacter* at a number of different pH levels. In both cases the curves obtained by plotting oxygen uptake against pH were flat-topped; the plateau of the curve for *Nitrosomonas* covered the range pH 8·5–8·8, and that for *Nitrobacter* pH 8·3–9·3. On each side of the optimum range both curves descended steeply. However, there is no reason to suppose that all strains have these same optima, and in fact Winogradsky & Winogradsky (1933) obtained six strains of *Nitrosomonas* with optima ranging from pH 6 to 9 and seven of *Nitrobacter* with optima ranging from pH 6·3 to 9·4; in each case the optimum was measured by the oxidation of ammonia or nitrite respectively in growing cultures. The optimum reaction in culture, however, is not necessarily related to the natural conditions in which the strain has been living. Heubült (1929) found that strains of *Nitrosomonas* isolated from soils with pH values ranging from 5·6 to 6·8 grew in a liquid medium of pH 7·1–7·2. Cultures of *Nitrosomonas*, isolated from an acid Uganda soil having a surface pH of about 5, actually grew better in culture at pH 7·0 than at pH 5·4 (Meiklejohn 1953c).

(4) *Nutrient requirements*

The nitrifiers are usually grown in media containing no organic matter, but only an ammonium salt or a nitrite, and a mixture of mineral salts. The composition of some of these media is given in Table 1. Ammonia or nitrite is presumably the source of nitrogen, as well as the source of energy, but nothing is known about nitrogen assimilation by any of these organisms. The amount of nitrogen assimilated is, of course, very small indeed compared with the amount of ammonia or nitrite oxidized in the reactions providing energy. No growth takes place in the absence of carbon dioxide, which appears to be the only source of carbon which the nitrifiers are capable of using. The other nutrient requirements have not yet been fully worked out, but they are known to include calcium, magnesium, phosphorus, iron and probably copper. As far as is known, potassium, sulphur and zinc are not required. Calcium is necessary for *Nitrosomonas europaea*, and it cannot be replaced by strontium. The amount of calcium giving maximum growth is *c*. 1 mg./l. (Kingma Boltjes, 1935). Magnesium, phosphate and iron are all essential for both *Nitrosomonas* and *Nitrobacter* (Bömeke, 1949), but the minimum requirements for growth are very small, *c*. 0·003 mg./l. of phosphorus (as

Table 1. *Media used for growing nitrifying organisms*

1, Omeliansky (1899); 2, Gibbs (1919); 3, Kingma Boltjes (1935); 4, Meiklejohn (1950).

Ingredient (g.)	Ammonia-oxidizers				Nitrite-oxidizers			
	1	2	3	4	1	2	3	4
$(NH_4)_2SO_4$	2·0	1·0	1·0	0·66	—	—	—	—
$NaNO_2$	—	—	—	—	1·0	1·0	1·0	0·5
K_2HPO_4	1·0	1·0	1·0	—	0·5	0·5	0·5	—
KH_2PO_4	—	—	—	0·14	—	—	—	0·14
NaCl	2·0	2·0	2·0	0·3	0·5	0·5	0·5	0·3
$MgSO_4$	0·5	0·5	0·5	0·14	0·3	0·3	0·3	0·14
$Fe_2(SO_4)_3$	0·4	Trace	0·01	—	0·4	Trace	0·4	—
$FeSO_4$	—	—	—	0·03	—	—	—	0·03
Na_2CO_3	—	—	—	—	1·0	1·0	1·0	—
$MgCO_3$	Excess	Excess	10·0	—	—	—	—	—
$CaCO_3$	—	—	—	10·0	—	—	—	10·0
Trace elements	—	—	—	+	—	—	—	+
H_2O, distilled	1 l.	1 l.	—	1 l.	1 l.	1 l.	—	1 l.
H_2O, tap	—	—	1 l.	—	—	—	1 l.	—

phosphate), and *c*. 0·0002 mg./l. of magnesium (Meiklejohn, 1952). The minimum amount of iron required for growth is very small, but the optimum amount is very high, corresponding rather to the amounts required by *Aspergillus* than to those required by other bacterial species. Pure cultures of *Nitrosomonas europaea* and *Nitrobacter winogradskyi* can grow on media previously treated with 8-hydroxyquinoline, and therefore containing infinitesimally small amounts of iron; but the nitrifiers grow much better if more iron is added. The stimulating effect of iron begins to appear at a concentration of 0·1 mg./l. for *Nitrosomonas* and 0·3 mg./l. for *Nitrobacter*, and the optimum amount for both species is *c*. 6 mg./l., which is very high for bacteria. Amounts of iron much higher than the optimum are tolerated by the nitrifiers; oxidation of ammonia or nitrite takes place at normal speed in the presence of 112 mg./l. of iron, though the process is slowed down in the presence of

560 mg./l. (*c.* 0·01 M). Manganese cannot replace iron, in fact it appears to be toxic in quite small amounts (Meiklejohn, 1953*a*). Since reagents which remove copper from the medium by chelation stop nitrification, it is probable that the nitrifiers require copper, but the amount needed is not yet known (Lees, 1948).

(5) *Inhibitory substances*

The only comprehensive study of the substances which depress growth and respiration of *Nitrosomonas* and *Nitrobacter* was carried out as long ago as 1916 by Meyerhof, using cultures from Winogradsky's laboratory, and working with the Warburg respirometer. This work may be summarized by saying that both species were poisoned by general cell poisons, and that they were especially sensitive to certain nitrogenous organic compounds, to lipoid-soluble substances and to relatively small quantities of certain electrolytes, in particular the ions of heavy metals. On the whole, *Nitrosomonas* was more sensitive than *Nitrobacter*. In the first place, potassium cyanide and narcotics such as urethane acted on the nitrifiers in the same concentrations as they act on other plant and animal cells. High concentrations of narcotics had a proportionately greater effect than low concentrations on oxygen uptake by *Nitrobacter*, and two urethanes applied together were more toxic than the sum of their separate toxicities. On the other hand, cyanide and a narcotic applied together had a total effect less than the sum of their separate effects.

The electrolytes to which the nitrifiers are sensitive include their own substrates, and in each case the substrate of the other species was found to be much more toxic than the organism's own substrate. Nitrite depressed both respiration and growth of *Nitrosomonas*, e.g. 0·1 M-sodium nitrite gave a 36 % depression of oxygen uptake, and 0·3 M caused complete inhibition. *Nitrobacter* was sensitive to the ammonium ion, but even more so to free ammonia, e.g. 0·001 M-ammonia at pH 9·5 gave a 70 % inhibition of oxygen uptake. Nitrate, like most other anions applied as the sodium salt, was only slightly toxic to both species, a concentration of the order 0·3 M being required before 100 % inhibition was approached. A large number of sodium salts were found to have about this level of toxicity, but borate and benzoate were more poisonous, and at 0·1 M each stopped the oxygen uptake of *Nitrosomonas* completely. Meyerhof pointed out that this effect may be connected with the lipoid-solubility of borates.

The cations least poisonous to the nitrifiers were those of the alkali and alkaline earth metals. The chlorides of sodium, potassium and magnesium completely inhibited oxygen uptake by *Nitrosomonas* at 0·5 M; sodium and potassium applied together had the same effect as either alone, but if magnesium was added the toxicity was lessened. Calcium, strontium and barium chlorides gave complete inhibition at 0·2 M, and caesium chloride at 0·05 M. Other metals were more poisonous; ferrous and ferric chlorides, and those of aluminium, zinc, copper, lead and manganese, gave substantial reductions in respiration at 0·01 M. Cobalt chloride gave a 60 % inhibition at 0·001 M, and nickel chloride complete inhibition at 0·0002 M. The most poisonous metals of all were mercury and silver. Mercuric chloride (which is lipoid-soluble) caused a progressive drop in respiration at 1×10^{-5} M, and silver chloride had a similar action at a concentration of only $2\cdot5 \times 10^{-6}$ M.

Organic substances which do not contain nitrogen were not very toxic to the nitrifiers. For instance, 0·3 M-glucose did not affect the respiration of *Nitrobacter* at all, and it required 0·8 M to give a 25 % inhibition. *Nitrosomonas* was slightly more sensitive to glucose, as 0·6 M-glucose depressed its oxygen uptake 40 %. Glycerol, mannitol, sodium acetate, butyrate and valerate were also easily tolerated.

On the other hand, nitrogenous organic substances were not readily tolerated by the nitrifiers, and some of them were very poisonous indeed. Amines were toxic to both species, aliphatic amines more so than aromatic, and the toxicity of amines increased with increasing lipoid-solubility. Both species tolerated high concentrations of urea, *Nitrosomonas* being the more sensitive, e.g. 0·1 M gave a 77 % inhibition with this species, but 0·5 M gave only 32 % with *Nitrobacter*. Asparagine also had little effect on *Nitrobacter* (0·15 M gave 20 % inhibition), but was much more toxic to *Nitrosomonas* (0·02 M gave 87 % inhibition).

Amino-acids were toxic to *Nitrosomonas* in concentrations of 0·025–0·01 M, and so was caffeine. Other alkaloids were more poisonous, nicotine being effective at 0·005 M, strychnine at 0·0005 M and quinine at 0·0002 M. Among substances of a comparable degree of toxicity, such as pyridine and pentamethylenediamine, hydroxylamine is especially interesting, since it is a possible intermediate in the oxidation of ammonia to nitrite. Meyerhof found that the respiration of *Nitrosomonas* was depressed 96 % by 0·0017 M-hydroxylamine. Of those tested, the most poisonous compounds included aniline and diphenylaniline (*o*- and *m*-nitroaniline and dimethylaniline were less toxic), guanidine, α-triphenylguanidine and aminoguanidine (nitroguanidine was much less toxic), α-naphthylamine, *p*-phenylenediamine and *p*-nitrosodimethylaniline. This last compound gave a 78 % inhibition of oxygen uptake by *Nitrosomonas* in a concentration of only 1×10^{-5} M.

Many workers have found that peptone is toxic to the nitrifiers, and media containing peptone are commonly used for a routine test of the purity of cultures of these organisms. Kingma Boltjes (1935) found that the most toxic kinds of peptone were those with the highest content of free amino-acids. Heubült (1929) found that 0·05% peptone (Merck) was enough to stop nitrite formation almost completely in pure cultures of *Nitrosomonas europaea*; like Meyerhof he found that urea, glucose, glycerol, sodium butyrate and sodium citrate were less toxic than peptone. The reasons for the toxicity of such a wide range of organic compounds are largely unknown, though Lees (1952) has shown that the toxicity of many compounds to the nitrifiers is proportional to their power to remove metals from the medium by chelation.

There are, however, no real grounds for the common text-book statement that organic matter as such is poisonous to the nitrifying bacteria. This statement is based on a paper by Winogradsky & Omeliansky (1899), in which they showed that nitrification in cultures was delayed or stopped by small concentrations of glucose, peptone, asparagine and sodium butyrate, and by large concentrations of urine, broth and sodium acetate. As none of these substances was generally thought to be poisonous, the paper attracted much attention at the time it appeared, and very sweeping conclusions were drawn from it. But the method used to measure nitrification was not very accurate, and the control cultures showed great variability. At all events, as far as glucose is concerned Kingma Boltjes (1935) found that *Nitrosomonas* would grow in a 4% solution, and that all the sugar could be recovered unchanged; in other words, glucose was tolerated by the nitrifier, but was not utilized. The poisonous substance in media containing glucose is a decomposition product formed by heat sterilization (Jensen, 1950). Ecological evidence shows that nitrification is associated with large quantities of natural organic matter, and there is no reason to think that inhibitory concentrations of any of the substances which have been shown to be very poisonous to nitrifiers in the laboratory are encountered by them in nature.

ARE THE NITRIFIERS OBLIGATE AUTOTROPHS?

So far there is no satisfactory experimental evidence that the nitrifying bacteria are facultative autotrophs, that is, capable of heterotrophic growth. Such evidence has been obtained for the sulphate-reducing bacteria, which were grown heterotrophically, then tested and found to be able to grow as autotrophs in pure culture (Butlin & Adams, 1947). Several statements have been made to the effect that nitrifiers can grow in media such as broth in the absence of ammonia or nitrite, but that this converts them irreversibly into a non-nitrifying form (Fremlin, 1903). As pure cultures of the nitrifiers are so easily overgrown by heterotrophic contaminants, any reports of this kind are open to the very strong suspicion that the bacteria which grew in broth and did not nitrify were not, and never had been, nitrifiers. It is astonishing that one of these reports came from no less a bacteriologist than Beijerinck (1914).

If the nitrifiers are obligate autotrophs, they must be capable of synthesizing all the growth factors they require; if they are not, it should be possible to find out which factors they cannot synthesize. Though the accumulation of nitrate in saltpetre sites is associated with the presence of large quantities of organic matter, it is possible that the only reason for this is that nitrogenous organic matter is a source of ammonia, the starting material for nitrification. On the other hand, several workers have thought that there might be some kind of organic stimulant for the nitrifiers in natural organic matter, and infusions of horse-dung and soil extract have been reported to hasten the oxidation of ammonia in cultures (Winogradsky & Omeliansky, 1899; Hes, 1937). It is possible, however, that the effects observed were due to the presence of some essential element such as iron or calcium in the extracts; at all events, the media used by Hes were deficient in iron. Meiklejohn (1953 d) found that yeast extract, thiamine and urine added to mixed impure cultures containing both *Nitrosomonas* and *Nitrobacter* had no effect in small doses, and delayed nitrification in larger doses. β-Indoleacetic acid in concentrations ranging from 0·01–2·5 mg./ml. of medium had no effect.

There is, however, one organic substance which has been found to have a specific effect in improving the growth of the nitrifiers. Kingma Boltjes (1935) found that much larger colonies of both *Nitrosomonas* and *Nitrobacter* were produced on agar plates, if 0·75 % of Nährstoff-Heyden, a preparation of egg albumen, was added to the agar. The Nährstoff-Heyden was apparently not assimilated by the nitrifiers, as no growth took place unless carbon dioxide was also present; it probably improved their growth because it poised the oxidation-reduction potential of the medium at a suitable level. Kingma Boltjes takes the view that the nitrifiers are not strictly autotrophic, but are 'chemomixotrophic', since their growth is improved by Nährstoff-Heyden and is better on media made with tap water than with distilled water, an effect which may be due to the presence of organic growth factors in tap water. On the other hand, it may of course be due to the tap water containing essential trace elements.

HAVE THE NITRIFIERS A NORMAL METABOLISM?

There appears to have been an impression among biologists that auto-trophic bacteria do not possess a normal metabolism, but that their enzyme systems are completely different from those found in all other types of living cell. This supposition has already been shown not to be true of *Thiobacillus thio-oxidans* (Umbreit, 1947). One of the differences between autotrophs and other bacteria was held to be that autotrophs were capable of oxidizing only their one specific energy-yielding sub-strate, sulphur for *T. thio-oxidans*, ammonia for *Nitrosomonas*, nitrite for *Nitrobacter*. Vogler (1942) showed that in the absence of sulphur, *Thiobacillus thio-oxidans* took up oxygen and gave off carbon dioxide, and that therefore it must synthesize an organic storage material which it can break down to keep it alive in the absence of sulphur. It has been shown by Bömeke (1939) that the nitrifiers also have an endogenous respiration. Although the oxygen uptake of growing cultures of nitrifiers can all be accounted for by assuming that the oxidation of ammonia or nitrite is the only reaction taking place, Bömeke found that very thick suspensions made by centrifuging old cultures of *Nitrosomonas* and *Nitrobacter* showed small but measurable oxygen uptakes in the absence of ammonia and nitrite respectively. He attributed this to the break-down of cell substance synthesized during the growth of the culture. He also found that the small oxygen uptake was slightly increased by the addition of the following organic substances: yeast-water, broth, soil extract, peptone, sodium acetate, sodium lactate and calcium glycero-phosphate. No increase occurred on the addition of asparagine or urea. Bömeke states that the increase in oxygen uptake which followed the addition of these organic substances was too small to be the result of contamination by a heterotroph.

A little more may be deduced about the enzyme systems involved in the oxidation of ammonia and nitrite, and the synthesis of cell substance. It is evident that enzymes containing a heavy metal, probably iron, but perhaps also copper, are involved. This is shown by the toxicity of potassium cyanide and narcotics, and the action of chelating agents. The very large optimum iron requirement of both *Nitrosomonas* and *Nitro-bacter* is an indication that both species possess a complete cytochrome system, but this has not yet been investigated. Vogler & Umbreit (1942) were able to show that *Thiobacillus thio-oxidans* stores the energy derived from the oxidation of sulphur in compounds containing energy-rich phosphate bonds. It has not yet been possible to demonstrate this in the case of the nitrifiers, though phosphate is undoubtedly essential for their growth.

Fat plays a very important part in the metabolism of *T. thio-oxidans*, as sulphur particles are made accessible by being dissolved in the fat globules contained in its cells (Umbreit, Vogel & Vogler, 1942). It is difficult to see what part fat can play in the metabolism of the nitrifiers, which act on water-soluble substrates, and so Meyerhof's observation that the toxicity of a variety of poisons to the nitrifiers is related to their lipoid solubility, though interesting, is puzzling.

ARE THE NITRIFIERS PRIMITIVE ORGANISMS?

No organic stimulant for the nitrifiers has yet been discovered, so that it may be assumed that they are able to synthesize their entire cell substance from carbon dioxide, if the energy and nitrogen source (ammonia or nitrite) and the essential inorganic nutrients are present. According to Lwoff (1944), the ability to synthesize all the necessary cell constituents is an indication of the primitive character of the organism. The earliest forms of life to appear on earth had complete synthetic powers, and, as they developed and were differentiated, there was a progressive loss of the ability to synthesize cell constituents. If synthetic power were the only criterion of primitive character, then the nitrifiers must be regarded as being among the first organisms to appear. But other considerations make it likely that they were, though early, not the first forms of life. It is true that ammonia, the starting-point for nitrification, is thought to have been present in the earth's atmosphere before life developed, but it is very doubtful whether the primordial atmosphere contained any oxygen (Oparin, 1938). If this really was the case, then the first living things, though autotrophs, must have been independent of oxygen, which the nitrifiers certainly are not.

REFERENCES

AMER, F. M. & BARTHOLOMEW, W. V. (1951). Influence of oxygen concentration in soil air on nitrification. *Soil Sci.* **71**, 215.

ARND, T. (1919). Zur Kenntnis der Nitrifikation in Moorböden. *Zbl. Bakt.* (2. Abt.), **49**, 1.

BEIJERINCK, M. W. (1914). Über das Nitratferment und über physiologische Artbildung. *Folia microbiol., Delft,* **3**, 91.

BOERHAAVE, H. (1753). *Elementa Chemiae,* trans. P. Shaw. London: Longman.

BÖMEKE, H. (1939). Beiträge zur Physiologie nitrifizierender Bakterien. *Arch. Mikrobiol.* **10**, 385.

BÖMEKE, H. (1949). Über die Ernährungs und Wachstumsfaktoren der Nitrifikationsbakterien. *Arch. Mikrobiol.* **14**, 63.

BOULLANGER, E. & MASSOL, L. (1904). Études sur les microbes nitrificateurs. *Ann. Inst. Pasteur,* **18**, 181.

BOUSSINGAULT, J. B. (1844). *Economie Rurale.* Paris: Déchet Jeune.

BUTLIN, K. R. & ADAMS, M. E. (1947). Autotrophic growth of sulphate-reducing bacteria. *Nature, Lond.* **160**, 154.

FREMLIN, H. S. (1903). On the culture of the Nitroso-bacterium. *Proc. Roy. Soc. B,* **71**, 356.

GIBBS, W. M. (1919). The isolation and study of nitrifying bacteria. *Soil Sci.* **8**, 427.

GRIFFITH, G. AP. & MANNING, H. L. (1949). A note on nitrate accumulation in a Uganda soil. *Trop. Agriculture, Trin.* **26**, 108.

HES, J. W. (1937). Zur Stoffwechselphysiologie von *Nitrosomonas. Rec. Trav. bot. Néerl.* **34**, 233.

HESSELMAN, H. (1917). Studien über die Nitratbildung in natürlichen Böden. *Medd. Skogsförsöksanst. Stockh.* **14**, 297.

HEUBÜLT, J. (1929). Untersuchungen über Nitritbakterien. *Planta,* **8**, 398.

IMSENECKI, A. A. (1946). Symbiosis between myxobacteria and nitrifying bacteria. *Nature, Lond.* **157**, 877.

JENSEN, H. L. (1950). Effect of organic compounds on *Nitrosomonas. Nature, Lond.* **165**, 974.

JENSEN, H. L. (1951). Notes on the microbiology of soil from northern Greenland. *Medd. Grønland,* **142**, 23.

KINGMA BOLTJES, T. Y. (1935). Untersuchungen über die nitrifizierenden Bakterien. *Arch. Mikrobiol.* **6**, 79.

LEES, H. (1948). The effects of zinc and copper on soil nitrification. *Biochem. J.* **42**, 534.

LEES, H. (1952). The biochemistry of the nitrifying organisms. I. The ammonia-oxidizing systems of *Nitrosomonas. Biochem. J.* **52**, 134.

LEES, H. & QUASTEL, J. H. (1946). The site of soil nitrification. *Biochem. J.* **40**, 815.

LWOFF, A. (1944). *L'Évolution Physiologique. Étude des Pertes de Fonctions chez les Microorganismes.* Paris: Hermann.

MEIKLEJOHN, J. (1950). The isolation of *Nitrosomonas europaea* in pure culture. *J. gen. Microbiol.* **4**, 185.

MEIKLEJOHN, J. (1952). Minimum phosphate and magnesium requirements of nitrifying bacteria. *Nature, Lond.* **170**, 1131.

MEIKLEJOHN, J. (1953*a*). Iron and the nitrifying bacteria. *J. gen. Microbiol.* **8**, 58.

MEIKLEJOHN, J. (1953*b*). The nitrifying bacteria: a review. *J. Soil. Sci.* **4**, 59.

MEIKLEJOHN, J. (1953*c*). The microbiological aspects of soil nitrification. *E. Afr. agric. J.* **19** (in the Press).

MEIKLEJOHN, J. (1953*d*). Some organic substances and the nitrifying bacteria. *Proc. Soc. appl. Bact.* **15**, 77.

MEYERHOF, O. (1916). Untersuchungen über den Atmungsvorgang nitrifizierender Bakterien. *Pflüg. Arch. ges. Physiol.* **164**, 352; **165**, 229; **166**, 240.

OMELIANSKY, V. (1899). Über die Isolierung der Nitrifikationsmikroben aus dem Erdboden. *Zbl. Bakt.* (2. Abt.), **5**, 652.

OPARIN, A. I. (1938). *The Origin of Life.* New York: Macmillan.

ROMELL, L. G. (1927). En nitritbakterien ur svensk Skogsmark. *Medd. Skogsför-söksanst. Stockh.* **24**, 57.

SCHLOESING, T. & MÜNTZ, A. (1877). Sur la nitrification par les ferments organisés. *C.R. Acad. Sci., Paris,* **84**, 301.

UMBREIT, W. W. (1947). Problems of autotrophy. *Bact. Rev.* **11**, 157.

UMBREIT, W. W., VOGEL, H. R. & VOGLER, K. G. (1942). The significance of fat in sulphur oxidation by *Thiobacillus thio-oxidans. J. Bact.* **43**, 141.

VOGLER, K. G. (1942). The presence of an endogenous respiration in the autotrophic bacteria. *J. gen. Physiol.* **25**, 617.

VOGLER, K. G. & UMBREIT, W. W. (1942). Studies on the metabolism of the autotrophic bacteria. 3. The nature of the energy storage material active in the chemosynthetic process. *J. gen. Physiol.* **26**, 157.

WAKSMAN, S. A. & MADHOK, M. R. (1937). Influence of light and heat upon the formation of nitrate in soil. *Soil Sci.* **44**, 361.

WARINGTON, R. (1878). On nitrification. *J. chem. Soc.* **33**, 44.

WARINGTON, R. (1879). On nitrification. Part 2. *J. chem. Soc.* **35**, 429.

WARINGTON, R. (1884). On nitrification. Part 3. *J. chem. Soc.* **45**, 637.

WARINGTON, R. (1891). On nitrification. Part 4. *J. chem. Soc.* **59**, 484.

WINOGRADSKY, S. (1890). Recherches sur les organismes de la nitrification. Parts 1–3. *Ann. Inst. Pasteur*, **4**, 213, 257, 760.

WINOGRADSKY, S. (1891). Recherches sur les organismes de la nitrification. Parts 4 and 5. *Ann. Inst. Pasteur*, **5**, 92, 577.

WINOGRADSKY, S. (1904). Die Nitrifikation. In *Handbuch der Technischen Mykologie* (ed. F. Lafar, 1904–6), **3**. Jena: Fischer.

WINOGRADSKY, S. & OMELIANSKY, V. (1899). Über den Einfluss der organischen Substanzen auf die Arbeit der nitrifizierenden Bakterien. *Zbl. Bakt.* (2. Abt.), **5**, 329, 377, 429.

WINOGRADSKY, S. & WINOGRADSKY, H. (1933). Études sur la microbiologie du sol. 7. Nouvelles recherches sur les organismes de la nitrification. *Ann. Inst. Pasteur*, **50**, 350.

THE BIOCHEMISTRY OF THE NITRIFYING BACTERIA

H. LEES

Department of Biological Chemistry, University of Aberdeen

The task of dealing with the biochemistry of the nitrifying bacteria, *Nitrosomonas* and *Nitrobacter*, is not an easy one. This is not because the subject is inherently difficult, but because it scarcely exists. The known biochemistry of the nitrifiers consists of a few isolated observations made on soils or cultures; not one single reaction sequence has been worked out in plan, let alone in detail. The reason for this unsatisfactory state of affairs, which is in such dismal contrast to the vast and growing knowledge about heterotrophic organisms, is not difficult to find. It lies almost entirely in the technical difficulties involved in obtaining a crop of live organisms sufficiently large for biochemical studies. Even under optimum conditions, as far as these are at present understood, growth of the nitrifiers in pure culture is always slow and gives a poor yield of organism. Ideally one would wish for a rapid and heavy growth, but if this aim could be only partially achieved, i.e. were it possible to produce a heavy crop slowly or a light crop quickly, then the supply of live organisms would be considerably greater than it is at present. In what follows, therefore, I shall try not only to survey what knowledge we have of the biochemistry of the nitrifiers (in so far as this has not been done by Dr Meiklejohn, see p. 68, this volume), but also to speculate a little on means whereby a more satisfactory growth of the organisms might be attained.

GENERAL BIOCHEMISTRY OF THE NITRIFYING BACTERIA

Living organisms seem to resemble one another biochemically far more than they differ. The patterns of molecular architecture and the sequence of molecular movements are, as far as we can surmise, much the same in *Escherichia coli* as they are in *Nitrosomonas* and *Nitrobacter*. It would seem very doubtful whether the metabolism of most biochemically significant compounds within the cell boundaries of *Escherichia coli* is much different from the metabolism of these same compounds within the cell boundaries of *Nitrosomonas* and *Nitrobacter*. Where these two autotrophs differ from the heterotroph is in:

(*a*) their ability to synthesize every compound required for growth and metabolism, the synthetic abilities of the heterotroph being rather less complete;

(*b*) their ability to utilize for growth and metabolism the free energy released by an inorganic oxidation; this may not be an absolute differentiation between autotrophs and heterotrophs because it is always possible that the free energy released by the very small inorganic oxidations carried out by some heterotrophs (Cutler & Mukerji, 1931; Isenberg, Schatz & Hutner, 1952) is available to the organisms;

(*c*) their almost complete inability to utilize compounds they require for growth and metabolism when these compounds are supplied extracellularly. This inability may be connected with a highly selective absorption mechanism or rather less probably with an exceptionally integrated metabolism relatively insensitive to an artificial increase in the concentration of one of the metabolites.

The biochemistry of the autotrophs in general and, for our present purpose, the biochemistry of *Nitrosomonas* and *Nitrobacter* in particular, thus presents us with a number of interesting questions. Is the composition of the organisms much the same as the composition of heterotrophs, e.g. are the same amino-acids found both in autotrophs and heterotrophs? Are they completely indifferent to all externally supplied organic compounds or do some organic compounds stimulate their growth? It may, I think, be justly assumed that the characteristic inorganic oxidations performed by these organisms are used to supply reducing power to mechanisms reducing carbon dioxide, much as the photolytic reaction is used in photosynthesis (Vishniac & Ochoa, 1951); by what steps, then, are these oxidations carried out? What is the nature of the ordinary 'heterotrophic' cell metabolism of *Nitrosomonas* and *Nitrobacter*?

The answers to all these questions are, unfortunately, largely obscure at the present time. Such answers as can be given form the substance of this paper.

HETEROTROPHIC METABOLISM IN NITRIFYING BACTERIA

Heterotrophic metabolism in the nitrifiers must, of course, exist; the maintenance of order in the labile pattern of the cell constituents necessarily involves the expenditure of energy (Schrödinger, 1944). Owing to the difficulty in obtaining sufficient cell material for study, heterotrophic metabolism in the nitrifiers has, however, been studied only once. Bömeke (1939), using centrifuged suspensions, examined the metabolism of the nitrifiers manometrically in the absence of inorganic substrates. A small endogenous uptake of oxygen was observed; slight stimulations were noted with a number of compounds, e.g. acetate and lactate. We have confirmed the existence of the 'rest respiration' noted by Bömeke (see section on Free-Energy Efficiency, p. 93), but the usual paucity of cell material discouraged us from further investigations.

GROWTH FACTORS AND THE NITRIFYING BACTERIA

Nitrosomonas and *Nitrobacter* can undoubtedly develop in an entirely inorganic medium; whether they achieve maximal growth rates under these conditions is not necessarily certain. If they can synthesize some essential compound only at a limited rate, then their rate of growth will be limited by the rate of synthesis. Moreover, if it be assumed that the required compound can be absorbed by the cells, addition of the compound to the inorganic medium should stimulate the growth rate. Fred & Davenport (1921), Murray (1923), Kingma Boltjes (1935) and Hes (1937) have all noted that various organic compounds augment the growth of the nitrifiers. These results are in harmony with the common observation that nitrification proceeds most readily in organic soils, in dung heaps, in sewage, and in mixed (rather than pure) cultures. Meiklejohn (1953), on the other hand, found no stimulation of nitrification in mixed cultures by soil extract, yeast extract, thiamine, or urine. However, Dr T. Hofman & I (unpublished) have noted an irregular reduction in the lag period of *Nitrosomonas* when biotin is added to the inorganic medium.

The subject merits further study because the discovery of a factor stimulating the growth of the nitrifiers, if such a factor exists, could well result in much larger crops of organism than we can obtain under current culture conditions.

COMPOSITION OF NITRIFYING BACTERIA

Hofman (1953) has determined the amino-acids and carbohydrates present in hydrolysates of *Nitrosomonas*. The amino-acid spectrum of the organism was quite normal, no unusual amino-acid being found. The results of the carbohydrate analyses were somewhat unexpected, however, since although glucose appeared to be absent, galactose, ribose, rhamnose, xylose and an unknown carbohydrate were all detected. As no enzymic hydrolysis of the carbohydrate fraction was carried out, deoxyribose did not appear on the carbohydrate chromatograms, although it was presumably present in the organisms.

Similar investigations are now being carried out with *Nitrobacter*, but final data are not yet available. Preliminary analyses have, however, indicated nothing unusual in the amino-acid composition of the organisms.

PRIMARY OXIDATION REACTIONS OF THE NITRIFYING BACTERIA

A. *Primary oxidation reaction of* Nitrosomonas

By the term 'primary oxidation reaction' is meant the inorganic oxidation that releases the free energy subsequently utilized by the organism for growth, i.e. the free energy that is utilized to reduce carbon dioxide to the oxidation level of the cell constituents. If the energy released by the primary oxidation reaction is to be transported by the normal type of high-energy phosphate bond systems to the mechanisms responsible for the assimilation of carbon dioxide, and if the overall utilization of the energy is to be efficient,* then the energy must be released by the primary oxidation reactions in steps of 10–15 kcal., i.e. in amounts capable of being transported by the known energy-transport systems. Moreover, if the reducing power generated by the primary oxidation reaction is to be effectively utilized for the reduction of carbon dioxide, the primary oxidation reaction must consist of, or be intimately coupled to, dehydrogenation reactions leading eventually to the mechanisms assimilating carbon dioxide. On both counts the oxidation of ammonia by *Nitrosomonas* must be a stepwise process, since the total loss of free energy during the completion of the reaction

$$NH_4^+ + 1\tfrac{1}{2}O_2 \rightarrow 2H^+ + H_2O + NO_2^-$$

is *c.* 66 kcal. at physiological concentrations of the reactants (Baas-Becking & Parks, 1927), while the generation of reducing power is equivalent to three pairs of hydrogen atoms.

The simplest possible reaction sequence to account for the oxidation is that advanced by Kluyver & Donker (1926) who proposed the following scheme:

(1) ammonia → hydroxylamine,
(2) hydroxylamine → hyponitrite (or dihydroxyammonia),
(3) hyponitrite → nitrite.

Let us now examine the evidence for the occurrence of each of these steps.

---

* These terms are widely used, widely understood, and do not normally confuse anyone accustomed to employing them. They are retained here despite recent criticisms of their accuracy (Gillespie, Maw & Vernon, 1953).

(1) *Ammonia → hydroxylamine*

This reaction almost certainly takes place (Hofman & Lees, 1953). It is specifically inhibited by micromolar concentrations of thiourea and allylthiourea; *Nitrosomonas* cells suspended in 10^{-5} M-thiourea or allylthiourea solutions were incapable of oxidizing ammonia but would oxidize hydroxylamine at an undiminished rate. It was also found that the oxidation of hydroxylamine could be prevented by hydrazine, and in the presence of this inhibitor cells metabolizing ammonia were found to accumulate hydroxylamine but failed to form any nitrite. There are some reasons for believing that the enzyme involved here is a copper enzyme. The oxidation of ammonia in soil and culture is generally inhibited by chelating agents (Lees, 1946, 1952), especially by the thioureas, which are known to have a particularly high affinity for copper. This is, of course, inferential reasoning. The enzyme oxidizing ammonia has not been isolated, and whether it actually contains copper is therefore not known (Quastel & Scholefield, 1951).

According to Martin, Buehrer & Caster (1942) the loss of free energy involved in the oxidation of ammonia to hydroxylamine is very small. It therefore follows that the free-energy efficiency of *Nitrosomonas* would not be markedly affected if this oxidation is not coupled to the energy-transfer systems. If we make the assumption that such coupling is in fact absent, the first step in the primary oxidation reaction of *Nitrosomonas* may be formulated:

$$2NH_4^+ + O_2 \rightarrow 2NH_2OH + 2H^+.$$

This formulation is attractive for several reasons. Firstly, the ionic species are correct. Within the pH range in which *Nitrosomonas* exhibits greatest activity (pH 6–8·6), the bulk of the ammonia (pK 9·5) will be cationic whereas the bulk of the hydroxylamine (pK 5·9) will be uncharged. The generation of a proton at this point should thus seem to be inevitable. Secondly, the enzyme catalyzing the reaction is depicted as dealing directly with molecular oxygen and this is characteristic of all the known copper enzymes, polyphenol oxidase, monophenol oxidase, tyrosinase and ascorbic acid oxidase. Thirdly, the results of Meyerhof (1917) suggest that whereas NH_4^+ cannot easily penetrate the cell wall, NH_3 can and does; it might therefore be expected that while *Nitrosomonas* cannot easily take in cations, it can and does easily take in neutral molecules. Engel (1941) believed that ammonia oxidation in *Nitrosomonas* took place on the cell surface and not within the cytoplasm. If this theory is correct, the first step in ammonia oxidation by *Nitrosomonas* would simply be, according to the formulation used here, a device whereby the nitrogenous substrate of the primary oxidation reaction is transformed from a cationic form (NH_4^+) almost incapable of penetrating the cell wall to a neutral form (NH_2OH) capable of doing so. Subsequent oxidation of the NH_2OH, which apparently involves effectively all the free-energy loss entailed in the oxidation of ammonia to nitrite (Martin *et al.* 1942), could then take place within the cytoplasm of the cells where suitable energy coupling mechanisms would be readily available.

(2) *Hydroxylamine → X (dihydroxyammonia? nitroxyl? hyponitrite?)*

Since the elucidation of the first stage of ammonia oxidation, this second stage has become the key to a complete understanding of the primary oxidation reaction sequence of *Nitrosomonas*. The end-product of this second stage is, indeed, the only unknown intermediate in the whole sequence. Various possibilities can be suggested:

(*a*) *Hydroxylamine → dihydroxyammonia.* This can be formulated most logically as:

$$NH_2OH \cdot H_2O - 2H \rightarrow NH(OH)_2.$$

Such a reaction may well occur; the difficulty of proving that it does so lies in the fact that dihydroxyammonia has never been prepared because it is so unstable. It is conceivable that it has a transient existence on an enzyme surface prior to its dehydrogenation to nitrite. But such an existence could only be inferred from other data; free dihydroxy-ammonia could never be isolated from, or added to, metabolizing systems of *Nitrosomonas*.

(*b*) *Hydroxylamine → nitroxyl.* This reaction would result from the direct dehydrogenation of hydroxylamine:

$$NH_2OH - 2H \rightarrow NOH.$$

However, proof or disproof of the operation of such a reaction presents the same difficulties as appeared in connexion with reaction (*a*), since nitroxyl is also unstable.

(*c*) *Hydroxylamine → hyponitrite.* Largely owing to the fact that hyponitrite is a relatively stable ion, the possibility that hydroxylamine is converted to hyponitrite is the one which has received most attention. Since all the known hydrogen carriers such as pyridine nucleotides can accommodate only two hydrogen atoms at a time, this reaction, if it occurs, must almost certainly be a two-stage process involving the inter-mediate formation of either nitroxyl:

$$NH_2OH - 2H \rightarrow NOH,$$
$$2NOH \rightarrow HON:NOH,$$

or dihydroxyhydrazine:

$$2NH_2OH - 2H \rightarrow (HO)NH.NH(OH),$$
$$(HO)NH.NH(OH) - 2H \rightarrow HON:NOH.$$

The immediate question is, however, does hyponitrite occur at all as an intermediate in the primary oxidation reaction of *Nitrosomonas*? Evidence on this point is conflicting. Mumford (1914) and Corbet (1935) both seem to have detected hyponitrite in nitrifying cultures; on the other hand, Tanator (quoted by Mellor, 1928, p. 289) claimed that hyponitrite can be formed by purely chemical reaction in solutions containing small concentrations of both hydroxylamine and nitrite, especially if chalk is present. Since all these conditions are likely to be satisfied in growing cultures of *Nitrosomonas*, it is quite possible that the hyponitrite observed by Mumford (1914) and Corbet (1934, 1935) in such cultures was not formed biologically. Moreover, Dr T. Hofman and I have failed completely to observe any nitrite formation in washed suspensions of *Nitrosomonas* to which hyponitrite had been added (concentrations of hyponitrite ranging from 10^{-4} to $2 \cdot 5 \times 10^{-3}$ M were tested). Jensen, H. L. (1953, personal communication) was similarly unsuccessful in parallel experiments. However, it should be remembered that both Jensen's experiments and our own were carried out on whole cells; it is therefore possible that these negative results were due to inability of the hyponitrite to penetrate the cell wall.

(3) *X (dihydroxyammonia? nitroxyl? hyponitrite?)* → *nitrite*

Since the substrate has not been identified, useful speculation about the reaction is almost impossible. It may be argued that hyponitrite is not the substrate because, firstly, it does not act as a substrate when added extracellularly; secondly, that its conversion to nitrite by dehydrogenation reactions would be a complicated process, and thirdly (though perhaps this is less relevant), it is not an intermediate in the reverse process of denitrification as studied in *Pseudomonas stutzeri* (Allen & van Niel, 1952) or in cell-free extracts of *Bacillus pumilis* (Taniguchi, Mitsui, Toyoda, Yamada & Egami, 1953). In favour of the participation of dihydroxyammonia it may be pointed out that the very transience of its existence harmonizes with the fact that the substrate of this stage has so far defied attempts at identification. Moreover, the dehydrogenation of dihydroxyammonia would give rise to nitrite directly without any of the complications, such as the splitting of a nitrogen-nitrogen bond, involved in the conversion of hyponitrite to nitrite.

B. *Primary oxidation reaction of* Nitrobacter

The primary oxidation reaction of *Nitrobacter*,

$$NO_2^- + \tfrac{1}{2}O_2 \rightarrow NO_3^-,$$

involves a free-energy loss of 17·5 kcal. at physiological concentrations of the reactants (Baas-Becking & Parks, 1927), while the generation of reducing power is equivalent to one pair of hydrogen atoms. On both these counts the reaction is almost certainly a one-stage process.

At first glance it seems that the reaction could not be a simple dehydrogenation and that linkage of the reaction to dehydrogenation mechanisms must be through some cycle involving catalase or peroxidase or indeed any enzyme capable of dealing with molecular oxygen. But this assumption may not be correct. Mellor (1928, p. 466) presents evidence to show that nitrous acid can add on water to give a dihydroxy molecule; if this addition takes place, then the resultant molecule might well be dehydrogenated to yield nitric acid:

$$HNO_2 + H_2O \rightarrow O:NH(OH)_2,$$

$$O:NH(OH)_2 - 2H \rightarrow HNO_3.$$

Nason & Evans (1953) have in fact shown that nitrate reductase preparations from *Neurospora* reduce nitrate to nitrite at the expense of TPNH which becomes oxidized to TPN+.* Whatever the mechanism of the reaction may be, this work of Nason & Evans does show that in *Neurospora* the reduction of nitrate can be linked, presumably in a fairly simple way, with hydrogen transport systems. One may reasonably suppose that, in *Nitrobacter*, equally simple links might exist coupling nitrite oxidation with hydrogen transport systems.

Practically nothing is known about the enzyme involved in nitrite oxidation in *Nitrobacter*. Quastel & Scholefield (1951) studied the effects of various inhibitors on nitrite oxidation in soil, and to these workers goes the credit of having discovered a highly selective inhibitor of the oxidation, namely, nitrourea. In so far as we have repeated on cultures of *Nitrobacter* the experiments carried out by Quastel & Scholefield on soils, we have found no marked discrepancy between their results and ours.

Early in our work on *Nitrobacter* we were interested, as indicated at the beginning of this section, in the possible role of catalase in nitrite oxidation. The organisms certainly had catalase activity sensitive to

---

* TPNH and TPN+ represent reduced and oxidised forms of triphosphopyridine nucleotide respectively.

poisoning by azide, cyanide and hydroxylamine; nitrite oxidation as measured in the Warburg apparatus was also sensitive to these same three poisons. However, when the inhibitions of catalase activity brought about by the three poisons were compared quantitatively with the inhibitions of nitrite oxidation, it was found that no correlation existed, e.g. azide was a strong poison of nitrite oxidation but only a weak one of catalase activity, whereas with hydroxylamine the situation was reversed (Fig. 1). It therefore seems likely that catalase plays only a minor part, if any, in the oxidation of nitrite by *Nitrobacter*.

THE pH OPTIMA OF THE NITRIFYING ORGANISMS

During the course of his studies on the biochemistry of the nitrifying organisms Meyerhof (1916, 1917) investigated the effect of pH on the

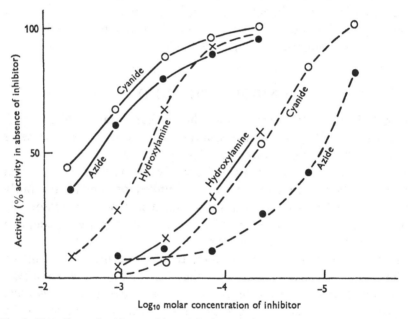

Fig. 1. The effects of azide, cyanide and hydroxylamine on catalase activity and nitrite oxidation in *Nitrobacter*. Catalase activity (—), nitrite oxidation (‑ ‑ ‑).

activity of *Nitrosomonas* and *Nitrobacter*. He found the pH optimum for ammonia oxidation by *Nitrosomonas* to be 8·6 with almost zero activity at pH 7·6 and 9·6. The corresponding figures for nitrite oxidation by *Nitrobacter* were pH 8·6 (the optimum) with pH 5·5 and 10 as the extreme limits. Somewhat different results were obtained by Meek & Lipmann (1922) and by Winogradsky & Winogradsky (1933).

For reasons that have already been fully discussed (Hofman & Lees, 1953) we have reinvestigated the pH-activity curves of both *Nitrobacter* and *Nitrosomonas*. For *Nitrosomonas* we found, in agreement with Meyerhof, that the pH optimum for ammonia oxidation was 8·6 and that the rate fell almost to zero at pH 9·6. On the other hand, ammonia oxidation proceeded, in our experiments (Hofman & Lees, 1953), with considerable speed at pH's as low as 6·5 (Fig. 2*a*); this agrees well with what is observed in soil where nitrification will usually continue quite rapidly at a pH as low as 5·5. For *Nitrobacter* we find (Fig. 2*b*) that the

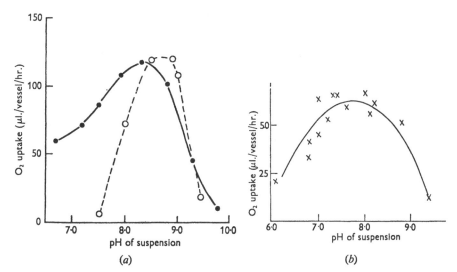

Fig. 2. Relation between pH of medium and rate of oxygen uptake by (*a*) *Nitrosomonas* and (*b*) *Nitrobacter*. Organisms suspended in phosphate-borate buffer of required pH and containing (*a*) 280 μg. NH_3-N/ml. and (*b*) 200 μg. NO_2-N/ml. (for experimental details, see Hofman & Lees, 1953). Dotted curve in (*a*) represents results of Meyerhof (1917) redrawn.

pH optimum for nitrite oxidation is 7·7 and not 8·6 as found by Meyerhof. Our own observations agree well with the results of soil studies carried out by Martin *et al.* (1942), in which it was shown that nitrate formation in alkaline soils treated with ammonium salts did not occur until the activities of *Nitrosomonas* had brought the pH to below a 'threshold value' of 7·7.

FREE-ENERGY EFFICIENCIES OF THE NITRIFYING BACTERIA

Baas-Becking & Parks (1927) calculated, from the experimental results obtained by other workers, the free-energy efficiencies of several autotrophs and came to the conclusion that these efficiencies were, in general, low. The principle of their calculations was as follows. The reduction of 1 g. of carbon from carbon dioxide to the oxidation level of carbon in the cell constituents consumes about 10,000 cal. of free energy (value obtained from combustion experiments with cells of autotrophs). The oxidation of 1 g. of substrate releases Z cal. of free energy, where Z has a value dependent on the substrate. Then, if analyses show that in a given culture of the autotroph A g. of substrate have been oxidized and B g. of carbon have been assimilated as cell material, then the free-energy efficiency of the autotroph is

$$\frac{B \times 10,000}{A \times Z} \times 100\%.$$

This calculation is perfectly valid so long as one remembers that the value obtained applies solely to the stage of growth at which the analyses were made. It is not necessarily justifiable to assume that the free-energy efficiency so calculated is constant for any one type of organism throughout all stages of growth of a culture of that organism. The amount of organic carbon (B) found in the culture at the time of sampling is not simply the amount that has been assimilated by the operation of the chemosynthetic mechanisms during growth of the culture; it is this amount *minus* the amount lost by respiration during the same period. The free-energy efficiency calculated by Baas-Becking & Parks is therefore in effect

$$\frac{(F-R) \times 10,000}{A \times Z} \times 100\%,$$

where $F =$ the total amount of carbon that has been assimilated during growth of the culture and $R =$ the total amount of carbon lost by respiration during growth of the culture. Now this expression cannot have a constant value unless $(F/A - R/A)$ remains constant. The value of F/A may well, perhaps, as a first approximation, be considered as constant, since it represents the degree of energy coupling between the primary inorganic oxidation reaction and the assimilatory mechanisms. No such assumption can be made about the value of R/A, which relates respiration and the primary inorganic oxidation reaction. For example, in the early stages of growth of a culture of *Nitrosomonas*, the primary inorganic oxidation reaction, carbon assimilation and respiration must all be proceeding simultaneously. However, the culture will eventually reach a stage beyond which further increase in nitrite concentration is impossible (this stage is usually attained when the nitrite concentration has risen to about 0·1 M). Consequently, both F and A will become constant and F/A will also become constant. But since at this stage the culture must contain many viable cells and respiration must therefore be going on, R must be increasing and the value of R/A must also be increasing. Therefore the value of $(F/A-R/A)$ must be decreasing. This value, on which the calculations of Baas-Becking & Parks were based, may thus very well depend on the stage of growth which the culture had attained at the time of sampling.

But if we ask ourselves what we mean when we talk about the free-energy efficiency of the autotrophs, I think the answer will probably be 'the efficiency implied by the ratio F/A, or at least, the efficiency implied by the maximal value of $(F/A - R/A)$, i.e. the intrinsic efficiency of chemosynthesis minimally complicated by respiratory factors'. In order to obtain maximum values for the free-energy efficiency, the determinations must be made when respiratory losses of carbon are minimal. Now it does not seem unreasonable to suppose that the ionic composition of the cytoplasm of autotrophic cells remains tolerably constant despite any changes in the ionic composition of the fluid medium. If this supposition is correct, and if the activities of the cells result in profound changes in the ionic composition of the fluid medium (e.g. the production of easily diffusible and doubtless toxic NO_2^- ions by *Nitrosomonas*), a relative increase in respiration might well be expected to occur as growth of the culture proceeded, because the organisms would be constrained to maintain within their cytoplasm a tolerably low concentration of nitrite against an ever-increasing concentration difference of nitrite between the cytoplasm and the external fluid medium. If nitrite is capable of diffusing across the cell wall, and there seems to be no good reason to suppose that it cannot, then a low cytoplasmic nitrite concentration could be maintained against an ever-increasing concentration of nitrite in the fluid medium only by the expenditure of energy in a mechanism excreting nitrite. In other words, the cells would be expected to have an ever-increasing 'osmotic respiration' consequent on the accumulation of the end-product of the primary inorganic reaction. Since the operation of the osmotic respiration would lower the overall free-energy efficiency as determined by the method used by Baas-Becking & Parks, we might expect that this overall free-energy efficiency would be higher in young cultures of *Nitrosomonas* (where the nitrite concentration of the fluid medium is small) than in older cultures (where the nitrite concentration is higher).

Whether the foregoing is or is not an accurate representation of the situation in growing cultures of *Nitrosomonas*, the fact remains that recent results (Hofman & Lees, 1952) suggest that the overall free-energy efficiencies of *Nitrosomonas* cultures fell as the cultures grew and accumulated nitrite. In the very early stages of growth, the overall free-energy efficiency was of the order of 40 % (this figure is difficult to determine accurately). In old cultures that had attained maximum growth and nitrite accumulation, the overall free-energy efficiency was 7 %. The sometimes rather conflicting determinations of overall free-energy efficiency of *Nitrosomonas* made by different workers (Winogradsky, 1890; Nelson, 1931; Hes, 1937; Bömeke, 1951) could all be reconciled with the recent determinations and with each other if due account were taken of the nitrite concentration in the culture when the determination was made. It is interesting to note in this connexion that the data of Vogler (1942) and of Vogler & Umbreit (1942) suggest that the intrinsic efficiency of the energetic coupling between sulphur oxidation and carbon dioxide assimilation in *Thiobacillus thio-oxidans* is about 90 %, whereas the overall free-energy efficiency of advanced cultures was 6·2 % (Waksman & Starkey, 1922). *T. thio-oxidans* resembles *Nitrosomonas* in so far as its activities result in changes in the ionic environment of the cells, and it may well be that in this organism too the operation of an 'osmotic respiration' accounts for the low overall free-energy efficiency of well-developed cultures.

Unfortunately this theory of an 'osmotic respiration' is, as far as *Nitrosomonas* is concerned, very difficult to test in practice. The 'rest respiration' of the organisms is low (Bömeke, 1939) and adequate amounts of cells cannot be easily obtained by any of the known culture methods. Warburg experiments on the effect of increasing nitrite concentrations on the rest respiration of the cells in the absence of ammonia must therefore be prolonged, with consequent risk of contamination by air-borne heterotrophs. Such experiments as we have carried out do, however, suggest that the rest respiration of the cells increases when the nitrite concentration of the suspension fluid is increased; but since the oxygen uptakes involved were 3–20 μl. O_2/vessel/5 hr. it would be unwise to place any great confidence in the actual numerical data. As negative evidence in support of the 'osmotic respiration' theory it is, however, worth noting that Baas-Becking & Parks found that those autotrophs whose activities did not result in changes in the ionic environment of the cells (the hydrogen bacteria and the methane bacteria) had overall free-energy efficiencies of *c.* 30 %, as compared with 6–9 % in other autotrophs. The overall free-energy efficiency of *Hydrogenomonas facilis* as studied by Schatz (1952) also seems to have been about 30 %.

The theory that 'osmotic respiration' is responsible for the fall in the overall free-energy efficiency of growing cultures of *Nitrosomonas*, and perhaps *Thiobacillus thio-oxidans*, is thus neither unreasonable nor inconsistent with available data, but it certainly cannot be said to be proved. If, however, it were proved, it would immediately suggest a way to improve crops of *Nitrosomonas*, since maximal crops could be expected if the nitrite formed by the organism were continuously removed from the medium. If this were done, e.g. by growing the organisms in a cellophan sac immersed in a regularly changed bath of dilute ammonium sulphate solution, the osmotic respiration would be kept at a minimal level and wastage of cell material much reduced.

CONCLUSIONS

The data now available, though limited, suggest that the general biochemistry of the nitrifying organisms is not radically different from that of the heterotrophs. If it appears to differ, this is only because the metabolism of the nitrifying organisms is heavily weighted towards synthesis of all essential compounds and towards the extraction of free energy from inorganic oxidations. The data are, however, fragmentary, largely because of the difficulty of obtaining sufficient cellular material for biochemical studies. Future developments in this field are likely to be slow unless more satisfactory methods of culturing the organisms are found.

REFERENCES

ALLEN, M. B. & VAN NIEL, C. B. (1952). Experiments on bacterial denitrification. *J. Bact.* **64**, 397.
BAAS-BECKING, L. G. M. & PARKS, G. S. (1927). Energy relations in the metabolism of autotrophic bacteria. *Physiol. Rev.* **7**, 85.
BÖMEKE, H. (1939). Beiträge zur Physiologie nitrifizierende Bakterien. *Arch. Mikrobiol.* **10**, 385.
BÖMEKE, H. (1951). *Nitrosomonas oligocarbogenes*, ein obligat autotrophes Nitritbakterium. *Arch. Mikrobiol.* **15**, 414.
CORBET, A. S. (1934). The formation of hyponitrous acid as an intermediate compound in the biological or photochemical oxidation of ammonia to nitrous acid. I. Chemical reactions. *Biochem. J.* **28**, 1575.
CORBET, A. S. (1935). The formation of hyponitrous acid as an intermediate compound in the biological or photochemical oxidation of ammonia to nitrous acid. II. Microbiological oxidations. *Biochem. J.* **29**, 1086.
CUTLER, D. W. & MUKERJI, B. K. (1931). Nitrite formation by soil bacteria other than *Nitrosomonas*. *Proc. Roy. Soc.* B, **108**, 384.
ENGEL, H. (1941). Arbeiten über Nitrifikationsbakterien. *Forschungsdienst Sonderh.* **16**, 144.
FRED, E. B. & DAVENPORT, A. (1921). The effect of organic nitrogenous compounds on the nitrate-forming organism. *Soil Sci.* **11**, 389.
GILLESPIE, R. J., MAW, G. A. & VERNON, C. A. (1953). The concept of phosphate bond energy. *Nature, Lond.* **171**, 1147.
HES, J. W. (1937). Zur Stoffwechselphysiologie von *Nitrosomonas*. Thesis, University of Groningen.

HOFMAN, T. (1953). The biochemistry of the nitrifying organisms. 3. Composition of *Nitrosomonas*. *Biochem. J.* **54**, 293.

HOFMAN, T. & LEES, H. (1952). The biochemistry of the nitrifying organisms. 2. The free-energy efficiency of *Nitrosomonas*. *Biochem. J.* **52**, 140.

HOFMAN, T. & LEES, H. (1953). The biochemistry of the nitrifying organisms. 4. The respiration and intermediary metabolism of *Nitrosomonas*. *Biochem. J.* **54**, 579.

ISENBERG, H. D., SCHATZ, A. & HUTNER, S. H. (1952). Oxidation of ammonia by *Streptomyces nitrificans*. *Bact. Proc.* p. 41.

KINGMA BOLTJES, T. Y. (1935). Untersuchungen über die nitrifizierenden Bakterien. *Arch. Mikrobiol.* **6**, 79.

KLUYVER, A. J. & DONKER, H. J. L. (1926). Die Einheit in der Biochemie. *Chem. Zelle*, **13**, 134.

LEES, H. (1946). Effect of copper enzyme poisons on soil nitrification. *Nature, Lond.* **158**, 97.

LEES, H. (1952). Biochemistry of the nitrifying organisms. 1. The ammonia-oxidizing systems of *Nitrosomonas*. *Biochem. J.* **52**, 134.

MARTIN, W. P., BUEHRER, T. F. & CASTER, A. B. (1942). Nitrification in alkaline desert soils. *Proc. Soil Sci. Soc. Amer.* **7**, 223.

MEEK, C. S. & LIPMANN, C. B. (1922). The relation of the reaction and salt content of the medium on the nitrifying bacteria. *J. gen. Physiol.* **5**, 195.

MEIKLEJOHN, J. (1953). The nitrifying bacteria. *J. Soil Sci.* **4**, 59.

MELLOR, J. W. (1928). *A Comprehensive Treatise on Inorganic and Theoretical Chemistry*, **8**. London: Longmans Green.

MEYERHOF, O. (1916). Untersuchungen über den Atmungsforgang nitrifizierender Bakterien. *Pflüg. Arch. ges. Physiol.* **164**, 352; **165**, 229.

MEYERHOF, O. (1917). Die Atmung des Nitritbilders und ihre Beeinflussung durch chemische Substanzen. *Pflüg. Arch. ges. Physiol.* **166**, 240.

MUMFORD, E. M. (1914). The mechanism of nitrification. *Proc. chem. Soc.* **30**, 36.

MURRAY, T. J. (1923). Food accessory factors and the nitrite bacteria. *Proc. Soc. exp. Biol., N.Y.* **20**, 301.

NASON, A. & EVANS, H. J. (1953). Triphosphopyridine nucleotide—nitrate reductase in *Neurospora*. *J. biol. Chem.* **202**, 655.

NELSON, D. H. (1931). Isolation and characterization of *Nitrosomonas* and *Nitrobacter*. *Zbl. Bakt.* (2. Abt.), **83**, 280.

QUASTEL, J. H. & SCHOLEFIELD, P. G. (1951). Biochemistry of nitrification in soil. *Bact. Rev.* **15**, 1.

SCHATZ, A. (1952). Uptake of carbon dioxide, hydrogen and oxygen by *Hydrogenomonas facilis*. *J. gen. Microbiol.* **6**, 329.

SCHRÖDINGER, E. (1944). *What is Life?* Cambridge University Press.

TANIGUCHI, S., MITSUI, H., TOYODA, J., YAMADA, T. & EGAMI, F. (1953). The successive reduction from nitrate to ammonia by cell-free bacterial enzyme systems. *J. Biochem., Tokyo*, **40**, 175.

VISHNIAC, W. & OCHOA, S. (1951). Photochemical reduction of pyridine nucleotides by spinach grana and coupled carbon dioxide fixation. *Nature, Lond.* **167**, 768.

VOGLER, K. G. (1942). Studies on the metabolism of the autotrophic bacteria. *J. gen. Physiol.* **26**, 103, 109.

VOGLER, K. G. & UMBREIT, W. W. (1942). Studies on the metabolism of the autotrophic bacteria. *J. gen. Physiol.* **26**, 159.

WAKSMAN, S. & STARKEY, R. L. (1922). On the growth and respiration of the sulphur-oxidizing bacteria. *J. gen. Physiol.* **5**, 285.

WINOGRADSKY, S. (1890). Recherches sur les organismes de la nitrification. *Ann. Inst. Pasteur*, **4**, 257, 760.

WINOGRADSKY, S. & WINOGRADSKY, H. (1933). Études sur la microbiologie du sol. *Ann. Inst. Pasteur*, **50**, 350.

THE NITROGEN METABOLISM OF THE BLUE-GREEN ALGAE (MYXOPHYCEAE)

G. E. FOGG AND MIRIAM WOLFE

Department of Botany, University College, London

INTRODUCTION

The blue-green algae (Myxophyceae or Cyanophyceae) are a class of micro-organisms distinguished chiefly by their apparent lack of an organized nucleus such as is found in other organisms, by the absence of any form of sexual reproduction, by their lack of flagella, and by their characteristic modes of cell division and movement. Although they share certain characteristics with the bacteria on the one hand (Pringsheim, 1949) and with the red algae (Rhodophyceae) on the other (Kylin, 1943), it seems unlikely that their affinity with these or any other groups can be anything but remote (Fritsch, 1945; Pringsheim, 1949). Their nutrition is characteristically phototrophic, but there is reason to believe that chemolithotrophs such as *Beggiatoa*, *Thiothrix* and *Achromatium*, and chemo-organotrophs such as *Vitreoscilla* should be included in the class (Pringsheim, 1949). There are thus indications that the blue-green algae show something of the variety of chemical activity characteristic of the bacteria, but, perhaps because of the difficulties of isolating and handling them in pure culture, comparatively few biochemical investigations have been made with these organisms. At present most information is available concerning their nitrogen metabolism, with which this review is concerned.

THE ASSIMILATION OF NITROGEN

Nitrogen may be assimilated by blue-green algae as the element, as nitrate, nitrite or ammonia, or in the form of various organic compounds.

Nitrogen fixation

Blue-green algae were among the first organisms to attract attention as 'fixers' of the free nitrogen of the atmosphere. Frank (1889) found that gains in combined nitrogen in soil cultures incubated in the light were associated with the development of these organisms, and Beijerinck (1901) obtained a copious growth of blue-green algae, consisting principally of *Anabaena catenula* and other *Anabaena* spp., in media initially free from combined nitrogen that were inoculated with a small amount of soil and incubated in the light. However, neither Frank's nor Beijerinck's observations constitute proof of nitrogen fixation by blue-green algae, since their cultures contained other kinds of micro-organism which may have been responsible for the observed fixation. It is perhaps worth noting that in the paper describing this work, which is well known as containing the first description of *Azotobacter*, Beijerinck's primary concern was with nitrogen fixation by blue-green algae. The isolation of the latter in pure bacteria-free culture, a matter of some difficulty because of the gelatinous sheaths with which they are generally invested, was eventually achieved by Pringsheim (1914). Pringsheim and his pupils Glade (1914) and Maertens (1914) were un-able to find evidence for nitrogen fixation in the species which they isolated, although these included several forms which have been subse-quently shown to possess this capacity. It is possible that the particular strains with which these investigators worked were not able to fix nitrogen, or that insufficient amounts of molybdenum for nitrogen fixation were present in the media, but neither of these explanations of the failure to detect this property seems convincing. As a result of this work with pure cultures it became generally believed that blue-green algae do not themselves fix nitrogen, and that where they develop in media deficient in combined nitrogen it is because they are in symbiotic association with bacteria possessing this capacity (see, for example, Jones, 1930). However, bacteria-free cultures of *Nostoc punctiforme* and *Anabaena variabilis* were isolated by Drewes (1928) and shown to be able to fix nitrogen. Later work has amply confirmed that the Myxophyceae include many nitrogen-fixing forms.

The methods first used for the isolation of pure cultures of blue-green algae were laborious and unreliable. However, treatment with ultra-violet light has been found to be a convenient and fairly certain means of obtaining pure cultures, since a suitable irradiation kills contami-nating bacteria without damaging the algae (Allison & Morris, 1930; Bortels, 1940; Gerloff, Fitzgerald & Skoog, 1950*a*; Fogg, 1951*a*). Particular attention has been paid to the checking of the purity of cultures (De, 1939; Fogg, 1942), and there can be little doubt that really pure cultures have been obtained. Conclusive proof of fixation by a given species requires the use of a reliable method of demonstrating an increase in combined nitrogen as well as the use of pure cultures. In the majority of studies the Kjeldahl method has been employed. Since cultures have generally been grown in media initially free from com-bined nitrogen the use of this method in this connexion is not open to objection (see Wilson, 1952). A precaution that was observed by

Beijerinck (1901), but not often by subsequent workers, is the removal of traces of nitrogenous impurities, such as ammonia and oxides of nitrogen, from the air supplied to the cultures. Since blue-green algae grow relatively slowly, considerable errors may arise if unpurified laboratory air is used, and suitable precautions to prevent contamination should always be taken in accurate work (Fogg, 1942). The use of heavy nitrogen as a tracer offers one of the most convincing means of proving nitrogen fixation, and this method has now been used to demonstrate fixation by *Nostoc muscorum* (Burris, Eppling, Wahlin & Wilson, 1942, 1943) and *Calothrix parietina* (Williams & Burris, 1952).

A list of blue-green algae which have been obtained in pure culture and examined for nitrogen-fixing properties is given in Table 1. In addition, there are various reports of nitrogen fixation based on inadequate evidence, and some of these have been discussed elsewhere (De, 1939; Fogg, 1951 *a*). Certain species belonging to the genera *Schizothrix* (Oscillatoriaceae), *Plectonema* (Scytonemataceae) (Watanabe, Nishigaki & Konishi, 1951) and *Gloeotrichia* (Rivulariaceae) (Williams & Burris, 1952; Fogg, G. E., unpublished) have been judged to be nitrogen-fixing on the basis of observations on impure cultures. It will be seen from Table 1 that orders other than the Nostocales have as yet been insufficiently examined for nitrogen-fixing species. While it appears that the property of nitrogen fixation is widely distributed in the Myxophyceae and is not peculiar to any one family or order, there is so far no unequivocal evidence for its possession by any members of the Chroococcales or Oscillatoriaceae. The high proportion of nitrogen-fixing species included in the Myxophyceae is striking (Williams & Burris, 1952); possibly it is a result of the appearance of the class at a period in the earth's development when less combined nitrogen was available than at the present time.

As in other nitrogen-fixing organisms, the assimilation of elementary nitrogen is suppressed when combined nitrogen is supplied to blue-green algae in a readily assimilable form such as nitrate or an ammonium salt (Fogg, 1942). All the nitrogen-fixing Myxophyceae grow best in neutral or alkaline media, and *Nostoc muscorum*, for example, does not fix nitrogen below about pH 5·7 (Allison & Hoover, 1935). Nitrogen fixation may occur during either phototrophic or chemotrophic growth. *Anabaena cylindrica* appears to be an obligate phototroph and cannot therefore fix nitrogen for more than a limited period in the dark (Fogg, G. E., unpublished), but *Nostoc muscorum* will grow and fix nitrogen slowly in the dark if provided with glucose (Allison, Hoover & Morris, 1937), and Winter (1935) found that endophytic strains of *N. punctiforme*

Table 1. *Distribution of the capacity to assimilate
elementary nitrogen in the Myxophyceae*

The classification used is that of Fritsch (1945). Not all of the families included in the class are mentioned. The references are as follows: (1) Allen, 1952; (2) Allison, Hoover & Morris, 1937; (3) Allison & Morris, 1930; (4) Allison & Morris, 1932; (5) Bortels, 1940; (6) Burris, Eppling, Wahlin & Wilson, 1942, 1943; (7) De, 1939; (8) Drewes, 1928; (9) Fogg, 1942; (10) Fogg, 1951a; (11) Fogg, G. E., unpublished; (12) Henriksson, 1951; (13) Singh, 1942; (14) Watanabe, 1951; (15) Williams & Burris, 1952; (16) Winter, 1935.

Order and family	Species able to fix nitrogen	Species not able to fix nitrogen
CHROOCOCCALES	—	*Chroococcus turgidus* (1)
		Chroococcus spp. (1)
		Coccochloris peniocystis
		(Gloeothece linearis) (15)
		Diplocystis (Microcystis)
		aeruginosa (15)
		Gloeocapsa membranina (15)
		G. dimidiata (15)
		Synechococcus cedrorum (1)
CHAMAESIPHONALES	—	—
PLEUROCAPSALES	—	—
NOSTOCALES		
Oscillatoriaceae	—	*Lyngbya aesturii* (1)
		Lyngbya spp. (1, 4)
		Oscillatoria spp. (1, 4, 11)
		Phormidium foveolarum (7)
		P. tenue (15)
		P. lividum (1)
		Phormidium spp. (1)
Nostocaceae	*Anabaena ambigua* (13)	*Anabaena variabilis* (1)
	An. cylindrica (5, 9)	*Anabaena* spp. (1)
	An. fertilissima (13)	*Aphanizomenon flos-aquae* (15)
	An. gelatinosa (7)	
	An. humicola (5)	
	An. naviculoides (7)	
	An. variabilis (3, 5, 7, 8)	
	Anabaena spp. (8)	
	Anabaenopsis sp. (14)	
	Aulosira fertilissima (13)	
	Cylindrospermum gorakporense (13)	
	C. licheniforme (5)	
	C. maius (5)	
	Nostoc paludosum (5)	
	N. punctiforme (5, 8, 16)	
	N. muscorum (2, 6, 15)	
	Nostoc spp. (1, 12, 14, 15)	
Rivulariaceae	*Calothrix brevissima* (14)	—
	C. parietina (15)	
Scytonemataceae	*Tolypothrix tenuis* (14)	*Plectonema notatum* (1)
		P. nostocorum (15)
STIGONEMATALES	*Mastigocladus laminosus* (10)	—

isolated from *Cycas* and *Gunnera* fixed more nitrogen in the dark than in the light, when supplied with fructose.

That traces of molybdenum are necessary for the achievement of maximum rates of nitrogen fixation was discovered by Bortels (1930) and has since been confirmed for *Azotobacter* spp. (Horner, Burk, Allison & Sherman, 1942), *Clostridium* spp. (Jensen & Spencer, 1947) and various species of Nostocaceae (Bortels, 1940; Fogg, 1949). For *Anabaena cylindrica* cultured in the usual medium free from combined nitrogen, the concentration of this element giving optimum growth in 18 days is of the order of 0·2 mg./l. or more (Wolfe, 1953). Fig. 1*a* shows the results of an experiment in which the basal medium was freed from traces of molybdenum by co-precipitation with iron and 8-hydroxyquinoline in the presence of acetic acid (Nicholas & Fielding, 1950). The trace element salts, which were added separately to the purified basal medium, were purified by three successive crystallizations, and the alga used as inoculum had been previously subcultured for three transfers in a medium containing ammonium chloride but no added molybdenum. The medium used in this experiment was found to contain not more than 0·5 μg./l. of molybdenum, but nevertheless an appreciable amount of growth occurred when the element was not deliberately supplied. To show that molybdenum is essential for nitrogen fixation by *An. cylindrica*, ethylenediamine tetra-acetic acid (versene) has been employed to remove from the medium the free molybdenum ions present as impurity (see Hutner, Provasoli, Schatz & Haskins, 1950). Using this method, and with the concentrations of other trace elements increased by suitable amounts, the results given in Fig. 1*b* were obtained, and these show that this element is in all probability essential for nitrogen fixation by the organism, since only small amounts of growth occurred in the absence of supplied molybdenum.

The properties of the enzyme system fixing nitrogen, which have so far only been examined in *Nostoc muscorum* among the Myxophyceae, appear to be very similar to those of *Azotobacter* and of the *Rhizobium*-legume system (Burris & Wilson, 1946). Fixation was found to occur at half the maximum rate when the partial pressure of nitrogen (pN_2) was *c*. 0·02 atm., and although this value was not determined with great precision, it is of the same order as those for *Azotobacter* (0·02 atm.) and *Trifolium pratense* (0·05 atm.). Like those of other nitrogen-fixing organisms, the enzyme system of *Nostoc* is specifically inhibited by hydrogen and by carbon monoxide. A point of difference which may be important is that in *Nostoc* the degree of inhibition of nitrogen fixation by carbon monoxide appears to depend on the pN_2, whereas in

Azotobacter and *Trifolium* the inhibition is non-competitive. The half-maximum rate of fixation occurs at a partial pressure of carbon monoxide (pCO) of *c.* 0·0006 atm. when the pN_2 is 0·1 atm. and at a pCO of 0·0014 atm. when the pN_2 is 0·75 atm. These values are intermediate between those found for *Trifolium* (0·0002 atm.) and *Azotobacter* (0·004–0·005 atm.).

In bacteria there appears to be some correlation between the presence of hydrogenase and the capacity to fix nitrogen (Lindstrom, Burris & Wilson, 1949), but this correlation does not hold for blue-green algae,

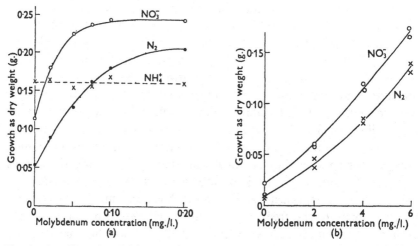

Fig. 1. The effect of molybdenum concentration on the growth of *Anabaena cylindrica*. (*a*) With gaseous nitrogen, potassium nitrate or ammonium chloride as nitrogen source. Cultures grown for 18 days (N_2, NO_3^-) or 14 days (NH_4^+). (*b*) With gaseous nitrogen or potassium nitrate as nitrogen source in the presence of ethylenediamine tetra-acetic acid (0·05 g./l.). Cultures grown for 14 days.

since neither *Nostoc muscorum* (Frenkel & Rieger, 1951) nor *Anabaena cylindrica* (Wolfe, 1953) appears to be able to utilize elementary hydrogen.

Using heavy nitrogen as a tracer it has been established that the nitrogen fixed by *Nostoc muscorum* appears first in ammonia, glutamic acid and aspartic acid (Wilson, 1952). Following exposure of the alga for 120 min. to an atmosphere containing nitrogen with 22·5 atoms % excess of ^{15}N, 0·46, 0·66 and 0·51 atom % excess of the tracer were found in ammonia (including amide-nitrogen), glutamic acid and aspartic acid respectively, as compared with 0·352 atom % excess for the total hydrolysate of the alga. This order of labelling is similar to that obtained when ammonia containing ^{15}N is supplied, and suggests that, as in other nitrogen-fixing organisms, ammonia is the first stable product of nitrogen fixation.

Because they combine capacities both for photosynthesis and for nitrogen fixation, blue-green algae are of considerable ecological importance in certain situations (De, 1939; Singh, 1942; Fogg, 1947; Watanabe *et al.* 1951; Williams & Burris, 1952). Little deliberate attempt has so far been made to exploit their special properties for economic purposes, but the use of blue-green algae in the reclamation of certain types of waste land has been shown to be possible (Singh, 1950). The isolation of a nitrogen-fixing blue-green alga which would grow rapidly in suspension might make possible an industrial photosynthetic process yielding valuable nitrogenous materials.

The assimilation of nitrate

Nitrate is readily utilized as a source of nitrogen by nearly all the blue-green algae which have been examined in pure culture (Pringsheim, 1914; Maertens, 1914; Glade, 1914; Fogg, 1942; Gerloff *et al.* 1950*b*, 1952; Allen, 1952). However, at least in the case of *Anabaena cylindrica*, it appears that the assimilation of nitrate requires that the organism should be adapted to this source of nitrogen. After being maintained on a medium without combined nitrogen *An. cylindrica* has been found to continue to fix nitrogen when subcultured into a medium containing nitrate (Fogg, G. E., unpublished). In one instance uptake of nitrate by an illuminated suspension of the same organism previously grown without this substance, began only after a lag period of about 9 hr. (Wolfe, 1953).

Nitrate assimilation by blue-green algae is accompanied by effects on gas exchange similar to those observed by Warburg & Negelein (1920) with *Chlorella*. In order to demonstrate this with *Anabaena cylindrica* it has been found desirable to use cells which have been starved of nitrogen by incubation in an atmosphere containing hydrogen at a sufficient partial pressure to inhibit nitrogen fixation, and also to carry out the experiment under conditions in which elementary nitrogen cannot be assimilated. Nitrate is then assimilated rapidly in the dark for 3–5 hr. After this, absorption may continue without assimilation so that nitrate accumulates within the cells. During assimilation, the respiratory quotient rises from the value of 0·9–1·0, characteristic of endogenous respiration in the absence of a source of nitrogen, to a value of the order of 2 (see Fig. 3*a*). However, whereas in *Chlorella* a similar rise in respiratory quotient during nitrate assimilation is due to an increased output of carbon dioxide (Warburg & Negelein, 1920), the rise in

Anabaena cylindrica has frequently, but not invariably, been found to be due to a decrease in oxygen consumption, the carbon dioxide output remaining about the same as in the absence of nitrate. Thus under certain circumstances it appears that in the latter organism nitrate competes with oxygen as hydrogen acceptor (Wolfe, 1953).

The implication of molybdenum in the assimilation of nitrate as well as free nitrogen was first noticed by Steinberg (1937) in work with *Aspergillus*. Since then it has been found with a variety of organisms that traces of molybdenum are necessary for the achievement of maximum rates of growth when nitrate is the source of nitrogen (for re-

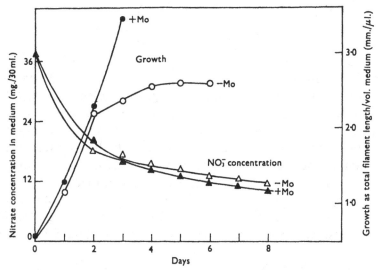

Fig. 2. The rate of growth and of nitrate absorption by *Anabaena cylindrica* with (+Mo) and without (−Mo) added molybdenum in the presence of ethylenediamine tetra-acetic acid.

ferences see Mulder, 1948; Hewitt, 1951; Wood, 1953). It will be seen from Fig. 1a, b that *Anabaena cylindrica* is no exception to the general rule. Both figures show that the rates of assimilation are greater at lower concentrations of molybdenum when the source of nitrogen is nitrate than when it is molecular nitrogen. It is the growth of the organism which is reduced in the absence of molybdenum, the rate of nitrate uptake itself being unaffected by molybdenum concentration within the limits investigated. This has been demonstrated for *An. cylindrica* in an experiment in which growth and the uptake of nitrate were followed in molybdenum-deficient and normal suspensions of the organism (Fig. 2). From the second day onwards, growth was much less in the absence of molybdenum than in its presence. Nevertheless, the amount of nitrate taken up was nearly the same in both cases. This is in agreement with the results of similar studies by Hewitt & Jones (1947) on nitrate assimilation by molybdenum-deficient cauliflowers, and by Sakamura & Maeda (1950) working with a yeast, *Hansenula anomala*. The addition of molybdenum to molybdenum-deficient cultures containing nitrate is followed immediately by an increased reduction of nitrate. This suggests that molybdenum participates directly in metabolism in the ionic form (Wolfe, 1953).

It is generally supposed that the first step in the assimilation of nitrate is its reduction to ammonia via nitrite and hydroxylamine (Wood, 1953), and the available evidence suggests that this is the path followed in the blue-green algae. Both nitrite and ammonia appear in the medium following the addition of nitrate to *Anabaena cylindrica* adapted to this source of nitrogen (Wolfe, 1953). On the other hand, oximes, which might be expected to occur if hydroxylamine were an intermediate in the reduction, have not been detected either in the medium or in the cells when this alga assimilates nitrate (Wolfe, 1953). Nitrite is a suitable source of nitrogen for a number of blue-green algae (Pringsheim, 1914). Thus Maertens (1914) found nitrite to be suitable for the growth of *Oscillatoria* spp. but not *Nostoc, Cylindrospermum* or *Calothrix* spp. At concentrations of up to 13·6 mg. nitrogen/l., nitrite is as effective as nitrate as a nitrogen source for *Microcystis aeruginosa* (Gerloff *et al.* 1952). The effect of higher concentrations of nitrite on *M. aeruginosa* is not recorded, but a concentration of 27 mg. nitrite-nitrogen/l. completely inhibits the growth of *Anabaena cylindrica* (Fogg, G. E., unpublished). The presence or absence of molybdenum appears to have no effect on the accumulation of nitrite in cultures of *An. cylindrica* assimilating nitrate (Wolfe, 1953). Presumably molybdenum is necessary for the growth of blue-green algae with nitrite as the source of nitrogen, but this does not appear to have been demonstrated. Small amounts of nitrate or nitrite have been reported to occur in cultures of *An. cylindrica* supplied with ammonium-nitrogen only (see Fogg, 1952, Table 11), indicating that the reduction of nitrate to ammonia may be partly or wholly reversible.

The assimilation of ammonia

Ammonium salts have been found to provide a readily assimilable source of nitrogen for all the blue-green algae which have been examined in pure culture (Pringsheim, 1914; Maertens, 1914; Walp, 1942; Fogg, 1949; Gerloff *et al.* 1952; Allen, 1952; Wilson, 1952). However, ammonium-nitrogen is only as favourable as nitrate for growth if the pH changes attendant upon its absorption are reduced to a minimum (Gerloff *et al.* 1952; Fogg, 1952). Concentrations of ammonium nitrogen as low as 5×10^{-5} M suppress the formation of heterocysts by *An. cylindrica*, an indication of an important effect on the course of metabolism (Fogg, 1949). An increase in rate of respiration following the supply of an ammonium salt to nitrogen-starved cells, similar to that observed by Syrett (1953*a*) with *Chlorella*, has been found to occur in *Anabaena cylindrica* (Wolfe, 1953).

Results such as those shown in Fig. 1*a* make it clear that normal growth with ammonia as a source of nitrogen can take place in the absence of the relatively high concentrations of molybdenum necessary for growth on nitrate or free nitrogen, but leave open the question of whether or not such growth is possible in the complete absence of molybdenum. An experiment in which the effects of the small amounts of molybdenum inevitably present as impurity have been reduced by chelation has not been carried out with *Anabaena* cultures supplied with an ammonium salt. However, there is evidence that molybdenum is necessary in minute amounts for the growth of higher plants supplied exclusively with ammonium nitrogen (for references see Wood, 1953), and perhaps a similar requirement could be demonstrated for blue-green algae if more refined methods were used.

The assimilation of organic nitrogen

The assimilation of nitrogen in organic combination by blue-green algae has been little studied. Allen (1952) has stated that the amino-acids present in casein hydrolysate form a good nitrogen source for all the species which she studied. Her work shows that species of Myxophyceae such as *Oscillatoria* and *Chroococcus* spp. are capable of assimilating amino-nitrogen, but in the absence of analytical data it is impossible to decide whether the nitrogen-fixing species of *Anabaena* and *Nostoc* with which she worked actually assimilated the amino-acids supplied in the medium. A few blue-green algae such as a *Lyngbya* sp. appear to be able to use urea as a source of nitrogen (Allen, 1952). The growth of *Anabaena cylindrica* at high light intensities is promoted by $2 \cdot 5 \times 10^{-4}$ M-asparagine, but $2 \cdot 5 \times 10^{-4}$ M-aspartate and 5×10^{-4} M-urea inhibit completely the growth of this species (Fogg, 1949). Most blue-green algae, including *An. cylindrica*, are able to utilize intact casein (Allen, 1952). The effect of molybdenum on the assimilation of organic nitrogen by blue-green algae has evidently not been investigated.

Most blue-green algae appear to be completely autotrophic in their nitrogen nutrition, but certain species, such as *Synechococcus cedrorum*, which grows on casein hydrolysate but not on inorganic sources of nitrogen (Allen, 1952), may have a requirement for organic nitrogenous compounds.

THE INTERMEDIARY METABOLISM OF NITROGEN AND THE ROLE OF MOLYBDENUM

Very few studies have been made on the intermediary metabolism of nitrogen in blue-green algae. However, the distribution of heavy nitrogen among various substances in *Nostoc muscorum*, following supply of the tracer as free nitrogen or ammonia, suggests that the mechanisms involved are similar to those which have been found in other organisms (Wilson, 1952). It would appear from these results that, as in higher plants, the principal means of entry of ammonia into metabolism is by amination of α-ketoglutaric acid to form glutamic acid, irrespective of whether the ammonia is assimilated directly or produced by fixation of free nitrogen. Aspartic acid, which does not become labelled so rapidly as glutamic acid, may possibly be formed by an analogous mechanism, but the amino-groups of other amino-acids are evidently derived from these two compounds by transamination.

Investigations carried out in this laboratory have been chiefly concerned with the role of molybdenum in the intermediary metabolism of nitrogen in *Anabaena cylindrica*. Previous studies of the metabolic role of molybdenum have been made with higher plants or fungi and have been reviewed by Hewitt (1951) and Wood (1953). Molybdenum deficiency produces alterations in the proportions of various nitrogenous compounds present in plants, decreasing the concentrations of most amino-acids (Hewitt, Jones & Williams, 1949; Agarwala & Williams, 1951) but increasing those of others such as arginine (Agarwala & Williams, 1951) and of amides (Sakamura & Maeda, 1950). The effects of deficiency are not confined to nitrogenous substances, since the ascorbic acid content of molybdenum-deficient plants has been found to be from 25 to 45 % of that of normal plants (Hewitt, Agarwala & Jones, 1950). These results do not appear to provide a basis for any comprehensive explanation of the role of molybdenum in metabolism.

A moderate deficiency of molybdenum in cultures of *An. cylindrica* supplied with nitrate or free nitrogen results in the production of cells having a low content of total combined nitrogen on a dry weight basis and having a characteristic orange-yellow colour due to a reduction in the content of the blue pigment, phycocyanin (Fogg, 1952). A comparison of the metabolism of such cells and those supplied with sufficient molybdenum for normal growth has yielded results which appear to

throw considerable light on the part played by this micro-nutrient
element in nitrogen metabolism (Wolfe, 1953).

The effects of the addition of potassium nitrate on the respiration and
the content of various nitrogenous substances in molybdenum-deficient
and normal cells of *An. cylindrica* are illustrated in Figs. 3 and 4. The
algal material, which was adapted to nitrate, was starved of nitrogen
overnight before use, and the experiments were conducted in the dark
in an atmosphere containing 80 % H_2 and 20 % O_2, so that both photosynthesis and nitrogen fixation were suppressed. Respiration was

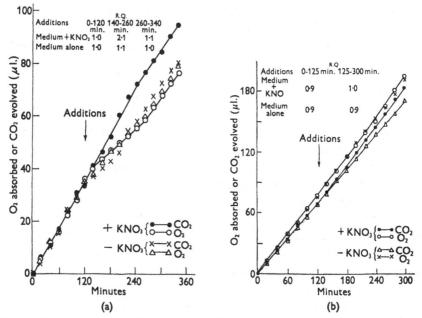

Fig. 3 *a, b*. The effect of the addition of nitrate (10 μmoles/ml.) on the respiration of *Anabaena cylindrica* after nitrogen starvation. (*a*) with 22·4 mg. dry weight of normal alga; (*b*) with 16·1 mg. dry weight of molybdenum-deficient alga.

measured manometrically by Warburg's 'direct' method, and samples
for analysis were obtained from suspensions incubated under similar
conditions.

The rate of endogenous respiration in molybdenum-deficient cells is
considerably higher than that in normal ones. In the particular experiments the results of which are shown in Fig. 3 *a, b*, the rates were 2·48 and
0·78 μl. O_2/mg. dry cells/hr. for deficient and normal cells respectively.
A similar increase in respiration rate of molybdenum-deficient *Hansenula*
has been noted by Sakamura & Maeda (1950). The initial respiratory
quotients, on the other hand, were of the same order, i.e. 0·9–1·0, in both

types of *Anabaena*. Following the addition of nitrate the respiratory quotient of normal cells rose to 2·1 then fell as the polysaccharide reserves of the cells became exhausted. The addition of nitrate to molybdenum-deficient cells caused the respiratory quotient to rise only slightly, from 0·9 to 1·0.

Both the total nitrogen content and the content of all organic nitrogenous fractions are less in molybdenum-deficient than in normal material. In both types of material nitrate is rapidly absorbed and accumulates within the cells (Fig. 4a, b). The rate of its subsequent dis-

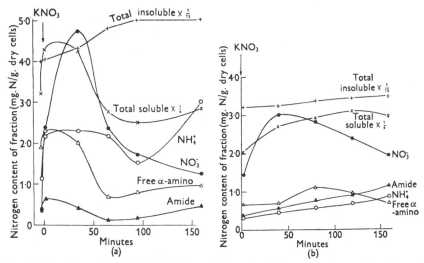

Fig. 4. The effect of the addition of nitrate (10 μmoles/ml.) on the nitrogen fractions of *Anabaena cylindrica* after nitrogen starvation. (*a*) normal cells; (*b*) molybdenum-deficient cells.

appearance is less in molybdenum-deficient than in normal cells. This decrease in the rate of nitrate reduction together with the increase in rate of endogenous respiration in molybdenum-deficient cells is sufficient to account for the small increase in respiratory quotient noted above. In normal cells ammonia at first accumulates following supply of nitrate, then falls as it is converted to organic nitrogen. The rise in ammonia and amide concentration towards the end of the experiment (Fig. 4a) is thought to be related to depletion of carbohydrate reserves. In the first 5 min. after the addition of nitrate there is a rapid increase of free α-amino-, amide- and soluble combined α-amino-nitrogen. Later, these fractions decrease in amount and there is a corresponding rise in the insoluble nitrogen of the cells. In molybdenum-deficient cells the amount of ammonia steadily rises following supply of nitrate. Amide-nitrogen likewise shows a steady rise in deficient cells but free α-amino-nitrogen, after showing a slight rise, falls to its original level. The rise in concentration of total soluble nitrogen in these cells is thus largely accounted for by increases in ammonia and amide-nitrogen. There is only a negligible increase in total insoluble nitrogen in molybdenum-deficient cells.

Addition of molybdate has been found not to affect the activity of nitrate reductase extracted from higher plants (Evans & Nason, 1953) or from fungi (Nicholas, Nason & McElroy, 1953), but it has been observed that in molybdenum-deficient fungi there is a reduction in the concentration of nitrate reductase to between one-fifth and one-thirtieth of that in normal material (Nicholas *et al.* 1953). It was pointed out that this decrease may arise either because of a general decrease in protein brought about by nitrogen deficiency or because molybdenum is a component of the nitrate reductase itself, but brief mention was made of results which indicate that the latter alternative is the correct one. It thus seems that molybdenum may be directly concerned in the reduction of nitrate to ammonia, as well as in other reactions, but the results of Sakamura & Maeda (1950) as well as those of Wolfe (1953) described above, make it clear that a moderate molybdenum deficiency does not entirely prevent the reduction of nitrate, whereas it blocks completely the incorporation of the ammonia produced by this reduction into amino-acids and proteins.

Estimations of carbohydrate fractions show that there is a rapid depletion of polysaccharide associated with the assimilation of nitrate in the dark by normal cells of *Anabaena* (Wolfe, 1953). Thus, in one instance over 65 % of the total polysaccharide disappeared in under 2 hr. when the cells were incubated in the presence of nitrate, whereas in its absence there was only a 13·8 % decrease in the same period. Mono- and di-saccharides showed relatively little change. These results are similar to those obtained by Syrett (1953b) for *Chlorella*. In molyb-denum-deficient cells of *Anabaena* the consumption of polysaccharide is much reduced, a decrease of 21 % being observed during a period of nitrate assimilation comparable with that in which there was a 65 % decrease in normal cells (see also Fig. 7).

These results lend support to the suggestion put forward by other workers in this field (Sakamura & Maeda, 1950; Agarwala, 1952) that molybdenum deficiency affects nitrogen metabolism through changes in carbohydrate metabolism. Further information on this has therefore been sought in experiments in which the effects of various carbon sources supplied exogenously have been determined. Addition of glucose, acetate, fumarate, succinate or citrate to normal cells of *An. cylindrica* assimilating nitrate stimulates respiration and, of these substances, fumarate, succinate and citrate have proved to be available for nitrate reduction, whereas glucose and acetate are scarcely utilized for this purpose (Wolfe, 1953). Addition of fumarate or succinate to molyb-

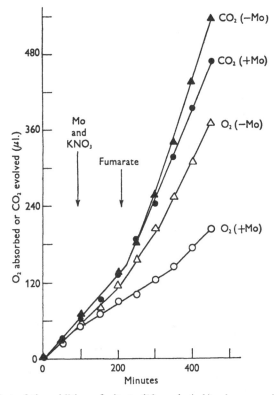

Fig. 5. The effect of the addition of nitrate ($10\,\mu$moles/ml.), alone or with molybdenum (6 mg./l.), and of the subsequent addition of fumarate, on the respiration of molybdenum-deficient cells of *Anabaena cylindrica* after nitrogen starvation. Each manometer contained $22\cdot7$ mg. dry weight of alga.

denum-deficient cells supplied with nitrate is followed by an increase in the respiratory quotient suggesting that the addition of either of these substances enables nitrate assimilation to proceed in the absence of molybdenum. This possibility has been investigated with results such as those summarized in Figs. 5–7.

The addition of nitrate to molybdenum-deficient cells results in an appreciable increase in respiratory quotient only when molybdenum is supplied simultaneously, the values in the instance shown in Fig. 5

being 1·8 for cells supplied with molybdenum and 1·2 for deficient cells. The addition of fumarate causes an increase in rate of respiration and respiratory quotient, the value of the latter indicating reduction of nitrate both when molybdenum is deficient and when it is supplied.

In the molybdenum-deficient cells, amide accumulates following the addition of nitrate (Fig. 6a), but in similar cells supplied with nitrate and molybdenum simultaneously, no such continued accumulation occurs (Fig. 6b). This finding confirms the idea derived from changes in respiratory quotients that no period of adaptation is required after the addition of molybdenum before normal nitrate assimilation can begin. The

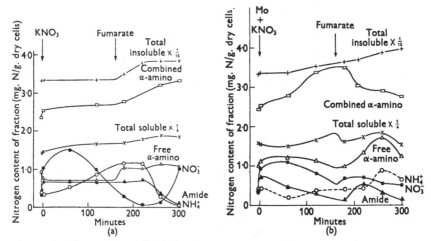

Fig. 6. The effect of the addition of nitrate (10 μmoles/ml.), alone or with molybdenum (6 mg./l.), and of the subsequent addition of fumarate (6·7 μmoles/ml.), on the nitrogen fractions of molybdenum-deficient cells of *Anabaena cylindrica* after nitrogen starvation. (a) without added molybdenum; (b) with added molybdenum.

addition of fumarate to cells supplied with molybdenum results in a stimulation of nitrate assimilation, a transitory increase in ammonium-, amide- and free α-amino-nitrogen, and a continued increase in total insoluble nitrogen. In cells not supplied with molybdenum (Fig. 6a) the addition of fumarate results in a fall to a low level of nitrate concentration, an increase in nitrite and free α-amino-nitrogen, and after an hour, a sharp fall in amide-nitrogen. An increase in total insoluble nitrogen shows that the addition of fumarate removes the block to protein synthesis.

The reduction of nitrate is accompanied by disappearance of polysaccharide (Fig. 7), the decrease being considerably greater in cells supplied with molybdenum. The addition of fumarate leads to synthesis of polysaccharide in both kinds of cells. That in cells not supplied with molybdenum takes place more rapidly but is followed by a sharp decrease, coinciding with the decrease in amide, evidently resulting from utilization in the synthesis of protein and conversion to disaccharide.

On the basis of their observations on *Hansenula*, Sakamura & Maeda (1950) suggested that molybdenum may be involved in an oxidation reaction concerned in nitrogen anabolism subsequent to the formation

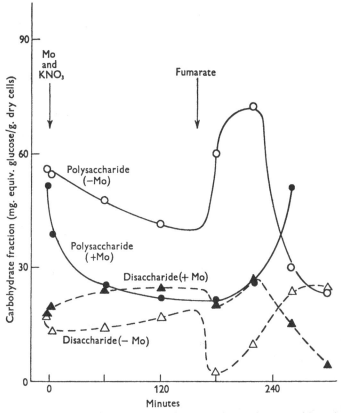

Fig. 7. The effect of the addition of nitrate (10μmoles/ml.), alone or with molybdenum (6 mg./l.), and of the subsequent addition of fumarate (6·7μmoles/ml.), on the carbohydrate fractions of molybdenum-deficient cells of *Anabaena cylindrica* after nitrogen starvation.

of amino-acids, but did not explain satisfactorily why such a reaction should be important when nitrate is supplied but not when ammonia is supplied. Hewitt (1951), in reviewing the effects of molybdenum deficiency in plants, emphasized that the metal may have a multiple role in metabolism but concluded that an interpretation of the information then available was not possible. However, a step towards an explanation of the metabolic function of molybdenum has been made recently by Wood (1953) who pointed out that investigations of the effect of this element on enzyme systems have shown that it has a specific effect on phosphatase action. It has been reported by several workers that traces of molybdenum inhibit, both *in vivo* and *in vitro*, the hydrolysis by phosphatases of a variety of organic phosphates such as glucose-6-phosphate, hexose diphosphate, glycerophosphate and adenosine triphosphate (for references see Wood, 1953). Wood suggested that, because of this, hexose phosphates and adenosine triphosphate are present in low concentration in molybdenum-deficient plants and that consequently the glycolytic and respiratory cycles provide hydrogen for nitrate reduction and energy for protein synthesis at reduced rates. Thus according to this view, molybdenum should be necessary for amino-acid and protein synthesis when ammonia is given as a source of nitrogen but 'since the energy requirements for these syntheses are considerably less than for nitrate reduction, onset of deficiency symptoms might be expected later in the life cycle than is the case with plants receiving nitrate only'.

This suggestion appears to us to afford an essential clue towards the solution of the problem but to be unsatisfactory in that it does not give a sufficient explanation of the fact that relatively large amounts of molybdenum are required for the assimilation of nitrate or free nitrogen, whereas extremely minute traces of the metal suffice for healthy growth on ammonia. If it is accepted that the important role of molybdenum is the inhibition of the enzymic hydrolysis of organic phosphates, then a consistent explanation of this major feature of the metabolic function of molybdenum and of the observations described above appears to be as follows. The increase in rate of hydrolysis of organic phosphates brought about by molybdenum deficiency will have two major effects, first, it will decrease the rate of glycolysis and secondly, it will increase the rate of oxidation of the products of glycolysis. The reduction in rate of glycolysis, for which we have clear evidence in *Anabaena* (see p. 112 and Fig. 7) will occur because the concentration of the immediate substrate, hexose diphosphate, and intermediates such as glycerophosphate, will be reduced by increased hydrolysis. Syrett (1953b) has found evidence which suggests that the rate of respiration in *Chlorella* is normally determined by the concentrations of substances such as adenosine diphosphate capable of acting as acceptors of the organic phosphate groups produced in aerobic respiration. Evidently the situation is similar in *Anabaena*, since molybdenum deficiency, which we suppose to increase the rate of production of phosphate acceptors, is characterized by a considerable increase in the rate of aerobic respiration (see p. 110). In the absence of nitrate a new position of equilibrium is evidently reached in molybdenum-deficient cells, in which the increased rate of hydrolysis of organic phosphates is compensated by the greater rate of production of these substances by aerobic respiration. In the presence of ammonium-nitrogen normal protein synthesis will be possible because of this maintenance of the supply of organic phosphates. In the presence of nitrate, however, such a balance cannot be maintained. The reduction of nitrate requires the production of hydrogen-donors in excess of the amount involved in respiration. This cannot be achieved by a greater rate of glycolysis, as it is in normal cells, because of the limitation imposed by molybdenum deficiency, and nitrate reduction can thus only occur at the expense of aerobic respiration. This results in a decrease in the rate of formation of organic phosphates so that the rate of glycolysis is still further decreased. Thus the concentrations of tricarboxylic acid cycle intermediates are reduced to the level at which none are available to provide the carbon skeletons for amino-acids, consequently the ammonia produced by the reduction of nitrate must accumulate or be added as amide groups to existing amino-acids. Provision of tricarboxylic acid cycle intermediates such as fumarate will make good this deficiency and, as reported above, enables the block caused by molybdenum deficiency to be by-passed so that protein synthesis can occur.

Nitrogen fixation, involving as it does the reduction of the element to the ammonia level, will affect the metabolic balance of molybdenum-deficient cells in the same way as nitrate reduction and will therefore require a relatively high molybdenum concentration for elaboration of its products to protein. If this interpretation is correct, the similar requirement for molybdenum in nitrate reduction and in nitrogen fixation does not mean that these two processes have any part of their pathways in common. With both *An. cylindrica* (Fig. 1) and *Azotobacter*, it has been observed that a higher concentration of molybdenum is required for maximum growth on free nitrogen than on nitrate, but the reason for this is not apparent.

THE PRODUCTS OF NITROGEN METABOLISM

Amino-acids

The results of various amino-acid analyses of blue-green algae are summarized in Table 2. In general agreement with these are results of qualitative analyses made by Watanabe (1951) and Fowden (1951a).

In the analysis of *Phormidium* sp. separation of amino-acids was achieved by precipitation and other chemical methods, whereas analyses of the other species were made by a photometric ninhydrin method after chromatographic separation. For all species except *Anabaena cylindrica* only about half the total nitrogen is accounted for, and it is possible that in the case of *Phormidium* sp. the fraction examined may not have been representative of the whole (Fowden, 1951 b). Thus it is unlikely that cystine, which is reported as absent in *Phormidium* sp. and *Gloeotrichia echinulata* by Mazur & Clarke (1938, 1942) but which is present in *Anabaena cylindrica* (Fowden, 1951 a), in *Tolypothrix tenuis* and *Anabaenopsis* sp. (Watanabe, 1951), and in *Oscillatoria* sp. (Wassink & Ragetli, 1952), is actually absent in any representatives of this group.

Table 2. *The amino-acids of the bulk proteins of various blue-green algae*

Amounts are given as g. amino-acid-N/100 g. protein-N.

	Phormidium sp. (Mazur & Clarke, 1938)	Diplocystis (Microcystis) aeruginosa (Williams & Burris, 1952)	Nostoc muscorum (Williams & Burris, 1952)	Calothrix parietina (Williams & Burris, 1952)	Anabaena cylindrica (Fowden, 1954)
Aspartic acid	0·9	4·6	7·9	5·8	6·9
Glutamic acid	4·4	6·5	6·0	8·4	5·6
Serine	—	3·3	2·6	4·0	2·4
Threonine	—	3·2	4·6	3·4	5·7
Glycine	1·6	4·9	5·0	4·5	5·5
Alanine	5·2	5·4	8·4	6·1	6·0
Valine	6·7	4·1	3·7	4·1	7·0
Leucine	2·1	4·2	4·2	} 7·5	6·2
Isoleucine	—	2·2	2·8		3·9
Phenylalanine	1·1	4·4	1·9	2·0	2·9
Tyrosine	1·8	—	0·8	1·0	1·6
Proline	7·0	3·2	2·9	2·9	5·0
Tryptophane	0·2	—	—	—	1·0
Methionine	2·0	1·7	0·4	1·2	1·2
Cystine	0·0	—	—	—	—
Arginine	9·2	—	—	—	11·7
Histidine	3·8	—	—	—	2·5
Lysine	0·0	—	—	—	6·6
Amide	—	—	—	—	8·0
Total	46·0	47·7	51·2	50·9	89·7

Taking the analysis of *Anabaena cylindrica* as the most reliable it would not appear from this that there is any major difference between the Myxophyceae and other algae such as *Chlorella vulgaris* (Fowden, 1951 b, 1954) in respect of any of the amino-acids listed in Table 2.

The Myxophyceae are, however, characterized by the production of an amino-acid which so far has not been found in any other group of organisms except the bacteria. This is α-ε-diaminopimelic acid, which occurs in *Anabaena cylindrica*, *Oscillatoria* sp. and *Mastigocladus laminosus* but not in representatives of other groups of algae, including

Chlorophyceae, Xanthophyceae, Bacillariophyceae, Euglenineae, Phaeophyceae or Rhodophyceae (Work & Dewey, 1953). Various unidentifiable spots appearing on paper chromatograms of preparations from blue-green algae (Fowden, 1951 *a*; Watanabe, 1951; Wassink & Ragetli, 1952) may perhaps prove to be due to amino-acids peculiar to this group.

Polypeptides

Extracellular nitrogenous products have been noted frequently in cultures of nitrogen-fixing blue-green algae (Allison & Morris, 1932; Winter, 1935; De, 1939; Fogg, 1942, 1951 *a*; Henriksson, 1951; Watanabe, 1951), and it appears that the liberation of such substances may be characteristic of most members of the class whether they fix nitrogen or not (Fogg, 1952). In *Anabaena cylindrica*, and probably in other species also, the substances concerned are principally polypeptides (Fogg, 1952), free amino-acids being present only in minute amounts (Watanabe, 1951; Fogg, 1952). In one sample of extracellular polypeptide from *An. cylindrica* the principal component amino-acids were found to be serine and threonine with smaller amounts of glutamic acid, glycine and tyrosine, and traces of alanine, valine and leucine. The chemical constitution of these polypeptides has not yet been established, but since they do not give the ninhydrin reaction it seems possible that they have a cyclic structure (Fogg, 1952).

The liberation of extracellular polypeptide is not due to autolysis and invariably appears to accompany growth (Fig. 8). The relative amounts of extracellular polypeptide produced vary with the cultural conditions, being, for example, increased in the later stages of growth by deficiency of iron. Molybdenum deficiency has been found to decrease only slightly the production of extracellular nitrogenous substances (Fogg, 1952), but in view of the differential effects of such deficiency on the proportions of different fractions of combined nitrogen found by Wolfe (1953) the interpretation of this observation may not be as straightforward as was previously supposed. From the physiological relations in their production it seems possible that extracellular polypeptides are released in the course of metabolism at the outer surface of the cell (Fogg, 1952). Because of their capacity to form complexes with various ions, extracellular polypeptides have important effects on the growth of the alga in culture and probably also in natural environments (Fogg & Westlake, 1954).

Proteins

While nothing is known of the proteins of the Myxophyceae in general, a moderate amount of information is available concerning phycocyanin, the chromoprotein from which the class derives its characteristic colour. Phycocyanin ($\lambda_{max.} = c.$ 620 mμ.) is the principal phycobilin pigment in blue-green algae, but it may be accompanied or replaced by phycoerythrin ($\lambda_{max.} = c.$ 565 mμ.) in certain species. These

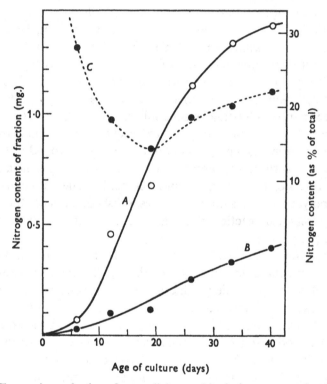

Fig. 8. Changes in production of extracellular combined nitrogen occurring during the growth of cultures of *Anabaena cylindrica*. *A*, intracellular combined nitrogen; *B*, extracellular combined nitrogen; *C*, extracellular combined nitrogen as a percentage of the total.

pigments are similar to but not identical with phycobilins found in the Rhodophyceae (Kylin, 1937).

The phycobilins contain a metal-free linear tetrapyrrolic chromophoric group in combination with a globulin-like protein (see Rabinowitch, 1945). Phycocyanin from *Aphanizomenon flos-aquae* has a molecular weight of about 208,000 but decomposes into smaller units as the pH is increased (Svedberg & Katsurai, 1929). Paper chromato-

graphic analysis has shown that the amino-acid composition of phycocyanin from *Oscillatoria* sp. is similar to that of bulk proteins of other
organisms except that no appreciable amount of arginine is present
(Wassink & Ragetli, 1952).

Light absorbed by phycocyanin in the living alga can be utilized in
photosynthesis (Blinks, p. 224, this volume). The amount of phycocyanin
in blue-green algae is considerably reduced in nitrogen starvation
(Boresch, 1913; Fogg, 1952) and may also vary according to the colour
and intensity of the light to which the alga is exposed (Fritsch, 1945;
Rabinowitch, 1945; Blinks, p. 224, this volume).

Cyanophycin

The conspicuous granules of cyanophycin, a material soluble in dilute
acids, having characteristic staining reactions, and commonly occurring
in the cells of Myxophyceae (Fritsch, 1945), have attracted considerable
attention from microscopists but little from chemists. Cyanophycin is
generally thought to be proteinaceous in nature (Fritsch, 1945), although
there is little definite evidence for this. It does, however, contain a large
proportion of arginine (Fogg, 1951*b*). Cyanophycin is evidently a
reserve product, since it disappears in material placed in the dark and is
absent from actively growing cells (Fritsch, 1945; Fogg, 1951*b*).

Nucleic acids

It is evident from their staining reactions that blue-green algae contain
nucleic acids (Poljansky & Petruschewsky, 1929; Fogg, 1951*b*), but no
investigation appears to have been made of either the chemistry or the
metabolism of these and related substances in such organisms. In view
of the anomalous nature of the nuclear apparatus of the Myxophyceae
such investigations might be expected to yield especially interesting
results.

CONCLUDING REMARKS

Our knowledge of the nitrogen metabolism of the Myxophyceae is as yet
fragmentary, but from the information which we have it does not appear
that it differs in any fundamental way from that of other micro-organisms.
The properties of the enzyme system fixing nitrogen which is present in
many of the species, the paths by which nitrate and ammonia enter into
metabolism, and the amino-acid composition of their bulk proteins are
all similar to those of organisms belonging to other classes and indicate
a substantial conformity with the metabolic pattern which appears to be
common to all forms of life. So far, studies with blue-green algae have
contributed little to our knowledge of this general pattern, but investiga-

tions on the effects of molybdenum deficiency in *Anabaena cylindrica* appear to provide a basis for a theory of the function of this element in nitrogen metabolism. This theory can also explain a variety of observations made with other types of micro-organisms and with higher plants.

However, the nitrogen metabolism of the blue-green algae has certain special features. Metabolic flexibility, such as enables *An. cylindrica* to utilize nitrogen sources ranging from the free element to casein, appears to be a characteristic of many primitive micro-organisms. Some peculiarities, for example, the capacity for nitrogen fixation and the production of the uncommon diaminopimelic acid, are shared with the bacteria, while another characteristic, the production of phycobilins, is also possessed by the Rhodophyceae but by no other group of organisms. Other metabolic products, as, for example, cyanophycin, may occur only in the Myxophyceae. Further investigation may be expected to reveal other unique features of the chemical activities of these organisms.

We wish to express our thanks to Mr P. J. Syrett for his helpful discussion of certain sections of this paper, and to the Council of the Royal Society for permission to reproduce Fig. 8. We are also grateful to the Department of Scientific and Industrial Research for a maintenance allowance made to one of us (M.W.), which enabled some of the investigations which have been described to be carried out.

REFERENCES

AGARWALA, S. C. (1952). The effect of molybdenum and nitrate status on the carbohydrate and nitrogen metabolism of cauliflower plants in sand culture. *Rep. agric. hort. Res. Sta., Bristol*, 1951, p. 70.

AGARWALA, S. C. & WILLIAMS, A. H. (1951). The effects of the inter-relationships between molybdenum and nitrogen supply on the free amino-acid status of cauliflower plants grown in sand culture. *Rep. agric. hort. Res. Sta., Bristol*, 1950, p. 66.

ALLEN, M. B. (1952). The cultivation of Myxophyceae. *Arch. Mikrobiol.* **17**, 34.

ALLISON, F. E. & HOOVER, S. R. (1935). Conditions which favour nitrogen fixation by a blue-green alga. *Proc. III Int. Congr. Soil Sci.* **1**, 145.

ALLISON, F. E., HOOVER, S. R. & MORRIS, H. J. (1937). Physiological studies with the nitrogen-fixing alga, *Nostoc muscorum*. *Bot. Gaz.* **98**, 433.

ALLISON, F. E. & MORRIS, H. J. (1930). Nitrogen fixation by blue-green algae. *Science*, **71**, 221.

ALLISON, F. E. & MORRIS, H. J. (1932). Nitrogen fixation by soil algae. *Proc. II Int. Congr. Soil Sci.* **3**, 24.

BEIJERINCK, M. W. (1901). Ueber oligonitrophile Mikroben. *Zbl. Bakt.* (2. Abt.), **7**, 561.

BORESCH, K. (1913). Die Färbung von Cyanophyceen und Chlorophyceen in ihrer Abhängigkeit vom Stickstoffgehalt des Substrates. *Jb. wiss. Bot.* **52**, 145.

BORTELS, H. (1930). Molybdäns als Katalysator bei der Biologischen Stickstoffbindung. *Arch. Mikrobiol.* **1**, 333.

BORTELS, H. (1940). Über die Bedeutung des Molybdäns für stickstoffbindende Nostocaceen. *Arch. Mikrobiol.* **11**, 155.

BURRIS, R. H., EPPLING, F. J., WAHLIN, H. B. & WILSON, P. W. (1942). Studies of biological nitrogen fixation with isotopic nitrogen. *Proc. Soil Sci. Soc. Amer.* **7**, 258.

BURRIS, R. H., EPPLING, F. J., WAHLIN, H. B. & WILSON, P. W. (1943). Detection of nitrogen fixation with isotopic nitrogen. *J. biol. Chem.* **148**, 349.

BURRIS, R. H. & WILSON, P. W. (1946). Characteristics of the nitrogen-fixing enzyme system in *Nostoc muscorum. Bot. Gaz.* **108**, 254.

DE, P. K. (1939). The role of blue-green algae in nitrogen fixation in rice-fields. *Proc. Roy. Soc.* B, **127**, 121.

DREWES, K. (1928). Über die Assimilation des Luftstickstoffs durch Blaualgen. *Zbl. Bakt.* (2. Abt.), **76**, 88.

EVANS, H. J. & NASON, A. (1953). Pyridine nucleotide-nitrate reductase from extracts of higher plants. *Plant Physiol.* **28**, 233.

FOGG, G. E. (1942). Studies on nitrogen fixation by blue-green algae. I. Nitrogen fixation by *Anabaena cylindrica* Lemm. *J. exp. Biol.* **19**, 78.

FOGG, G. E. (1947). Nitrogen fixation by blue-green algae. *Endeavour*, **6**, 172.

FOGG, G. E. (1949). Growth and heterocyst production in *Anabaena cylindrica* Lemm. II. In relation to carbon and nitrogen metabolism. *Ann. Bot., Lond.* N.S. **13**, 241.

FOGG, G. E. (1951a). Studies on nitrogen fixation by blue-green algae. II. Nitrogen fixation by *Mastigocladus laminosus* Cohn. *J. exp. Bot.* **2**, 117.

FOGG, G. E. (1951b). Growth and heterocyst production in *Anabaena cylindrica* Lemm. III. The cytology of heterocysts. *Ann. Bot., Lond.* N.S. **15**, 23.

FOGG, G. E. (1952). The production of extracellular nitrogenous substances by a blue-green alga. *Proc. Roy. Soc.* B, **139**, 372.

FOGG, G. E. & WESTLAKE, D. F. (1954). The importance of extracellular products of algae in freshwater. *Proc. Int. Ass. Limn.* **12** (in the Press).

FOWDEN, L. (1951a). Amino-acids of certain algae. *Nature, Lond.* **167**, 1030.

FOWDEN, L. (1951b). The composition of the bulk proteins of *Chlorella. Biochem. J.* **50**, 355.

FOWDEN, L. (1954). A comparison of the compositions of some algal proteins. *Ann. Bot. Lond.* (in the Press).

FRANK, B. (1889). Ueber den experimentellen Nachweis der Assimilation freien Stickstoffs durch erdbodenbewohnende Algen. *Ber. dtsch. bot. Ges.* **7**, 34.

FRENKEL, A. W. & RIEGER, C. (1951). Photoreduction in algae. *Nature, Lond.* **167**, 1030.

FRITSCH, F. E. (1945). *The Structure and Reproduction of the Algae*, **2**. Cambridge University Press.

GERLOFF, G. C., FITZGERALD, G. P. & SKOOG, F. (1950a). The isolation, purification, and culture of blue-green algae. *Amer. J. Bot.* **37**, 216.

GERLOFF, G. C., FITZGERALD, G. P. & SKOOG, F. (1950b). The mineral nutrition of *Coccochloris peniocystis. Amer. J. Bot.* **37**, 835.

GERLOFF, G. C., FITZGERALD, G. P. & SKOOG, F. (1952). The mineral nutrition of *Microcystis aeruginosa. Amer. J. Bot.* **39**, 26.

GLADE, R. (1914). Zur Kenntnis der Gattung *Cylindrospermum. Beitr. Biol. Pfl.* **12**, 295.

HENRIKSSON, E. (1951). Nitrogen fixation by a bacteria-free, symbiotic *Nostoc* strain isolated from *Collema. Physiol. Plant.* **4**, 542.

HEWITT, E. J. (1951). The role of the mineral elements in plant nutrition. *Annu. Rev. Plant Physiol.* **2**, 25.

HEWITT, E. J., AGARWALA, S. C. & JONES, E. W. (1950). Effect of molybdenum status on the ascorbic acid content of plants in sand culture. *Nature, Lond.* **166**, 1119.

HEWITT, E. J. & JONES, E. W. (1947). The production of molybdenum deficiency in plants in sand culture with special reference to tomato and brassica crops. *J. Pomol.* **23**, 254.

HEWITT, E. J., JONES, E. W. & WILLIAMS, A. H. (1949). Relation of molybdenum and manganese to the free amino-acid content of the cauliflower. *Nature, Lond.* **163**, 681.

HORNER, C. K., BURK, D., ALLISON, F. E. & SHERMAN, M. S. (1942). Nitrogen fixation by *Azotobacter* as influenced by molybdenum and vanadium. *J. agric. Res.* **65**, 173.

HUTNER, S. H., PROVASOLI, L., SCHATZ, A. & HASKINS, C. P. (1950). Some approaches to the study of the role of metals in the metabolism of microorganisms. *Proc. Amer. phil. Soc.* **94**, 152.

JENSEN, H. L. & SPENCER, D. (1947). The influence of molybdenum and vanadium on nitrogen fixation by *Clostridium butyricum* and related organisms. *Proc. Linn. Soc. N.S.W.* **72**, 73.

JONES, J. (1930). An investigation into the bacterial associations of some Cyanophyceae. *Ann. Bot., Lond.* **44**, 721.

KYLIN, H. (1937). Über die Farbstoffe und die Farbe der Cyanophyceen. *K. fysiogr. Sällsk. Lund Förh.* **7** (12), 1.

KYLIN, H. (1943). Verwandtschaftliche Beziehungen zwischen den Cyanophyceen und den Rhodophyceen. *K. fysiogr. Sällsk. Lund Förh.* **13** (17), 1.

LINDSTROM, E. S., BURRIS, R. H. & WILSON, P. W. (1949). Nitrogen fixation by photosynthetic bacteria. *J. Bact.* **58**, 313.

MAERTENS, H. (1914). Das Wachstum von Blaualgen in mineralischen Nährlösungen. *Beitr. Biol. Pfl.* **12**, 439.

MAZUR, A. & CLARKE, H. T. (1938). The amino-acids of certain marine algae. *J. biol. Chem.* **123**, 729.

MAZUR, A. & CLARKE, H. T. (1942). Chemical components of some autotrophic organisms. *J. biol. Chem.* **143**, 39.

MULDER, E. G. (1948). Importance of molybdenum in the nitrogen metabolism of micro-organisms and higher plants. *Plant & Soil,* **1**, 94.

NICHOLAS, D. J. D. & FIELDING, A. H. (1950). Use of *Aspergillus niger* as a test organism for determining molybdenum available in soils to crop plants. *Nature, Lond.* **166**, 342.

NICHOLAS, D. J. D., NASON, A. & McELROY, W. D. (1953). Effect of molybdenum deficiency on nitrate reductase in cell-free extracts of *Neurospora* and *Aspergillus.* *Nature, Lond.* **172**, 34.

POLJANSKY, G. & PETRUSCHEWSKY, G. (1929). Zur Frage über die Struktur der Cyanophyceenzelle. *Arch. Protistenk.* **67**, 11.

PRINGSHEIM, E. G. (1914). Kulturversuche mit chlorophyllfürenden Mikro-organismen. iii Mitteilung. Zur Physiologie der Schizophyceen. *Beitr. Biol. Pfl.* **12**, 49.

PRINGSHEIM, E. G. (1949). The relationship between bacteria and Myxophyceae. *Bact. Rev.* **13**, 47.

RABINOWITCH, E. I. (1945). *Photosynthesis and Related Processes,* **1**. New York: Interscience.

SAKAMURA, T. & MAEDA, K. (1950). On the assimilation of nitrate nitrogen by *Hansenula anomala. J. Fac. Sci. Hokkaido Univ.* ser. v, **7**, 79.

SINGH, R. N. (1942). The fixation of elementary nitrogen by some of the commonest blue-green algae from the paddy field soils of the United Provinces and Bihar. *Indian J. agric. Sci.* **12**, 743.

SINGH, R. N. (1950). Reclamation of 'Usar' lands in India through blue-green algae. *Nature, Lond.* **165**, 325.

STEINBERG, R. A. (1937). Role of molybdenum in the utilisation of ammonium and nitrate nitrogen by *Aspergillus niger*. *J. agric. Res.* **55**, 891.

SVEDBERG, T. & KATSURAI, T. (1929). The molecular weights of phycocyan and of phycoerythrin from *Porphyra tenera* and of phycocyan from *Aphanizomenon flos-aquae*. *J. Amer. chem. Soc.* **51**, 3573.

SYRETT, P. J. (1953a). The assimilation of ammonia by nitrogen-starved cells of *Chlorella vulgaris*. Part I. The correlation of assimilation with respiration. *Ann. Bot., Lond.* N.S. **17**, 1.

SYRETT, P. J. (1953b). The assimilation of ammonia by nitrogen-starved cells of *Chlorella vulgaris*. Part II. The assimilation of ammonia to other compounds. *Ann. Bot., Lond.* N.S. **17**, 21.

WALP, L. (1942). The effect of nitrate and ammonium assimilation on cell prolifera-tion of *Nostoc muscorum*. *Growth*, **6**, 173.

WARBURG, O. & NEGELEIN, E. (1920). Über die Reduktion der Salpetersäure in grünen Zellen. *Biochem. Z.* **110**, 66.

WASSINK, E. C. & RAGETLI, H. W. (1952). Paper chromatography of hydrolysed *Oscillatoria* phycocyanin. *Proc. K. Akad. Wet. Amst.* C, **55**, 462.

WATANABE, A. (1951). Production in cultural solution of some amino-acids by the atmospheric nitrogen-fixing blue-green algae. *Arch. Biochem. Biophys.* **34**, 50.

WATANABE, A., NISHIGAKI, S. & KONISHI, C. (1951). Effect of nitrogen-fixing blue-green algae on the growth of rice plants. *Nature, Lond.* **168**, 748.

WILLIAMS, A. E. & BURRIS, R. H. (1952). Nitrogen fixation by blue-green algae and their nitrogenous composition. *Amer. J. Bot.* **39**, 340.

WILSON, P. W. (1952). The comparative biochemistry of nitrogen fixation. *Advanc. Enzymol.* **13**, 345.

WINTER, G. (1935). Über die Assimilation des Luftstickstoffs durch endophytische Blaualgen. *Beitr. Biol. Pfl.* **23**, 295.

WOLFE, M. (1953). The effect of molybdenum upon the nitrogen metabolism of *Anabaena cylindrica*. Ph.D. Thesis, University of London.

WOOD, J. G. (1953). Nitrogen metabolism of higher plants. *Annu. Rev. Plant Physiol.* **4**, 1.

WORK, E. & DEWEY, D. L. (1953). The distribution of α-ε-diaminopimelic acid among various micro-organisms. *J. gen. Microbiol.* **9**, 394.

AMMONIA AND NITRATE ASSIMILATION
BY GREEN ALGAE (CHLOROPHYCEAE)

P. J. SYRETT

Department of Botany, University College, London

INTRODUCTION

Green algae, like higher plants and some bacteria, can live autotrophically, obtaining their carbon from carbon dioxide by photosynthesis and their nitrogen from either nitrate or ammonia. We now know much about carbon assimilation from work with *Chlorella* and *Scenedesmus*, but nitrogen assimilation has received less attention. Yet unicellular algae are particularly favourable objects for the study of nitrogen assimilation. Since the cells live naturally in an aqueous environment, ions in solution can be directly supplied to them and their small size (diam. $3–10\mu$) makes diffusion effects unimportant. They can be grown quickly and fairly uniformly in pure culture, and since the cells are green and can photosynthesize, the relationship between nitrogen assimilation and photosynthesis can be studied. Some of them, e.g. *Chlorella* and *Scenedesmus*, grow well heterotrophically if glucose is supplied and consequently nitrogen assimilation can be studied in the presence of adequate amounts of carbohydrate; this allows the effects of heavy metal deficiencies on nitrate assimilation to be studied, although these deficiencies reduce the rate of photosynthesis (Noack & Pirson, 1939; Alberts-Dietert, 1941).

From well before 1900 information has been accumulating about the availability of nitrate and ammonia nitrogen for algal growth (see review by Ludwig, 1938), and Pringsheim (1946) has concluded that while some colourless algae cannot utilize nitrate, all green forms can use either nitrate or ammonia in the light. But although information of this kind is available for many different species, work on the physiology of nitrogen assimilation, like that on carbon assimilation, has been confined chiefly to a few species of *Chlorella*, *Scenedesmus* and *Ankistrodesmus*.

In this paper work with these organisms is considered. The principal topics discussed are the relationship of ammonia and nitrate assimilation to respiration in darkness, and the effects of anaerobiosis and light on assimilation.

GROWTH IN AMMONIA AND NITRATE MEDIA

Chlorella grows well in light in an inorganic medium containing either ammonium or nitrate ions, but unless the ammonium medium is buffered, the pH decreases rapidly as the ammonia is assimilated; in a nitrate medium the pH tends to rise from initial values of 3·8–5·2 to a final value of about 7 (Pratt & Fong, 1940). Such changes of pH may restrict growth in an ammonium medium but not in a nitrate one. However, *Chlorella* grows well in the presence of quite high concentrations of phosphate, so the medium can be buffered with 0·067 or 0·033 M phosphate (Hopkins & Wann, 1926; Pearsall & Loose, 1937; Noack & Pirson, 1939).

It has been reported (Noack & Pirson, 1939; Alberts-Dietert, 1941) that the autotrophic growth of *Chlorella* in an adequately buffered medium is the same whether nitrate or ammonia is the nitrogen source, but in these experiments growth may have been limited by the rate of carbon assimilation. In cultures shaken in darkness with 1·5 % glucose, Noack & Pirson again found growth was the same with either compound, but Alberts-Dietert in similar experiments found better growth in the ammonium cultures. Österlind (1949) compared the autotrophic growth of *Scenedesmus quadricauda* in nitrate and ammonia media buffered with bicarbonate-carbon dioxide and found better growth in the nitrate media over the pH range 6·8–11·2. But in these experiments growth was not followed over a period of time and the culture tubes were neither shaken continuously nor aerated, so there may have been differences in the diffusion of the ions to the cells. It is possible that nitrate is the better source of nitrogen at low pH values and that ammonia is better at higher ones (around neutrality), but no critical comparison of the two ions as nitrogen sources for algal growth appears to have been carried out under conditions where it is certain that growth is not limited by changes in pH, the rate of photosynthesis, or the efficiency of aeration.

When both ammonium and nitrate ions are supplied together most workers agree that *Chlorella* assimilates much more ammonia than nitrate (Pearsall & Loose, 1937; Pratt & Fong, 1940; Cramer & Myers, 1948; Schuler, Diller & Kersten, 1953). Urhan (1932) found that *Chlorella* and *Scenedesmus* assimilated both ions equally from ammonium nitrate in light, but these results were obtained with cultures which were not shaken.

AMMONIA ASSIMILATION IN DARKNESS

The study of ammonia assimilation has been aided by the use of cells starved of nitrogen (Urhan, 1932; Pirson, 1937; Myers, 1949; Syrett, 1953 *a*, *b*). After cells of *Chlorella vulgaris* have been allowed to photosynthesize for 16 hr. in a nitrogen-free medium they can assimilate added ammonia in darkness much more rapidly than normal cells, and whereas the rate of assimilation by normal cells is increased by adding glucose that of nitrogen-starved cells is not (Fig. 1 *a*, *b*). Presumably the nitrogen-starved cells contain sufficient carbon reserves to saturate the enzyme systems involved in assimilation.

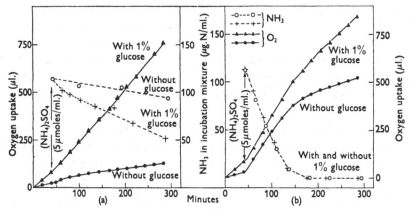

Fig. 1. The effect of glucose on the oxygen uptake and ammonia assimilation by *Chlorella vulgaris* at 25° and pH 6·1. (*a*) normal cells (4·75 mg. dry weight/ml.); (*b*) nitrogen-starved cells (7·65 mg. dry weight/ml.). (From Syrett, 1953 *a*.)

The rapid assimilation of ammonia by nitrogen-starved cells is accompanied by the rapid disappearance of acid-hydrolysable polysaccharide in the cells (Syrett, 1953 *b*), and it is possible that fat is also metabolized. When a large quantity of ammonia is added assimilation apparently continues until the carbon reserves of the cells are exhausted. If glucose is added, more ammonia can be assimilated (Syrett, 1953 *a*).

The products of ammonia assimilation by *C. vulgaris* have been studied (Syrett, 1953 *b*; Syrett & Fowden, 1952). The addition of ammonia is followed by an immediate increase of soluble organic nitrogen within the cells due to the formation of free and combined α-amino-nitrogen and amide-nitrogen (Fig. 2*a*). The amide-nitrogen reaches a maximum value after about 1 hr. and then decreases. Insoluble nitrogen shows no increase until some 20 min. after the addition of the ammonia. These changes are similar to those which follow the feeding of ammonia to other organisms, e.g. nitrogen-starved *Torula utilis* (Roine, 1947). The rapid formation of amides, chiefly glutamine, and the increase in glutamic acid (observed on chromatograms), indicate that glutamic acid may be a primary product of ammonia assimilation by *Chlorella* as it is in other organisms. It is of interest that Millbank (1953) has demonstrated the presence of a glutamic acid transaminase in *C. vulgaris*.

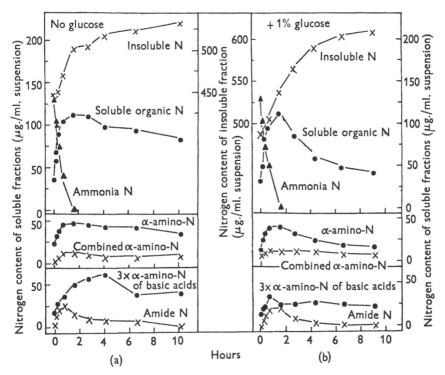

Fig. 2. The products of ammonia assimilation by nitrogen-starved cells of *Chlorella vulgaris* at 25° and pH 6·1. 5 μmoles $(NH_4)_2SO_4$ per ml. added at zero time. Soluble nitrogen extracted with hot water. (*a*) glucose absent; 9·3 mg. dry weight cells/ml.; (*b*) 1% glucose present; 7·05 mg. dry weight cells/ml. (From Syrett & Fowden, 1952.)

In one experiment (Fig. 2*a*) the ammonia added was completely assimilated in 90 min., but much of it remained in the cells as soluble organic nitrogen. When a similar experiment was carried out with 1 % glucose added the ammonia was completely assimilated to insoluble nitrogen (Fig. 2*b*). Apparently the cells with a limited carbon reserve could convert all the added ammonia to soluble organic nitrogen, but more carbon was required to complete the assimilation to protein. These cells with a high content of soluble organic nitrogen show a peculiar feature; rather more than half of the soluble organic nitrogen is nitrogen of the basic amino-acids. Chromatograms showed the presence of arginine, lysine, ornithine and histidine, arginine being the most abundant. It is of interest that when cells with a limited carbon reserve assimilate large amounts of ammonia they convert much of it to basic amino-acids, i.e. to compounds with a low C/N ratio.

THE RELATIONSHIP BETWEEN AMMONIA ASSIMILATION AND RESPIRATION

Cramer & Myers (1948) observed that the addition of ammonium sulphate to cells of *Chlorella pyrenoidosa* respiring glucose in the dark was followed by a decrease in the rate of carbon dioxide production; the rate of oxygen consumption remained constant and the respiratory quotient (R.Q.) decreased, e.g. in one experiment from 1·42 to 1·18. The change in R.Q. was very sharp when nitrogen-starved cells were used (Myers, 1949). Cramer & Myers (1948) pointed out that the high R.Q. before ammonia was added indicated that glucose was being assimilated to more reduced cellular materials, possibly fat. The decrease in R.Q. after the addition of ammonia showed that less reduced products, e.g. protein, were then being formed.

When ammonia is added to nitrogen-starved cells of *C. vulgaris* changes in the rate of respiration are very pronounced. The rate of oxygen uptake may increase by 600 %, but the rate of carbon dioxide production does not increase so much and the R.Q. drops from 1·2 to 0·75 (Fig. 3). When all the ammonia has been assimilated the respiration rate returns to its original value. The low R.Q. observed immediately after the addition of ammonia is to be expected if dicarboxylic amino-acids and amides are synthesized, since these compounds are more oxidized than the cellular carbohydrates from which they are formed. The sudden increase of respiration rate after the addition of ammonia to nitrogen-starved cells is not confined to *Chlorella*. It has been observed with the bacterium *Serratia marcescens* (McLean & Fisher, 1947, 1949), yeast (Yemm, E. W. & Folkes, B. F., unpublished) and excised barley roots (Willis, 1951), and the phenomenon may be a general one.

When the respiration rates of nitrogen-starved and normal cells are compared (Fig. 1, Table 1) one sees that the addition of ammonia to nitrogen-starved cells is followed by a rate of respiration as high as that of normal cells respiring glucose. In some way the addition of ammonia to nitrogen-starved cells allows the rapid metabolism of the carbon reserves which the cells contain. Polysaccharide disappears rapidly, and if this is the source of the respiratory carbon dioxide, about 25 % is respired and the rest presumably converted to organic nitrogen compounds (Syrett, 1953 *b*).

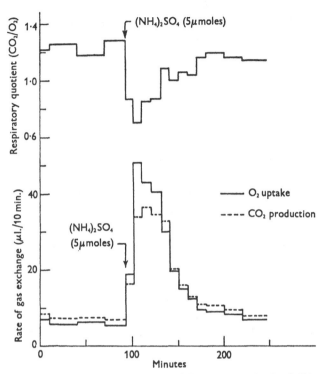

Fig. 3. The effect of ammonia on the respiration of nitrogen-starved cells of *Chlorella vulgaris* at 25° and pH 6·2. Each manometer contained 14·7 mg. dry weight cells. (From Syrett, 1953 *a*.)

Table 1. *The effect of ammonia on the rate of oxygen uptake by* Chlorella vulgaris *in the presence and absence of glucose at 25° and* pH 6·1. (From Syrett, 1953 *a*)

Cells	Glucose added (w/v)	Q_{O_2}	
		NH_4^+ absent	0·0044 M-NH_4^+ present
Nitrogen-starved	—	2·7	16·1
Nitrogen-starved	1 %	7·8	16·7
Normal	—	3·0	2·3
Normal	1 %	12·5	17·5

A possible explanation of the great stimulation of respiration following the addition of ammonia can be proposed if it is assumed that the metabolism of *Chlorella* involves phosphorylated compounds such as adenosine triphosphate (ATP) and adenosine diphosphate (ADP) and an organic acid cycle. When ammonia is added it is assimilated to amino-acids and amides, and the carbon skeletons for their synthesis probably come from the organic acids of the cycle. Yemm, E. W. & Folkes, B. F. (unpublished) have pointed out that the conversion of organic acids to nitrogen compounds will increase their rate of formation from carbo-hydrate together with the rates of the consequent oxidations and decarb-oxylations. Hence the rate of respiration may be expected to increase.

The concentration of phosphorylated compounds may also be important. Johnson (1941), Lardy (1952) and others have suggested that the availability of phosphate acceptors may control metabolic rates. Lardy & Wellman (1952) showed that the addition of phosphate acceptors to a preparation of rat-liver mitochondria greatly increased the rate of oxygen uptake. If there is little synthesis in nitrogen-starved cells of *Chlorella* and hence little utilization of energy-rich phosphate com-pounds, the concentration of phosphate acceptors may be low and this may limit the rate of respiration. When ammonia is added glutamine is synthesized, and this is known to involve the utilization of ATP (Speck, 1947; Elliott, 1951). Possibly other reactions in ammonia assimilation also involve utilization of ATP. Thus the increase in respiration rate may be due to an increase in ADP concentration because of these reactions. The addition of glucose to nitrogen-starved cells may bring about a smaller increase in respiration rate because much of it is converted to polysaccharide (Syrett, 1951), and this conversion presumably also utilizes ATP.

Some evidence that such a system is involved can be obtained by the use of 2:4-dinitrophenol. This uncouples phosphorylation from oxida-tion (Loomis & Lipmann, 1948), and a suitable concentration should therefore increase the respiration rate of nitrogen-starved cells just as ammonia does (cf. Lardy & Wellman, 1952). Experiments show that this is so, and a concentration of dinitrophenol which stimulates respiration reduces the rate of ammonia assimilation as one would expect if it uncouples phosphorylation (Table 2). Similar results have been obtained and a similar interpretation proposed by Yemm, E. W. & Folkes, B. F. (unpublished) from experiments with nitrogen-starved yeast.

Direct evidence for a change in the phosphate compounds in the cells after addition of ammonia has been sought, and it was found that the amount of total soluble phosphorus decreased as soon as ammonia was added (Table 3). Attempts to identify the phosphorus compounds involved have been unsuccessful.

If there is a control system which slows down the rate of breakdown of carbohydrate it probably operates by decreasing the rate of organic acid metabolism rather than the rate of conversion of carbohydrates to organic acids. This is shown by the fact that acetate affects nitrogen-

Table 2. *The effect of* 2:4-*dinitrophenol on the respiration and ammonia assimilation of nitrogen-starved cells of* Chlorella vulgaris *at* 25° *and* pH 6·1

Cells suspended in 0·0025 M-MgSO₄ and 0·1 M-phosphate. In this experiment and those following, respiration was measured by Warburg's 'direct' method and a correction for carbon dioxide retention applied as described by Johnson (Umbreit, Burris & Stauffer, 1949). Each manometer contained 16·5 mg. dry weight cells. Initially, 1·5 μmoles (NH₄)₂SO₄ added and the Q_{NH_3} calculated from ammonia determinations at 15 and 30 min.

	Before ammonia addition				After ammonia addition					
2:4-D.N.P. (M)	Q_{O_2}	% Control	Q_{CO_2}	% Control	Q_{O_2}	% Control	Q_{CO_2}	% Control	Q_{NH_3}	% Control
0	2·9	100	3·3	100	16·5	100	13·8	100	21·8	100
1·8 × 10⁻⁴	7·2	246	7·5	228	14·7	89	13·4	97	15·2	70
2·2 × 10⁻⁴	8·3	285	8·7	264	14·5	88	13·3	96	11·9	54

Table 3. *Changes in the total soluble phosphorus compounds of nitrogen-starved cells of* Chlorella vulgaris *following the addition of ammonia at* 25° *and* pH 7·2

Cells suspended in 0·011 M-KHCO₃; gas phase 5 % CO₂—95 % air. Soluble phosphorus compounds extracted by the method of Wassink, Wintermans & Tjia (1951). Values given below are mean values for duplicate flasks; individual values were within 2 % of the mean.

	(μg./g. dry weight cells)	
	Total soluble P	Change
Initial	857	—
Final:		
Control (0·02 M-Cl⁻ as HCl-KCl)	860	+3
Expt. (0·02 M-NH₄Cl)	706	−151

Table 4. *A comparison of the effects of acetate and glucose on nitrogen-starved cells of* Chlorella vulgaris *suspended in* 0·067 M-*phosphate at* 25° *and* pH 6·1

Additions	Rate of O₂ uptake (μl./10 min.)
None	5·5
Acetate (0·0044 M)	17·5
Glucose (0·0044 M)	18·2
Ammonia (0·0013 M)	28·0
Acetate (0·0044 M) + ammonia (0·0013 M)	36·0
Glucose (0·0044 M) + ammonia (0·0013 M)	34·0

starved cells in the same way as glucose; only when ammonia as well as acetate is added, is the maximum rate of oxygen consumption attained (Table 4).

THE ANAEROBIC ASSIMILATION OF AMMONIA
BY NITROGEN-STARVED CELLS

Cells of *C. vulgaris* which have been starved of nitrogen for 16 hr. can assimilate about 1·2–1·5 μmoles NH_3/mg. dry cells in air, whereas anaerobically they can assimilate only about 0·08 μmole NH_3/mg. dry cells. The quantity of ammonia assimilated aerobically can be increased if glucose is added, but the addition of glucose has no effect on the total quantity of ammonia assimilated anaerobically. The anaerobic ammonia assimilation is much slower: 0·13 μmole/mg. dry cells/hr. against *c*. 1·0 μmole/mg. dry cells/hr. aerobically.

The anaerobic assimilation of ammonia is accompanied by little or no increase in carbon dioxide production, but when the cells are suspended in bicarbonate buffer it is clear that much acid is produced. The amount of carbon dioxide liberated from the bicarbonate shows that approximately three hydrogen ions are produced for each ammonium ion which is assimilated. Some of the acid production is doubtless due to the replacement of assimilated ammonium ions by hydrogen ions, but this can only account for one hydrogen ion per ammonium ion, and the remainder of the acid production must be metabolic. Preliminary experiments indicate that lactic acid is produced. Gaffron (1939) and Michels (1940) have shown that *Chlorella* produces acid from glucose anaerobically and Michels identified lactic acid. A little acid is produced when ammonia is assimilated aerobically, but it can all be accounted for by the replacement of ammonium ions by hydrogen ions.

Ammonia appears to be assimilated anaerobically to α-amino-nitrogen, and chromatograms show that only glutamic acid increases significantly. Further work is in progress to confirm these results and work out quantitative relationships.

NITRATE ASSIMILATION

If suitable carbon compounds are present some green algae can assimilate nitrate in darkness, but others are able to do so only in the light (Pringsheim, 1946). This is of interest, since Burström (1945) has suggested that in higher plants there may be two mechanisms for nitrate assimilation: one which occurs in darkness and another which only occurs in illuminated green tissues.

There is little doubt that nitrate must be reduced before being assimilated into organic compounds, but opinions differ about how far the reduction must proceed. Most of the evidence agrees with the view that it is reduced to ammonia before being assimilated, but Virtanen &

Rautanen (1952) have reviewed evidence which suggests that assimilation may occur at the hydroxylamine level of reduction.

Nitrite and hydroxylamine have often been suggested as intermediates in nitrate reduction. Many bacteria reduce nitrate to nitrite and small amounts of nitrite can be detected in the tissues of higher plants, particularly under anaerobic conditions (Nance, 1950). The evidence for the formation of hydroxylamine in plant tissues is open to question, but that for the presence of oximes is less doubtful (Wood, 1953).

There are several reports of nitrite production by green algae. Warburg & Negelein (1920) found that *C. pyrenoidosa* in a mixture of nitric acid and sodium nitrate produced ammonia under aerobic conditions. When the oxygen concentration was decreased, less ammonia was produced but nitrite accumulated. The production of ammonia was very sensitive to cyanide (96 % inhibition by 10^{-5} M-HCN), but nitrite production was said to be unaffected, and Warburg & Negelein concluded that nitrite was not an intermediate in the normal reduction of nitrate to ammonia. However, the evidence in support of their statement that nitrite production is insensitive to cyanide is rather indirect. Mayer (1950) found that *C. vulgaris* produced both ammonia and nitrite from sodium nitrate in the light, but in darkness less ammonia and nitrite were formed (see Table 9). When glucose was added to the illuminated cultures, much less nitrite could be detected and the rate of nitrogen assimilation increased. No hydroxylamine could be detected.

Kessler (1953) has shown that nitrite is formed from nitrate by *Ankistrodesmus* in darkness. The accumulation of nitrite is greater the lower the pH, partly because the rate of nitrite assimilation decreases with increasing acidity. The reduction of nitrate to nitrite by *Ankistrodesmus* is very sensitive to cyanide. Nitrite added to *Ankistrodesmus* is assimilated, and if one compares the gas exchanges which occur when nitrite and nitrate are added they are consistent with the view that nitrite is an intermediate of nitrate reduction. The same is true when ammonia, nitrite and nitrate are added to nitrogen-starved cells of *Chlorella* (Fig. 4). Algae are able to use nitrite for growth provided that the concentration is not too high or the pH too low (Ludwig, 1938; Kessler, 1953). Thus the evidence for nitrite as an intermediate in nitrate reduction by algae is fairly strong.

Hydroxylamine does not serve as a nitrogen source for the growth of *Chlorella* (Ludwig, 1938). When a small quantity of ammonia, nitrite or hydroxylamine is added to nitrogen-starved cells of *C. vulgaris* the rate of respiration increases and the compounds disappear (Fig. 4). The gas exchanges following the addition of ammonia, nitrite or nitrate are

P. J. SYRETT

consistent with their all being assimilated to the same nitrogen compounds (see p. 139). If hydroxylamine is assimilated in the same way, the gas exchange following its addition should be intermediate between those following the addition of nitrite and ammonia. It is, however, quite different, which suggests that hydroxylamine is assimilated differently. Free hydroxylamine is thus unlikely to be an intermediate in nitrate reduction.

THE THERMODYNAMICS OF NITRATE REDUCTION

The reduction of nitrate to nitrite and of nitrite to ammonia are endergonic reactions if they are written as:

$$H^+ + NO_3^- \rightarrow HNO_2 + \tfrac{1}{2}O_2 \quad \Delta F = 22\cdot3 \text{ kcal.} \tag{1}$$

$$H^+ + H_2O + HNO_2 \rightarrow NH_4^+ + 1\tfrac{1}{2}O_2 \quad \Delta F = 59\cdot3 \text{ kcal.} \tag{2}$$

$$2H^+ + H_2O + NO_3^- \rightarrow NH_4^+ + 2O_2 \quad \Delta F = 81\cdot6 \text{ kcal.} \tag{3}$$

(ΔF calculated for $1\cdot0$ M-NO_3^- and $1\cdot0$ M-NH_4^+; $pO_2 = 0\cdot2$ atm.; pH = $7\cdot0$; data of Burström (1945) corrected to pH $7\cdot0$.)

In illuminated green cells it seems probable that light energy can be used for nitrate reduction (see p. 144). In the dark there are two possibilities. Respiratory energy might be used in some way to split water molecules so that reducing hydrogen together with oxygen were produced; the oxygen would be consumed by respiration:

$$
\begin{array}{l}
\textit{Respiratory energy} \\
4H_2O \left\{ \begin{array}{l} 2O_2 + 2(CH_2O) \rightarrow 2CO_2 + 2H_2O \\ 8H + NO_3^- + 2H^+ \rightarrow NH_4^+ + 3H_2O \end{array} \right.
\end{array} \tag{4}
$$

A second, more likely, possibility is that the reducing hydrogen comes from an organic molecule via a hydrogen-carrier. Nason & Evans (1953; Evans & Nason, 1953) have shown that *Neurospora* and higher plants contain a nitrate reductase which catalyses the reduction of nitrate to nitrite using reduced triphosphopyridine nucleotide as a hydrogen-donor. The oxidation of an organic compound by nitrate could be brought about through such a coenzyme coupling, and the overall reaction would occur with a large decrease of free energy, e.g. if glucose were oxidized:

$$
\left. \begin{array}{l}
\tfrac{1}{3}(C_6H_{12}O_6) + 2H_2O \rightarrow 8H + 2CO_2 \\
NO_3^- + 2H^+ + 8H \rightarrow NH_4^+ + 3H_2O
\end{array} \right\} \quad \Delta F = -155\cdot0 \text{ kcal.} \tag{5}
$$

(Glucose $0\cdot01$ M, $pCO_2 = 0\cdot03$ atm.: other concentrations as above.)

Carbohydrates need not be completely oxidized to carbon dioxide. Such incomplete oxidation presumably occurs when pyruvic acid accumulates in *Fusarium* on a medium containing nitrate (Wirth & Nord, 1942) or when acetaldehyde and carbon dioxide are formed by wheat roots reducing nitrate anaerobically (Nance & Cunningham, 1951).

The chief difference between the two possibilities is the immediate source of the hydrogen used for nitrate reduction; it may come directly from water or directly from an organic compound (in which case it may come indirectly from water). There is no apparent reason why both processes should not be anaerobic, since there is no net uptake of oxygen.

NITRATE ASSIMILATION IN DARKNESS AND RESPIRATION

The above equations show that if the respiratory substrate contains hydrogen and oxygen in the same proportions as water and is completely respired to carbon dioxide, then two molecules of carbon dioxide should be formed for each nitrate ion reduced. This is true whichever method of nitrate reduction occurs in the dark. The carbon dioxide so produced was called 'extra' carbon dioxide by Warburg & Negelein (1920), since it is not associated with an increased uptake of oxygen. If the respiratory substrate is more reduced less 'extra' carbon dioxide is expected.

The production of 'extra' carbon dioxide accompanying nitrate assimilation has been observed with a variety of organisms (Burström, 1945) and was first studied by Warburg & Negelein (1920) with *Chlorella pyrenoidosa*. This organism had an R.Q. of 1·0 in a nitrate-free medium, but when suspended in a nitrate mixture the rate of carbon dioxide production increased by 110 % while the oxygen uptake increased by only 40 %, the R.Q. being 1·5–1·6. Cramer & Myers (1948) found that the R.Q. of *C. pyrenoidosa* in a nitrogen-free glucose medium was 1·3; when ammonia was added the R.Q. fell to 1·2, but with nitrate it increased to 1·6. The R.Q. increases to a still higher value when nitrate is added to nitrogen-starved *Chlorella* (Table 5). Kessler (1953) has shown that the R.Q. of *Ankistrodesmus* increases when either nitrite or nitrate is added.

Warburg & Negelein attempted to relate the quantity of 'extra' carbon dioxide to the amount of nitrate reduced. They studied nitrate reduction by cells suspended in a mixture of 0·01 M-HNO$_3$ and 0·1 M-NaNO$_3$; this mixture has a pH of 2 and was used to obtain a high concentration of unionized nitric acid molecules and hence a rapid rate of nitrate reduction. The low pH, however, makes the general applicability of the results doubtful. In this mixture ammonia was produced and its rate of production increased with time. Warburg & Negelein suggested that any ammonia produced by nitrate reduction was at first assimilated and only later did it accumulate. The rate of carbon dioxide production decreased with time, and during the third or fourth hour of the experiments approximately one molecule of ammonia was formed for each two molecules of 'extra' carbon dioxide. Warburg & Negelein suggested that the equation

$$HNO_3 + H_2O + 2C = NH_3 + 2CO_2$$

now applied, the carbon coming from intracellular carbon compounds. This equation does not imply that the hydrogen for reduction comes from water, since it might just as well come from the carbon compound. Neither does the fact that an 'extra' CO_2/NH_3 ratio of 2 was found indicate that any particular mechanism of reduction is involved. This value is expected, whatever the mechanism, if carbohydrate is the respiratory substrate and is completely respired while nitrate is reduced to ammonia (Burström, 1945, and compare equations (4) and (5) above).

In fact, however, as Burström points out, the evidence that the ratio 'extra CO_2'/$NH_3 = 2$ is not very convincing from the experiments of Warburg & Negelein, since the ratio was much higher during the early part of the experiments when little ammonia accumulated; the ratio approached 2 after 3–4 hr., but there is no evidence that it remained constant since the experiments were then discontinued.

The 'extra' carbon dioxide production associated with *nitrite* reduction has been studied by Kessler (1953) with *Ankistrodesmus*. If an equation is written for nitrite reduction similar to that of Warburg & Negelein for nitrate it can be seen that 1·5 molecules of carbon dioxide are expected for each nitrite ion reduced:

$$NO_2^- + H_2O + 2H^+ + 1\tfrac{1}{2}C \rightarrow NH_4^+ + 1\tfrac{1}{2}CO_2.$$

In two experiments Kessler found 1·55 and 1·38 molecules 'extra' carbon dioxide per nitrite molecule reduced. These figures agree well with the expected values, but in these experiments some of the ammonia produced by nitrite reduction must have been assimilated. This would also affect the gaseous exchange, and as no allowance appears to have been made for this, the true volume of 'extra' carbon dioxide must be somewhat greater than that quoted by Kessler.

The measurement of the 'extra' carbon dioxide associated with nitrate reduction can also be investigated with nitrogen-starved cells. Warburg & Negelein showed in their experiments that no ammonia could be detected in the medium when nitrogen-starved cells were used, presumably because it was all assimilated. If the carbon dioxide production is measured when a quantity of nitrite or nitrate is assimilated by nitrogen-starved cells and compared with the carbon dioxide production when the same quantity of ammonia nitrogen is assimilated, the difference should be a measure of the 'extra' carbon dioxide associated with nitrite or nitrate reduction provided that the products of nitrogen assimilation are the same in all experiments.

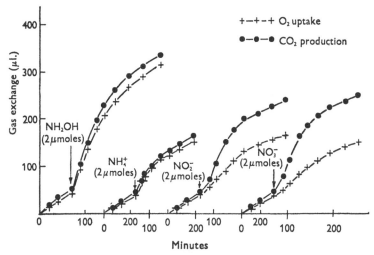

Fig. 4. The effect of ammonia, hydroxylamine, nitrite and nitrate on the respiration of nitrogen-starved cells of *Chlorella vulgaris* at 25° and pH 6·2. Each manometer contained 13·5 mg. dry weight cells suspended in 0·067 M-phosphate, 0·0017 M-KCl and 0·0017 M-MgSO₄.

Table 5. *The effect of ammonia, nitrite and nitrate on the respiration of nitrogen-starved cells of* Chlorella vulgaris

Experimental details as in Fig. 4.

Additions (2 μmoles)	Total gas exchange (μl.)		'Extra' CO₂ (μl.)	'Extra' CO₂ / N added (μmoles/ μg. atom)	Maximum rate of gas exchange (μl./10 min.)		R.Q.
	O_2	CO_2			O_2	CO_2	
NH_4^+	−73	71	—	—	−26·7	21·7	0·81
NO_2^-	−80	140	69	1·54	−10·0	21·3	2·13
NO_3^-	−71	166	95	2·12	−9·3	23·0	2·47

The results of such an experiment are shown in Fig. 4 and Table 5. The nitrogen compounds added (containing 2 μg. atom nitrogen) were completely assimilated during the experiment. It is not easy to interpret this type of graph as one does not know how much of the rapid gaseous exchange which occurs after the additions is connected with the nitrogen assimilation, i.e. there is doubt whether one should allow for a basic metabolism during this period or not (Syrett, 1953 a). However, if one assumes that all the gaseous exchange during the period of rapid metabolism is associated with nitrogen assimilation it can be seen that the volumes of 'extra' carbon dioxide are approximately 1·5 moles/mole nitrite and 2·0 moles/mole nitrate assimilated. However, because of the uncertainty of the method of calculation not too much weight can be attached to these values.

It has already been pointed out that a given quantity of nitrogen-starved cells can only assimilate a limited quantity of ammonia; when the intracellular carbon reserves are exhausted assimilation stops. This being so, one expects that the same quantity of cells would assimilate

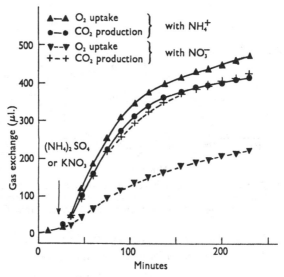

Fig. 5. The effect of a large quantity (40 μmoles) of ammonia or nitrate on the respiration of *Chlorella vulgaris* at 25° and pH 5·7. Each manometer contained 16·7 mg. dry weight cells suspended in 0·067 M-phosphate and 0·0017 M-MgSO₄.

much less nitrate-nitrogen because more of the carbon reserves are lost as carbon dioxide. This was confirmed by an experiment in which the gas exchange was followed (Fig. 5) and when rapid metabolism ceased, the remaining ammonia and nitrate were determined. It was found that the ammonia assimilated was 19·9 μmoles and the nitrate 6·2 μmoles.

It can be seen from Fig. 5 that the curves for carbon dioxide production in this experiment are almost identical. Not only is the maximum rate of production the same whether nitrate or ammonia is added, but the total volume produced and the point at which the rate of production begins to fall are the same. This means that the carbon reserves are used at the same rate in both experiments, and the same quantity of carbon must be converted to organic nitrogen compounds. Since only one-third as much nitrogen is assimilated in the nitrate experiment, the C/N ratio of the products must be some three times higher. Moreover, the products of nitrate assimilation must be more oxidized (see Table 6). Preliminary experiments show that the cells receiving ammonia produce more soluble organic nitrogen and that this contains appreciable quantities of basic amino-acids, particularly arginine and amides. Much smaller amounts of these compounds are formed in the cells receiving nitrate. Such differences are consistent with the conclusions deduced from the gas exchanges.

The agreement of the observed 'extra' carbon dioxide production with that expected on the assumption that the products of ammonia and nitrate assimilation are the same, indicates that this assumption is correct

Table 6. *Oxidation level of the products when a large quantity* (40 μ*moles*) *of ammonia or nitrate is assimilated by nitrogen-starved cells of* Chlorella vulgaris

	Nitrogen source	
	Nitrate	Ammonia
Nitrogen assimilated (μg. atom)	6·2	19·9
H oxidized by nitrate during reduction to ammonia level (μg. atom)	49·6 ($8 \times 6·2$)	—
Oxygen uptake (μmoles)	7·3	17·8
H oxidized by oxygen (μg. atom)	29·2 ($4 \times 7·3$)	71·2 ($4 \times 17·8$)
Total H oxidized (μg. atom)	78·8	71·2
H oxidized/N assimilated	12·7	3·6

Substantially the same amount of carbon is metabolized in both cases; it therefore follows that the products of nitrate assimilation must be more oxidized than the products of ammonia assimilation.

when small quantities are assimilated (see also p. 147). But when larger quantities are added the products of ammonia assimilation are likely to have a lower C/N ratio and a higher degree of reduction. Spoehr & Milner (1949) found that cells of *Chlorella pyrenoidosa* grown in an ammonia medium were more reduced than those grown with nitrate.

Whether ammonia, nitrite or nitrate is added to nitrogen-starved cells the rate of carbon dioxide production is much the same (Figs. 4, 5; Table 5). This suggests that this rate of decarboxylation may be close to the maximum possible. Reducing hydrogen is made available in the cells by reactions coupled to decarboxylations (e.g. in the Krebs cycle). If these reactions limit the rate of release of reducing hydrogen the rate of reduction of nitrite or nitrate will also be limited. This may explain why nitrate is assimilated more slowly than ammonia (see Fig. 4).

The rapid increase in the rate of carbon dioxide production following the addition of nitrate can be explained by the hypothesis that the concentration of phosphate acceptors limits the respiration rate of nitrogen-starved cells. If the hydrogen released by cellular reactions is oxidized by nitrate rather than oxygen, less phosphorylation is expected, since the decrease in free energy is smaller. At the same time the assimilation of the products of nitrate reduction will utilize high-energy compounds. Both these factors will lead to an increase in the concentration of phosphate acceptors.

THE EFFECT OF ANAEROBIOSIS ON NITRATE ASSIMILATION

Equations (4) and (5) (p. 136) suggest that nitrate reduction should be an anaerobic process, and work with both bacteria and higher plants shows that the rate of nitrate reduction increases as the oxygen tension is lowered (e.g. Stickland (1931) with *Escherichia coli*, Nance (1948) and Nance & Cunningham (1951) with wheat roots).

Table 7. *A comparison of the aerobic and anaerobic assimilation of ammonia, nitrite and nitrate by nitrogen-starved cells of* Chlorella vulgaris *at 25° and* pH 6·0

Cells suspended in 0·0017 M-MgSO₄ and 0·067 M-phosphate and incubated for *c.* 90 min. Ammonia, nitrite and nitrate were determined by the methods of Conway (1947). The mean values from duplicate experiments are given.

Additions (3 µg. atom N)	Nitrogen assimilated (%)	
	Gas phase air	Gas phase N₂
(NH₄)₂SO₄	100	66·4
NaNO₂	96·5	35·1
KNO₃	71·1	0·5

Different results are obtained with algae. Warburg & Negelein (1920) found that *Chlorella pyrenoidosa* in a mixture of sodium nitrate and nitric acid produced less ammonia in 0·01 % oxygen than in air. Nitrite was formed at low oxygen tensions, but if the sum of ammonia- and nitrite-nitrogen was taken as a measure of nitrate reduction this decreased as the oxygen tension was lowered. Kessler (1953) found that *Ankistrodesmus* in a nitrate medium produced less nitrite anaerobically and added nitrite was assimilated more slowly than in air. Here again nitrate reduction and nitrite assimilation were slower under anaerobic conditions.

Similar results are obtained with nitrogen-starved cells of *Chlorella vulgaris* which assimilate appreciable quantities of ammonia anaerobically. In short experiments about 35 % of the added nitrite is assimilated but no nitrate (Table 7). From these experiments it is uncertain whether anaerobiosis is stopping nitrate *reduction* since nitrate *absorption* may be an active process dependent on aerobic respiration. This is so with wheat roots which reduce nitrate more quickly anaerobically when it is in the roots but absorb it more slowly than in air (Nance, 1948). Experiments to investigate this possibility with *Chlorella* have as yet been inconclusive.

THE EFFECT OF LIGHT ON AMMONIA AND NITRATE ASSIMILATION

Ammonia assimilation

When normal cells of *C. vulgaris* are illuminated in air, the rate of ammonia assimilation is doubled, but the rapid assimilation by nitrogen-starved cells is hardly affected (Fig. 6). The increased ammonia assimilation by normal cells can be interpreted as a response to a better supply of carbohydrate resulting from photosynthesis, but it is probable that light has other effects as well. Myers (1949) found that photosynthesizing *C. pyrenoidosa* assimilated ammonia 2·5 times more quickly than similar cells in darkness in a glucose medium. Here, intermediate products of photosynthesis may combine with ammonia so that it is assimilated more quickly (cf. the rapid appearance of ^{14}C in alanine and glutamic acid in photosynthesis experiments with *Chlorella* (Calvin *et al.* 1951)). Pearsall & Bengry (1940), however, found that a very small light intensity caused an increase of 60 % in the exponential growth rate of *Chlorella* cultures in a glucose medium. Since further increases of light intensity had little effect even if bicarbonate was added, light increased the growth rate independently of photosynthesis. As the medium contained ammonium nitrate the cells must have assimilated chiefly ammonia for growth. Further work is required to elucidate this effect of light.

Nitrate assimilation

Since light accelerates ammonia assimilation one expects that it will increase the rate of assimilation of the products of nitrate reduction and it undoubtedly does so. But as Warburg & Negelein (1920) clearly showed, light also increases the rate of nitrate reduction.

Two views can be held about the acceleration of nitrate reduction by green cells in the light. The first is that reduction in the light is the same process as reduction in darkness but that it takes place at an increased rate, possibly because light increases the rate of nitrate absorption, as suggested by Warburg & Negelein, or because a better supply of organic hydrogen-donors is available. On the other hand, it has been suggested (see Rabinowitch, 1945) that illuminated chloroplasts can bring about a direct photochemical reduction of nitrate. Evidence that this is so

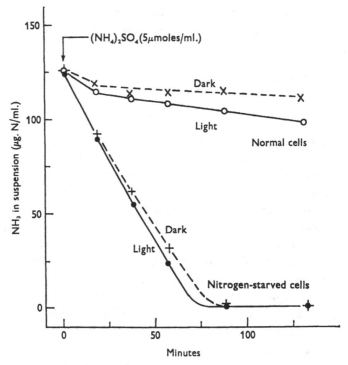

Fig. 6. The effect of light on ammonia assimilation by normal and nitrogen-starved cells of *Chlorella vulgaris* at 25° and pH 6·1. Normal cell suspension contained 7·69 mg. dry weight cells/ml. and nitrogen-starved cell suspension contained 7·54 mg. dry weight cells/ml. Light intensity 280 ft.c. (From Syrett, 1953a.)

has been obtained by Evans & Nason (1953). They showed that the leaves of several plants contain a nitrate reductase which reduces nitrate to nitrite if TPNH is present.* Vishniac & Ochoa (1951) showed that illuminated chloroplast preparations can reduce TPN+ with the evolution of oxygen,* and Evans & Nason were able to obtain a reduction of nitrate to nitrite by illuminated chloroplast grana of soya-bean leaves when nitrate reductase and TPN+ were added. Little nitrite was formed in darkness or if TPN+ was omitted and none at all without the

* See footnote, p. 91.

chloroplast grana. Here then the reactions which occur are presumably:

$$H_2O + TPN^+ \xrightarrow{\text{light}} TPNH + H^+ + \tfrac{1}{2}O_2$$
$$H^+ + TPNH + NO_3^- \longrightarrow NO_2^- + TPN^+ + H_2O$$

$$\overline{NO_3^- \xrightarrow{\text{light}} NO_2^- + \tfrac{1}{2}O_2}$$

The dependence of the above reactions on TPN^+ may explain the failure of Holt & French (1948) to find oxygen evolution when spinach chloroplasts were illuminated with nitrate in the absence of TPN^+.

Several workers have studied the effect of light on the reduction or assimilation of nitrate by green algae, but it is not possible to conclude with certainty how light acts. The first investigation was by Warburg & Negelein (1920) with *C. pyrenoidosa*. They found that, on illumination, the 'extra' carbon dioxide produced in the dark in their nitrate mixture was replaced by an 'extra' oxygen production. Ammonia was formed just as it was in the dark, but the rate of formation was 3–4 times faster in the light experiments. Warburg & Negelein pointed out that if oxygen and ammonia were the end-products of nitrate reduction in light one expected the equation

$$HNO_3 + H_2O = NH_3 + 2O_2.$$

In some experiments they found the O_2/NH_3 ratio to be 2 as expected, but in others it was higher and they postulated that ammonia assimilation accounted for this. Warburg & Negelein suggested that nitrate was reduced in the light by the same reactions as in darkness, the 'extra' oxygen in the light being produced from the dark 'extra' carbon dioxide by photosynthesis:

Dark	$HNO_3 + H_2O + 2C = NH_3 + 2CO_2$	(6)
Light	$2CO_2 = 2C + 2O_2$	(7)

$$\overline{HNO_3 + H_2O = NH_3 + 2O_2}$$

To explain the higher rate of formation of ammonia in light, they assumed that reaction (6) was normally limited by the rate of entry of nitrate into the cell, and that the rate of entry was increased when the cells were exposed to light. In support of this idea they showed that in the presence of 0·013 % phenylurethane, which almost completely inhibits photosynthesis (reaction (7)), the cells produced *carbon dioxide* in the light and at a rate 2–3 times greater than that observed in the dark. No measurements of ammonia were made. Rabinowitch (1945, p. 539) has suggested that the results could be interpreted equally well in terms of a direct photochemical reduction of nitrate, the urethane inhibiting the stage of the photochemical reaction which liberates oxygen, so that the primary product of photochemical oxidation is reduced by organic hydrogen-donors with the liberation of carbon dioxide. Warburg (1948) has also suggested recently that the direct photochemical reduction of nitrate may occur. Since the experiments of Warburg & Negelein, other workers have shown that the gas exchange in the light alters when nitrate is added. As in the dark experiments the changes are particularly noticeable when nitrogen-starved cells are used. Pirson & Wilhelmi (1950) added nitrate to markedly nitrogen-deficient cells of *Ankistrodesmus falcatus* which had an *assimilatory quotient* (CO_2 consumed/O_2 produced) of 1·0 before the addition of nitrate, indicating that carbohydrate was the product of photosynthesis. When nitrate was added the rate of oxygen production increased, but carbon dioxide assimilation decreased and the assimilatory quotient fell to 0·57.

Table 8. *Photosynthesis and respiration of nitrogen-deficient* Chlorella pyrenoidosa *in Knop's medium* (after Myers, 1949)

	Gas output (μl./μl. cells/hr.)		
	Oxygen	Carbon dioxide	$-CO_2/O_2$
Nitrate present, in light	39·0	−28·7	0·74
Nitrate present, in dark	−2·0	3·2	1·6
Nitrate absent, in light	31·6	−31·4	0·99
Nitrate absent, in dark	−1·4	1·4	1·0

Similar results were obtained by Myers (1949) with nitrogen-deficient cells of *Chlorella pyrenoidosa* (Table 8). Here again the addition of nitrate was followed by an increased oxygen production and a decreased carbon dioxide consumption in light.

These results can be interpreted in two ways: one can assume, as Warburg & Negelein did, that reaction (6) takes place more quickly in light and that some, but not all, of the 'extra' carbon dioxide is assimilated. Thus the oxygen evolution will increase, but the net carbon dioxide uptake will fall because of the unassimilated 'extra' carbon dioxide produced. Myers found that the overall rate of oxygen production in the light was 7·4 μl./hr. greater when nitrate was added and the rate of carbon dioxide absorption 2·7 μl./hr. less. This could be accounted for if 10·1 (7·4 + 2·7) μl. 'extra' CO_2/hr. were produced by reaction (6) in the light, and of this, 7·4 μl./hr. were assimilated by reaction (7) with the production of an equal volume of oxygen. However, in the dark, reaction (6) produced only 1·2 (3·2−2·0) μl. 'extra'

CO_2/hr., consequently if this interpretation is correct, reaction (6) must have proceeded eight times more rapidly in the light. The other possibility is that nitrate is reduced photochemically. If this is so, the oxygen production will increase and carbon dioxide consumption may decrease because carbon dioxide reduction and nitrate reduction are competing photochemical reactions (Cramer & Myers, 1948; Pirson & Wilhelmi, 1950).

Cramer & Myers (1948; Myers & Cramer, 1948) have studied the effect of nitrate assimilation on the assimilatory quotient of cells of *C. pyrenoidosa* with differing C/N ratios. They normally grew the cells at low light intensity in a nitrate medium, and these cells had an elementary composition which could be expressed as $C_{5.7}H_{9.8}O_{2.3}N_{1.0}$. If such cells are allowed to photosynthesize in a nitrate medium at low light intensity (40 ft.c.), one expects that the final product of carbon and nitrate assimilation will be new cells of the same composition. Since the elementary composition of the cells was known, Cramer & Myers were able to write the overall metabolic equation they expected as

$$1.0NO_3^- + 5.7CO_2 + 5.4H_2O \rightarrow C_{5.7}H_{9.8}O_{2.3}N_{1.0} + 8.25O_2 + 1.0OH^-.$$

Thus the predicted assimilatory quotient was $5.7/8.25 = 0.69$. The observed value was 0.68. A similar equation could be written for ammonia if it was assumed that cells of the same composition were formed. The predicted quotient was then 0.91 and the observed value 0.94.

When the gaseous exchange of these cells was measured at high light intensity (600 ft.c.) in a nitrate medium the assimilatory quotient was 0.88, not 0.68 as at the low light intensity. This means that less nitrate was assimilated in relation to the carbon dioxide assimilation, i.e. relatively more non-nitrogenous products were synthesized. A similar value (0.91) was obtained at low light intensity after the cells were starved of carbon for 3 days, and here again the synthesis of non-nitrogenous compounds predominated when the cells were illuminated. On the other hand, cells which were grown at high light intensity (300 ft.c.) had a quotient of 0.32 when placed in nitrate at low light intensity (45 ft.c.); this indicates that much nitrate was reduced.

Thus the effect of nitrate on the gas exchanges depends very much on the C/N balance of the cells. If the cells are rich in carbon compounds but nitrogen-deficient, nitrate is rapidly reduced and the assimilatory quotient is low. If the cells have plenty of nitrogen, relatively little nitrate is reduced, and the high assimilatory quotient shows that the synthesis of carbon compounds predominates.

The effect of light on nitrite formation

Kessler (1953) showed that nitrite accumulated in *Ankistrodesmus* cultures in darkness when nitrate was present. The accumulation was particularly pronounced at pH 2·7. In the light, nitrite did not accumulate, and if dark cultures were illuminated the nitrite present rapidly disappeared. Mayer (1950), however, found that cultures of *Chlorella vulgaris* formed more nitrite in the light than in darkness (Table 9). The difference between these results and those of Kessler can be explained

Table 9. *The formation of ammonia and nitrite from nitrate (initial concn.* 0·02 M) *by* Chlorella vulgaris *suspended in* 0·067 M-*phosphate at* pH 6·0 (after Mayer, 1950)

	In dark		In light	
Time (hr.)	Nitrite-N	Ammonia-N	Nitrite-N	Ammonia-N
24	4·6	—	46	—
96	3·8	1·4	70	12·6
144	5·2	12·9	73	29·2

μg. N/40 × 10^6 cells

by the difference in the experimental conditions. Kessler used thin suspensions, photosynthesis was rapid and conditions were favourable for nitrogen assimilation. Mayer used fairly dense cell suspensions which were not shaken continuously and photosynthesis did not compensate for respiration. Thus conditions were unfavourable for nitrogen assimilation. The fact that so much more nitrite was formed in the light shows, once more, the stimulatory effect of light on nitrate reduction.

CONCLUSION

It is clear that there are large gaps in our knowledge of the assimilation of ammonia and nitrate by algae. There is good evidence that nitrate is reduced to nitrite, but the intermediate compounds of nitrite assimilation are unknown. Ammonia can be formed, but whether it is a normal intermediate is uncertain. This problem is linked to another which requires further study, namely, the differences between the organic products of nitrate and ammonia assimilation. There is no doubt that assimilation is closely connected with the respiratory and photosynthetic mechanisms, but the exact nature of the coupling is obscure. In particular more information is needed about the way in which light acts.

I should like to thank the Editor of the *Annals of Botany* for permission to reproduce Fig. 1, 3 and 6 and Table 1, and the Editor of *Physiologia Plantarum* for permission to reproduce Fig. 2. I should also like to thank Dr L. Fowden for his helpful criticism of the manuscript of this paper.

The respiration experiments mentioned here were carried out with a Warburg apparatus purchased with a grant from the Dixon Fund of the University of London.

REFERENCES

ALBERTS-DIETERT, F. (1941). Die Wirkung von Eisen und Mangan auf die Stickstoffassimilation von *Chlorella*. *Planta*, **32**, 88.

BURSTRÖM, H. (1945). The nitrate nutrition of plants. *LantbrHögsk. Ann.* **13**, 1.

CALVIN, M., BASSHAM, J. A., BENSON, A. A., LYNCH, V. H., OUELLET, C., SCHOU, L., STEPKA, W. & TOLBERT, N. E. (1951). Carbon dioxide assimilation in plants. In *Carbon Dioxide Fixation and Photosynthesis, Symp. Soc. exp. Biol.* **5**, 284.

CONWAY, E. J. (1947). *Microdiffusion Analysis and Volumetric Error.* London: Crosby Lockwood and Son.

CRAMER, M. L. & MYERS, J. (1948). Nitrate reduction and assimilation in *Chlorella*. *J. gen. Physiol.* **32**, 93.

ELLIOTT, W. H. (1951). Studies in the enzymic synthesis of glutamine. *Biochem. J.* **49**, 106.

EVANS, H. J. & NASON, A. (1953). Pyridine nucleotide-nitrate reductase from extracts of higher plants. *Plant Physiol.* **28**, 233.

GAFFRON, H. (1939). On the anomalous respiratory quotients of algae from sugar cultures. *Biol. Zbl.* **59**, 288.

HOLT, A. S. & FRENCH, C. S. (1948). Oxygen production by illuminated chloroplasts suspended in solutions of oxidants. *Arch. Biochem.* **19**, 368.

HOPKINS, E. F. & WANN, F. B. (1926). Relation of hydrogen-ion concentration to growth of *Chlorella* and to the availability of iron. *Bot. Gaz.* **81**, 353.

JOHNSON, M. J. (1941). The role of aerobic phosphorylation in the Pasteur effect. *Science*, **94**, 200.

KESSLER, E. (1953). Über den Mechanismus der Nitratreduktion von Grünalgen. *Flora, Jena*, **140**, 1.

LARDY, H. A. (1952). The role of phosphate in metabolic control mechanisms. In *The Biology of Phosphorus*, p. 131. Ann. Arbor: Michigan State College Press.

LARDY, H. A. & WELLMAN, H. (1952). Oxidative phosphorylations: role of inorganic phosphate and acceptor systems in control of metabolic rates. *J. biol. Chem.* **195**, 215.

LOOMIS, W. F. & LIPMANN, F. (1948). Reversible inhibition of the coupling between phosphorylation and oxidation. *J. biol. Chem.* **173**, 807.

LUDWIG, C. A. (1938). The availability of different forms of nitrogen to a green alga. *Amer. J. Bot.* **25**, 448.

MCLEAN, D. J. & FISHER, K. C. (1947). The relation between oxygen consumption and the utilisation of ammonia for growth in *Serratia marcescens*. *J. Bact.* **54**, 599.

MCLEAN, D. J. & FISHER, K. C. (1949). The extra oxygen consumed during growth of *Serratia marcescens* as a function of the carbon and nitrogen sources and of temperature. *J. Bact.* **58**, 417.

MAYER, A. M. (1950). Problems of the assimilation of nitrogen by *Chlorella vulgaris*. Ph.D. Thesis, University of London.

MICHELS, H. (1940). Über die Hemmung der Photosynthese bei Grünalgen nach Sauerstoffentzug. *Z. Bot.* **35**, 241.

MILLBANK, J. W. (1953). Demonstration of transaminase systems in the alga *Chlorella*. *Nature, Lond.* **171**, 476.

MYERS, J. (1949). The pattern of photosynthesis in *Chlorella*. In *Photosynthesis in Plants* (ed. J. Franck and W. E. Loomis). p. 349. Ames: Iowa State College Press.

MYERS, J. & CRAMER, M. L. (1948). Metabolic conditions in *Chlorella*. *J. gen. Physiol.* **32**, 103.

NANCE, J. F. (1948). The role of oxygen in nitrate assimilation by wheat roots. *Amer. J. Bot.* **35**, 602.

NANCE, J. F. (1950). Inhibition of nitrate assimilation in excised wheat roots by various respiratory poisons. *Plant Physiol.* **25**, 722.

NANCE, J. F. & CUNNINGHAM, L. W. (1951). Evolution of acetaldehyde by excised wheat roots in solutions of nitrate and nitrite salts. *Amer. J. Bot.* **38**, 604.

NASON, A. & EVANS, H. J. (1953). Triphosphopyridine nucleotide-nitrate reductase in *Neurospora*. *J. biol. Chem.* **202**, 655.

NOACK, K. & PIRSON, A. (1939). Die Wirkung von Eisen und Mangan auf die Stickstoffassimilation von *Chlorella*. *Ber. dtsch. bot. Ges.* **57**, 442.

ÖSTERLIND, S. (1949). Growth conditions of the alga *Scenedesmus quadricauda* with special reference to the inorganic carbon sources. *Symb. bot. upsaliens.* **10** (3), 1.

PEARSALL, W. H. & BENGRY, R. P. (1940). Growth of *Chlorella* in relation to light intensity. *Ann. Bot., Lond.* N.S. **4**, 485.

PEARSALL, W. H. & LOOSE, L. (1937). The growth of *Chlorella vulgaris* in pure culture. *Proc. Roy. Soc.* B, **121**, 451.

PIRSON, A. (1937). Ernährungs- und stoffwechselphysiologische Untersuchungen an *Fontinalis* und *Chlorella*. *Z. Bot.* **31**, 193.

PIRSON, A. & WILHELMI, G. (1950). Photosynthese-Gaswechsel und Mineralsalz-ernährung. *Z. Naturf.* 5B, 211.

PRATT, R. & FONG, J. (1940). Studies on *Chlorella vulgaris*. III. Growth of *Chlorella* and changes in the hydrogen-ion and ammonium-ion concentrations in solutions containing nitrate and ammonium nitrogen. *Amer. J. Bot.* **27**, 735.

PRINGSHEIM, E. G. (1946). *Pure Cultures of Algae*, p. 36. Cambridge University Press.

RABINOWITCH, E. I. (1945). *Photosynthesis and Related Processes*, **1**, 538. New York: Interscience.

ROINE, P. (1947). On the formation of primary amino-acids in the protein synthesis in yeast. *Ann. Acad. Sci. fenn.* ser. A, **2**, Chem. no. 26.

SCHULER, J. F., DILLER, V. M. & KERSTEN, H. J. (1953). Preferential assimilation of ammonium-ion by *Chlorella vulgaris*. *Plant Physiol.* **28**, 299.

SPECK, J. F. (1947). The enzymic synthesis of glutamine. *J. biol. Chem.* **168**, 403.

SPOEHR, H. A. & MILNER, H. W. (1949). The chemical composition of *Chlorella*; effect of environmental conditions. *Plant Physiol.* **24**, 120.

STICKLAND, L. H. (1931). The reduction of nitrates by *Bacterium coli*. *Biochem. J.* **25**, 1543.

SYRETT, P. J. (1951). The effect of cyanide on the respiration and the oxidative assimilation of glucose by *Chlorella vulgaris*. *Ann. Bot., Lond.* N.S. **15**, 473.

SYRETT, P. J. (1953*a*). The assimilation of ammonia by nitrogen-starved cells of *Chlorella vulgaris*. Part I. The correlation of assimilation with respiration. *Ann. Bot., Lond.* N.S. **17**, 1.

SYRETT, P. J. (1953*b*). The assimilation of ammonia by nitrogen-starved cells of *Chlorella vulgaris*. Part II. The assimilation of ammonia to other compounds. *Ann. Bot., Lond.* N.S. **17**, 21.

SYRETT, P. J. & FOWDEN, L. (1952). The assimilation of ammonia by nitrogen-starved cells of *Chlorella vulgaris*. III. The effect of the addition of glucose on the products of assimilation. *Physiol. Plant.* **5**, 558.

UMBREIT, W. W., BURRIS, R. H. & STAUFFER, J. F. (1949). *Manometric Techniques and Tissue Metabolism*, p. 19. Minneapolis: Burgess.

URHAN, O. (1932). Beiträge zur Kenntnis der Stickstoffassimilation von *Chlorella* und *Scenedesmus*. *Jb. wiss. Bot.* **75**, 1.

VIRTANEN, A. I. & RAUTANEN, N. (1952). Nitrogen assimilation. In *The Enzymes* (ed. J. B. Sumner and K. Myrbäck), **2**, part 2, p. 1089. New York: Academic Press.

VISHNIAC, W. & OCHOA, S. (1951). Photochemical reduction of pyridine nucleotides by spinach grana and coupled carbon dioxide fixation. *Nature, Lond.* **167**, 768.

WARBURG, O. (1948). *Schwermetalle*, 2nd ed. p. 184. Berlin: Werner Saenger.

WARBURG, O. & NEGELEIN, E. (1920). Über die Reduktion der Salzpetersaure in grunen Zellen. *Biochem. Z.* **110**, 66.

WASSINK, E. C., WINTERMANS, J. F. G. M. & TJIA, J. E. (1951). Phosphate exchanges in *Chlorella* in relation to conditions for photosynthesis. *Proc. Acad. Sci. Amst.* **54**, 2.

WILLIS, A. J. (1951). Synthesis of amino-acids in young roots of barley. *Biochem. J.* **49**, xxvii.

WIRTH, J. C. & NORD, F. F. (1942). Essential steps in the enzymatic breakdown of hexoses and pentoses. Interaction between dehydrogenation and fermentation. *Arch. Biochem.* **1**, 143.

WOOD, J. G. (1953). Nitrogen metabolism of higher plants. *Annu. Rev. Plant Physiol.* **4**, 1.

MECHANISM OF PHOTOSYNTHESIS

HANS GAFFRON

*Institute of Radiobiology and Biophysics (Fels Fund),
University of Chicago, Chicago, Illinois*

The problem of photosynthesis has lately attracted widespread attention. The engineer in search of methods for a better utilization of solar energy; the food economist wishing to harvest fifty instead of five tons of dry organic matter per acre per year; the biochemist to whom the availability of radioisotopes, such as ^{14}C, and the possibility of working with cell-free extracts promises a rapid extension of his research field; finally, the scientist who has heard about the long-standing feud concerning quantum yields and other very difficult questions—they all would like to share, if possible, the specialist's knowledge. At the present time, there exist so many detailed reviews on the subject that this wish should have been satisfied. These reviews, which have appeared at the rate of four to six per year, cover the literature as completely as anyone could desire, and one or other of these references to original papers should be easily available to any interested reader.

It is not the purpose of this article to provide another review. In the time preceding the era of isotopes and mass-spectrometers, there existed an overall picture of photosynthesis which served as a guide for the research student. The question we intend to examine is whether in the light of the many new discoveries the old picture has become obsolete and should be changed. Obviously to-day's reviewer cannot always repeat what has been said many times before; he proceeds under the assumption that the reader is familiar with the answers to those questions which, by explicit or tacit consent of the specialists, are considered as settled. In a sense this creates among the readers coming from nearby scientific fields the notion that there is no guiding overall picture of the photosynthetic process which would be worth knowing, no core of knowledge which could serve as a general basis for a discussion. Certainly, it would be easier to judge the importance of a new discovery, or to infer the role of a certain enzyme, or to evaluate the merits of a controversial issue, if there were an established principle, a picture summarizing the essence of our knowledge in this field which would be equally familiar to biologists, biochemists, physiologists and physicists.

Looking back, one can readily see that ten years ago there was excellent agreement among the students of photosynthesis as to what constituted a fact and how it should best be interpreted. The views expressed in the older articles by van Niel, Franck, Rabinowitch and Gaffron were very much alike, and the accompanying explanatory diagrams and equations were practically interchangeable. It was, therefore, easy to summarize the then prevalent ideas about the mechanism of photosynthesis in a diagram which is reprinted here (Fig. 1), together with the following overall equations:

Photosynthesis in all green plants

$$CO_2 + 4H—OH \xrightarrow{\text{light}} (CH_2O) + 3H_2O + O_2.$$

Photoreduction in bacteria and algae

$$CO_2 + 4H—OH + 2AH_2 \xrightarrow{\text{light}} (CH_2O) + 5H_2O + 2A.$$

Photoreaction with chloroplasts

$$2Q + 4H—OH \xrightarrow{\text{light}} 2QH_2 + 2H_2O + O_2.$$

Chemoreduction in most living cells

$$CO_2 + 4H_2A + O_2 \longrightarrow (CH_2O) + 3H_2O + 4A.$$

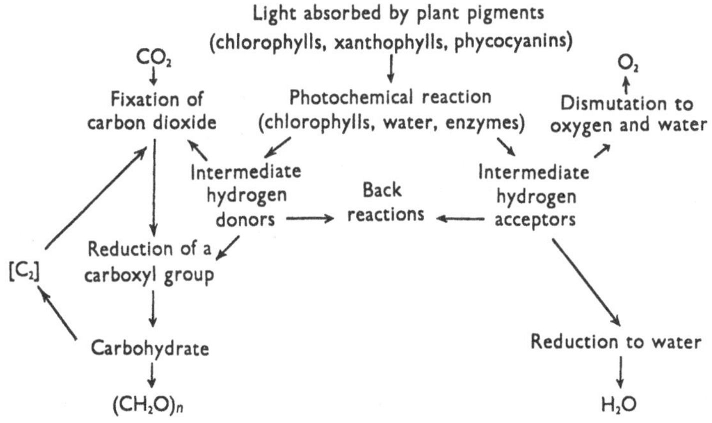

Fig. 1. Schematic representation of photosynthesis based on views held about ten years ago (1943). (Gaffron, 1946. Copyright. New York–London: Interscience Publishers.)

A new addition to the scheme is the expression (C_2) on the left-hand side of the diagram to indicate that a special cycle exists providing for a C_2 fragment which functions as an acceptor for carbon dioxide. The scheme left ample freedom for devising alternate mechanisms to explain the specific functions of each part. Yet it definitely excluded certain other notions. For instance, the scheme is incompatible with the idea that carbon dioxide reacts directly with the chlorophyll complex; or that there is a way by which one light quantum could cause the immediate evolution of one molecule of free oxygen; or that the oxygen originates from carbon dioxide; or that back reactions with molecular oxygen are driving the mechanism for carbon dioxide reduction; or that in photo-reduction the hydrogen of the hydrogen-donor is transferred directly to the carbon being reduced. These and other assumptions and hypotheses were regarded as having been refuted by experimental evidence available at that time. Some of these views are being revived to-day, often without any reference to, or consideration of, the experiments which showed why they had to be abandoned. Advance during recent years has been, for the most part, due to the introduction of new methods. There are the isotopes which have led to chemical investigations and to important studies with the mass spectrometer which were not possible before. The gas exchange during photosynthesis was formerly followed mainly by manometry. To determine oxygen, we now use a platinum or dropping mercury electrode, or measure the quenching of the phosphorescence of a dye adsorbed on silica gel. For carbon dioxide, we use the glass electrode or a gas analyser based on measurements of heat conductivity, or infrared absorption analysis. Under these circumstances, it should actually be easy to settle certain controversial points if it is found that identical results have been obtained with quite different methods. It should be possible to avoid hypotheses which explain only one single type of observation, while completely disregarding other observations which do not fit. Unfortunately, the field is beset not only with opposing beliefs but even with unresolved contradictions pertaining to factual observations. In the following pages some of the difficulties will be pointed out, the emphasis being on quantum yields and back reactions. A great deal of the experimental information mentioned has been received in the form of private communications, and specific references to journals will not be made.

I. LIGHT ABSORPTION AND INTERNAL CONVERSION

It has been definitely confirmed that the so-called accessory pigments (carotenoids, phycobilins) which occur in the plastids or grana of photosynthetic plants and bacteria can contribute towards the photo-chemical reduction of carbon dioxide. No plant, however, has been found which has active plastids containing only carotenoids and phycobilins without chlorophyll. Studies of energy transfer from one dyestuff to the other have shown that the blue light absorbed by the

carotenoids, and the green and orange-red light absorbed by the phycobilins is utilized only after a transfer of the energy to the chlorophyll
molecules. Two methods have been used in these studies: (1) the comparison of action spectra with absorption spectra, and (2) measurements
of the fluorescence intensity of chlorophyll *a* as a function of the intensity of the incident radiation absorbed by the other dyestuffs. The
discovery that chlorophyll *b* has to be relegated to the group of accessory
pigments was perhaps surprising to the plant physiologists but is in
better agreement with the behaviour of chlorophyll *in vitro*. The energy
of the photon absorbed by chlorophyll *b* is transferred for further pro-
cessing to chlorophyll *a*. It is assumed that only chlorophyll *a* is capable
of converting the energy of the light quanta into chemical energy.

The photochemistry of isolated chlorophyll is, therefore, again in the
limelight. The yield observed during oxidations or reductions with
chlorophyll as sensitizer (quantum yield near 100 %), as well as that of
photosynthesis at low light intensities (quantum yield 25–75 % or even
100 %), requires that the light energy be held in storage long enough for
specific chemical steps to follow. There has been some evidence for the
existence of a so-called metastable or triplet state with a lifetime perhaps
a million times longer than that of the initial electronic excitation of the
pigment molecule, but so far its existence has been demonstrated only
in other organic dyestuffs and not for chlorophyll within the living cell.
It might be expected that a small fraction of the light energy is re-
emitted from the metastable state either via the normal fluorescence
band or directly at longer wavelengths in the infrared. The search for
such a luminescence appearing in the infrared has led nowhere, and the
hope of finding a long-lived fluorescence or phosphorescence at the site
of the normal fluorescence band around 6800 Å has met with insur-
mountable difficulties because of the discovery by Strehler and Arnold
of a chemiluminescence of chlorophyll in living plants. The present
explanation of this chemiluminescence assumes the occurrence of some
back reactions inside the photochemical mechanism which excite the
chlorophyll complex and cause the emission of light. The yield is
exceedingly small. The effect lasts long enough to cover up any phosphorescence of the type one could expect to arise from the metastable
state. On the other hand, studies of the behaviour of chlorophyll *in
vitro*, particularly the reversible bleaching of solutions of chlorophyll in
pure organic solvents, have shown that there must exist an activation
state which lasts about one-thousandth of a second. During this time
the colour of the chlorophyll is markedly changed. The reaction is
perfectly reversible, and the photochemists believe this to be sufficient

evidence for the formation of a metastable state. If so, it is logical to assume that it represents the very first step in the conversion of light energy to chemical energy. And this again means that right here, at the beginning of photosynthesis, a certain price has to be paid in terms of energy to ensure the time needed for an efficient use of the excitation energy by chemical reactions. Consequently the photochemical mechanism does not pass on the full 41 kcal. of a mole quantum of light at 6800 Å. but an amount which is somewhat smaller. We should keep this in mind, since it constitutes one of the items enumerated in an energy budget for photosynthesis (Table 1).

To start the chain of chemical reactions after the absorption of light, we have hardly any other choice than to picture it as the transfer of radicals (H or OH) or of an electron to or from the chlorophyll complex. This view is based on the numerous observations about the capacity of fluorescent dyes to initiate either oxidations or reductions when irradiated in the presence of suitable acceptors.

The final outcome must be the severing of an O—H bond in a water molecule. Schematically, this is most often depicted as a reaction of the type

$$H-OH + X, Y \xrightarrow{h\nu} XH^* + YOH^*,$$

where an asterisk (*) is used to denote an 'active radical'. Linschitz resorted to temperatures as low as that of liquid nitrogen to observe in more detail what happens to the chlorophyll molecule under the influence of light. Chlorophyll and chlorophyll derivatives dissolved in pure organic solvents and frozen solid become bleached when exposed to intense illumination. After the light is turned off, the green colour returns. If the organic solution contains not only rather inert solvents but some molecules which can serve as an acceptor for hydrogen atoms, e.g. quinone, the light-induced bleaching of the chlorophyll persists even after the light has been turned off. In this case the chlorophyll regains its colour when the frozen mixture is brought back to room temperature. The interpretation is that any semiquinone formed in the light will be stable at low temperatures but will surrender again the hydrogen it just received in the light when the temperature rises. This does not yet prove that chlorophyll in the living cell works in this manner.

However, J. H. Smith's observations of a photochemical reduction of protochlorophyll (itself a green pigment) to chlorophyll in living cells seem to fit exactly into this pattern. At normal temperatures practically all the protochlorophyll is rapidly transformed in the light. At $-70°$ C. light still promotes the formation of chlorophyll (as shown by the analysis after thawing), except that the yield now rarely exceeds 50 % of the substrates originally present. These so far unique and important experiments strongly support the view that chlorophyll-like pigments can initiate biological reductions by a direct non-enzymic photochemical step. The variation of the yield with temperature suggests that the initial photoproducts (radicals) undergo dismutation into fully reduced products with the aid of an enzyme. Various schemes are possible. The following example need not be correct:

$$(1) \quad PChl + AH_2 \xrightarrow{\text{light}} PChl.H^* + AH^*,$$

$$(2) \quad 2PChl.H^* \xrightarrow{E_1} PChl + PChlH_2(= Chl),$$

$$(3) \quad 2AH^* \xrightarrow{E_2} AH_2 + A,$$

$$(4) \quad AH^* + PChl \xrightarrow{E_3} PChlH^* + A.$$

This scheme requires that at $-70°$ C. chlorophyll formation proceeds no further than step (1), with the rest to follow in the dark if and when the temperature allows. It is certainly possible that in photosynthesis chlorophyll sensitizes a hydrogen transfer not from its own molecule but from a molecule closely associated with it (J. H. Smith's pigment-protein complex, 'holochrome'). Under these circumstances, the chlorophyll itself would never show a change in colour.

At the moment we proceed under the assumption that chlorophyll when irradiated produces organic radicals due to the transfer of hydrogen atoms from water to a situation, XH^*, where the hydrogen is bound more loosely, i.e. with a higher energy than before. From here it may either return to its original place or initiate further reduction steps which by loss of free energy become practically irreversible. Corresponding steps lead from the original YOH radical to oxygen. This irreversibility, in other words, the absence of back reactions, can be inferred from the fact that photosynthesis remains a very efficient process even at low light intensities.

II. PHOTOCHEMICAL DEHYDROGENATION OF WATER AND QUANTUM EFFICIENCY

As our scheme indicates, the crucial photochemical reaction in photosynthesis is the decomposition of water. On this there has been general agreement for a long time. Ever since this became common knowledge, chemical engineers have been dreaming of reproducing this reaction *in vitro*, and thus of bringing about a photolysis of water with the liberation of hydrogen and oxygen as free gases. Actually, such a reaction would surpass somewhat the achievement of plants. The green plants do indeed release free oxygen, but no experimental trick, such as the use of special inhibitors, has so far induced them to evolve free hydrogen at the same time. On the other hand, we have learned that under anaerobic conditions plants as well as photosynthetic bacteria can produce free hydrogen photochemically, yet this happens only under conditions where no trace of oxygen is liberated. Whether this 'either—or' situation is inherent in the mechanism or merely a chance result of evolution might be worth investigating.

When, about thirty years ago, it became clear to plant physiologists that in the visible region of the spectrum more than one light quantum is necessary to bring about a complete reduction of carbon dioxide to carbohydrate, it was a new and fundamental insight. Nowadays, this statement appears trivial, but we still do not know by what specific mechanism the plant accumulates the energy of several discreet quanta. A succession of single hydrogen transfers is as far as the present hypotheses have gone. If one knew for sure the smallest number of light quanta which just sufficed to achieve the reduction of one molecule of carbon dioxide and the evolution of one molecule of oxygen, it would help to formulate a mechanism for the photochemical process. This seemingly straightforward experimental task appeared to have been finished at the time our guiding scheme was drawn up. There was convincing agreement between the results of different laboratories obtained with different methods as well as with different organisms that the minimum number was about 8. The results of earlier measurements, pointing to about half this number, were believed to be in error. At least they could easily be ascribed to the absence of stationary conditions in plant metabolism during the experimental period.

The question of whether 4 or 8 is the minimum number of quanta has recently been revived by the Warburg school. These workers claim that they can find conditions under which green plants definitely evolve one molecule of oxygen for each three (lower limit 2·8) quanta absorbed. Consequently measurements of the quantum efficiency have also been resumed in other laboratories. With the exception of those scientists who co-operated with Warburg, no one has been able to show that these values can be reproduced under conditions which satisfy the experimenter that a perfectly normal photosynthetic reaction is going on. On the contrary, measurements with the platinum electrode gave a lower limit of 6 with an average of 8. The manometric method, this time carefully rechecked and bolstered with all sorts of controls, continued to deliver values very little lower than before, namely, 10. A shift to other organisms, that is, studies with blue or red marine algae, or some strain of a green sulphur bacterium not so far studied, also produced no quantum number smaller than 10. This is the more interesting because in the case of the sulphur bacteria, thermodynamic reasoning reveals that less than 1 quantum would be required to promote the reaction. When looking for a way out of this impasse it is best not to resort to a scrutiny of methods but to view the problem from other angles.

A purely theoretical approach to the problem of photosynthetic quantum requirements is based on the laws of thermodynamics and general kinetics of photochemical reactions. It can be demonstrated, as Franck has repeatedly emphasized, that a process consisting of a complex set of photochemical partial reactions will never proceed with 100 % efficiency, that is, use all the light energy absorbed for photochemical purposes. If one assumes, for instance, as do Warburg and Burk, that the ideal theoretical quantum requirement should be 2·7 quanta, i.e. 112/41, because 112 kcal. is the heat of combustion of a CH_2O group and 41 kcal. the energy of a mole quantum of red light, one ignores the law of conservation of momentum (and even the second law of thermodynamics). Some dissipation of light energy into thermal energy is unavoidably connected with its utilization in photochemical reactions. Thus with less than 4 light quanta per molecule of carbon dioxide reduced, photosynthesis becomes impossible; i.e. photosynthetic efficiencies above 71 % should be discounted. But even a quantum requirement of 4 seems improbable, because all estimates of the energy losses turn out to be so high that even 4 quanta seem to provide too little energy. Wohl, in 1935, was the first to come to this conclusion. He pointed out that the photosynthetic efficiency remains constant in the region of low and even very low intensities (where photosynthetic rates are slower than those of respiration). From this he deduced a certain minimum stability of the photosynthetic intermediates and found that the energy of activation which so evidently prevents back reactions cannot be negligible. A corresponding amount of energy has to be provided by the light. Adding to these the losses incurred during the spontaneous evolution of oxygen, etc., Wohl arrived at a figure of about 180 kcal. In this way the minimum number of quanta required increases to 5. While this line of argument has been used, analysed and amplified often in discussions, Wohl's conclusions have not been modified very much, as may be seen from Table 1, which is Franck's version of an energy budget of photosynthesis. If similar estimates are made for a mechanism of the type now proposed by Warburg and Burk, the energy requirements seem to add up even higher.

If we want to approach the problem of efficiency and quantum numbers from a biochemical point of view, we have to make certain assumptions as to what the photochemical steps should look like. Let us assume the excited chlorophyll molecules always cause the same kind of hydrogen transfers, which result in the formation of a reducing radical XH and oxidizing radical YOH. With this premise, we obviously have no choice but to construct a reaction scheme on the basis of either 4 or an exact multiple of 4 such transfers. Many investigators, at different times and places and using different methods, have determined the efficiency of photosynthesis in the stationary phase, and because they were all unsuccessful in finding an evolution of oxygen with less than 8 quanta per molecule, we now have in the literature a so-called 8-quanta theory. Such a theory sets the task of distributing two quanta among the reactions leading to the final separation of one O—H bond. A 4-quanta

Table 1. *Franck's energy budget of photosynthesis*

	4 quanta (kcal.)	2·7 quanta (kcal.)
Energy income for process with:		
(a) Singlet-singlet transition at 41 kcal.	164	111
(b) Singlet-triplet transition at 36 kcal.	144	97
Energy expenditure:		
Free energy stored per CH_2O	117	117
Losses in photochemical steps at 7 kcal.	28	19
Losses for evolution of one O_2	20	20
Losses for radical transmutation at 1·5 kcal.	6	4
Losses for carboxylation with one CO_2 + phosphorylation	13	13
Total energy demand	184	173

theory, corresponding to the old results of Warburg and Negelein, would at first glance appear much simpler. Unfortunately, the challenge to interpreting either the alleged 4- or the 8-quanta process on the basis of a stoicheiometric transfer of 4 hydrogen atoms has been made less attractive by the observation that recent results in both series (i.e. 4- and 8-quanta experiments) indicate decidedly lower values. Burk and Warburg bettered their score of 4 by bringing about the evolution of 1 oxygen molecule with 2·8 quanta as the limit; the average being 3. In their newest experiments, therefore, photosynthesis showed an efficiency of about 100 % (see Table 1). Brackett and Olson, using a platinum electrode, and Stauffer with a dropping mercury electrode, added a few more measurements to the already long list of determinations showing a lower limit of 8 quanta. Although the agreement with the earlier results was good, some of the experiments gave values as low as 6, in agreement with a few scattered results mentioned by other workers many years ago. The latter had been ignored at that time because nothing much could be said about such numbers. Now the biochemistry of photosynthesis has progressed somewhat, and we have the wherewithal to explain these results in terms which should satisfy the biochemists as well as the physicists. Twenty years ago it was pointed out that respiration and photosynthesis might be linked by way of some common intermediates. With Calvin's discovery that phosphoglyceric acid is a photosynthetic intermediate, we can hardly avoid the conclusion that there must be conditions under which photosynthesis provides intermediate substrates for respiration, and vice versa. Whenever this occurs, the energy requirement for the uptake of carbon dioxide and for the production of oxygen may change considerably. Any cross-vein on the level of phosphoglyceric acid will lead to a diminution of the work

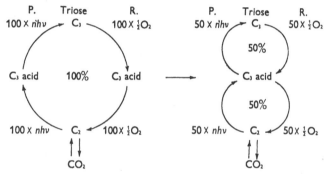

Fig. 2. Schematic explanation of the Kok effect. Compensation of respiration in the light via a full cycle consisting of independent pathways requires twice as much energy as the compensation of respiration via two half-cycles sharing a common pool of intermediates. (P = photosynthesis; R = respiration).

necessary to balance respiration, as shown in Fig. 2. Carbon dioxide is no longer reduced all the way to carbohydrate, nor the latter completely broken down again by way of respiration. Instead, we have two compensating half-cycles. The lower one lifts carbon dioxide only to the level of an intermediate, and the upper one constantly reoxidizes triose or sugar only to the level of the same intermediate. Under conditions of complete interference, the energy required for the sum of both processes is one-half that of normal photosynthesis. This example is quite arbitrary. It happens to ensure a constant ratio of 1:1 between oxygen and carbon dioxide. More likely than not the quotients of the two gases will not remain unity for long if such interferences take place. That they occur in reality is indicated by quite a few observations which we shall discuss later. We do not know yet whether the conditions under which Warburg and Burk have found 100 % efficiency favour metabolic cross-linkages, but it is certainly much more satisfactory to try to explain their wondrous results on the basis of cell physiology than to discuss the reliability of their manometric measurements.

Turning to those experiments which we have called the 8-quanta series, it should be mentioned that there is no necessity for insisting on 8 as the theoretical lower limit. Values of 7 or even 6 become acceptable if we give up the idea that the light reaction consists invariably of a transfer of a hydrogen atom leading directly to the permanent reduction of either a carboxyl or an aldehyde group. Several times, suggestions have been made that some energy might be used to create energy-rich phosphate groups attached to the organic intermediates. The organic compounds would be able to serve as substrates or acceptors for the actual reduction only after they have been phosphorylated and brought to a certain energy level. The example at hand is, of course, the reduction of phosphoglyceric acid to triose. If this occurred in a light reaction, it would first require the formation of an energy-rich acyl phosphate. Such a mixed photochemical mechanism, in combination with interfering respiratory reactions, might well account for any degree of overall efficiency corresponding to a number of quanta from 5 upwards. We shall return to the topic of phosphorylation during photosynthesis in connexion with the problem of back reactions. All this can be summed up in the following statements. Quantum yield measurements are no aid for devising a special photosynthetic mechanism. The overall efficiency in the continuous production of carbohydrate is hardly better than 35 %.

III. THE ACCEPTORS FOR H AND OH

The diagram in Fig. 1 indicates that the consequence of one or two photochemical steps should be the formation, on the one hand, of a hydrogen-donor and, on the other, of an oxidized substance, precursor to oxygen or hydrogen-acceptor. Calvin and Barltrop are at the moment elaborating a hypothesis to the effect that the sulphur atoms in thioctic acid may accept the water molecule, and they have given reasons why this reaction may proceed.

The main problem concerns the next step, the dismutation of two molecules of III to one fully reduced and one oxidized molecule. What is the probability of the back reaction forming water as compared with that of progress in the desired direction? The observations of Spikes and co-workers on the kinetics of the Hill reaction in chloroplasts are an indication of the complexity of the enzymic reaction providing for the stabilization of the photoproduct derived from the reaction with water.

It is evident that molecular oxygen cannot be formed in one step; most probably there are four; consequently, it has been common usage to speak of intermediate photoperoxides. Because the efficiency of photoreduction in bacteria and algae using molecular hydrogen is by no means better than that of oxygen evolution in green plants, it is convenient to let the mechanism of photoreduction branch off at this point. This also helps to explain observations which seem to show that photoreduction in plants interferes or competes with the evolution of oxygen. At recent scientific meetings, one question was in the foreground of attention; namely, how is hydrogen transferred from the original donor, water, to the organic intermediates which are known to be on the pathway of conversion of the carbon of carbon dioxide into sugar? Oddly enough, experimental evidence seems to support two quite different interpretations.

We have, on the one hand, the biochemists to whom it is a foregone conclusion that there must exist a coenzyme, i.e. a hydrogen carrier which, after having been reduced photochemically, brings about the reduction of carbon dioxide in a series of enzymic reactions very similar to those which have been studied in animal and plant tissues in the dark. The attitude of the biochemists at the present time is: why argue the point on the basis of a vague theoretical premise when, in fact, the question has been all but settled? Their attitude is based, of course, on the beautiful work with chloroplasts and coenzymes by Ochoa & Vishniac, Tolmach, Arnon, and Conn & Hendley. These experiments of Ochoa and others are important, mainly, for two reasons. Firstly, they show that in these systems chloroplasts, if they react at all, will react photochemically with coenzymes, and secondly, that chloroplasts which can reduce substances with the oxidation-reduction potential of the pyridine nucleotides must have retained their photochemical 'reducing power' virtually unimpaired. The range of the so-called 'Hill reactions' has been extended. But all this is as yet no proof that in the living cell the chlorophyll would of necessity function in that manner.

Let us hear the other side. Those familiar with the chemical reactions sensitized by illuminated dyestuffs hold that for the primary photochemical steps no help from enzymes is needed. The energy contained in the light-activated dye will promote unspecifically all those chemical changes in substances in contact with the dye for which the excitation energy is high enough. To achieve specificity during these very first steps, the substance X, Y must be present in a concentration high enough to crowd out all other possible reactions. The concentration of coenzymes or of thioctic acid is much too small for this purpose, while the concentration of photosynthetic intermediates may easily be high enough. As mentioned above, the chlorophyll molecules themselves might act as substance X or Y if in one way or another the bleaching of the dye connected with such reactions can be avoided. Because the concentrations of the favourite coenzymes are so inconveniently small, the idea of a photosynthetic unit is being revived again.

The non-specificity of the photochemical steps is attested by the great variety of reductions we are able to perform with chloroplast preparations. The excited chlorophyll complex will reduce indiscriminately nearly anything thrown in its path, e.g. quinones, dyestuffs, oxygen, coenzymes of the pyridine nucleotide or of the cytochrome type, ferric salts, the so-called methaemoglobin factor, and perhaps many other substances. Whether the reduction 'sticks', that is, leads to something which can be isolated or seen in a reduced form, is another matter and depends on the rate of back reactions, and this again depends on the radicals formed initially. If there are reactions to follow, the unstable radicals may not react back, and it is here that the biological specificity begins. For example, a C_3 acid intermediate may receive one hydrogen atom by direct photochemical reduction but lose it again if no specific enzyme comes to the rescue. If the chloroplasts are put to work on a complex enzyme system, such as those known to promote reversible carboxylations, it would be expected that the hydrogen would be received and delivered by the coenzyme in preference to the other components. Nature designed the enzymes with lower activation energies for just such purposes. Both the coenzymes triphosphopyridine nucleotide (TPN) and diphosphopyridine nucleotide (DPN), as well as thioctic acid, are found in normal quantities in green leaves, it is true, but enough reactions of another kind which need these coenzymes, some of them perhaps involved in photosynthesis, do occur in a green cell so that there is no difficulty in explaining their presence.

However persuasive these arguments may sound to a physicist, we feel that the heaviest weight in favour of this view has been provided by the low-temperature experiments on protochlorophyll formation described above (pp. 156–7).

As is often the case when reasonable arguments are brought up to bolster opposing views, there may be a compromise solution waiting. It is obviously possible that besides the chlorophyll molecules themselves some other part of the protein complex serves as the X and Y. The com-

plex may carry in addition a dismutation catalyst or hydrogen transferring enzyme. In the case of the artificial carbon dioxide fixing system
containing extracted chloroplasts, the added coenzymes would receive
the hydrogen secondhand, as it were, while in photosynthesis the built-in
enzyme would react directly and specifically with the organic intermediates. Such an arrangement must of necessity dispel all misgivings
arising from kinetic considerations. That it might be true is strongly
hinted at by the outcome of more recent experiments with chloroplast
preparations.

IV. FIXATION OF CARBON DIOXIDE. C_2 COMPLEX. INFLUENCE OF CARBON DIOXIDE CONCENTRATION

The scheme shown in Fig. 1 presupposes that carbon dioxide cannot be
reduced unless it appears first in a carboxyl group. The discovery by
Calvin, Benson and co-workers that phosphoglyceric acid is an intermediate amply confirmed this. Their sensitive method of radiochromato-
graphy enabled them to observe the results of extremely short periods of
photosynthesis. If radioactive tracer carbon dioxide is given to photosynthesizing plants for a few seconds before they are killed and analysed,
the tracer carbon is found in the carboxyl group of glyceric acid. This
was confirmed in other laboratories, and for several years there has been
complete agreement that the C_3 acid, if not free then in an active state,
is the first stepping stone in the pathway of the conversion of carbon
from carbon dioxide into sugar. One question, however, has remained
open; namely, whether in addition to this very special carboxylation,
photosynthesis makes use of some other carboxylation reactions, particularly those which are known to occur in the general metabolism of
nearly all living cells. Further research with radioactive carbon has
indicated that an unknown C_2 fragment must be the precursor to glyceric
acid, and that this C_2 compound or fragment is regenerated from products or intermediates of photosynthesis in a rapid cycle. This cycle is
thus a characteristic of the mechanism of photosynthesis. Opinions
were divided, however, as to the pathway along which this regeneration
could take place. Leaning heavily on the knowledge provided by comparative biology, the scientists at Berkeley made many efforts to prove
the existence of at least one more carboxylation. This would have made
it possible to account for the cycle in terms of well-explored enzymic
carbon-fixation reactions, such as the fixation of carbon dioxide in malic
acid. A great number of possible reaction sequences were written down
and published, but none of them could truly account for the observed
facts. Calvin and his co-workers as well as Fager and his colleagues

found that the percentage of total carbon dioxide fixed in phosphoglyceric acid during the shortest 'tagging periods' was always too high
in comparison with what one could expect to obtain if the cycle func-
tioned not with one but with two carboxylations. Fager's group later
observed that 90 % or more of the carbon dioxide fixed in the dark
immediately after an illumination enters the carboxyl group of glyceric
and pyruvic acids. They thought, therefore, that this pointed to the
existence of only one carboxylation reaction which mattered, and that
this was the step leading to phosphoglyceric acid. The cycle, they concluded, did not contain a dicarboxylic acid, and they proposed instead
something simpler, a regeneration of the C_2 fragment on the level of
carbohydrate (Fig. 3).

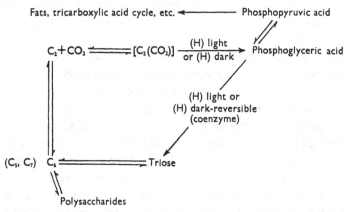

Fig. 3. Scheme of Fager and co-workers for fixation and assimilation of carbon dioxide.

Two quite different types of experiments have provided so much
support for this idea that it now appears to be the correct one. A rapid
transformation of a hexose molecule into molecules containing only two
carbon atoms, that is, dioses, was until recently hardly more than a
speculation. At the present moment, however, the free interconversion
of sugars (as studied by Dische, Horecker, Gibbs, Racker and others) by
way of an exchange of various pieces of their carbon chains, has become
a most intriguing chapter in the enzyme chemistry of carbohydrates.
It is, therefore, quite appropriate to make use of these findings to
explain the workings of the cycle leading to the precursor. Benson
showed that C_5 and C_7 sugars, which he identified as ribulose and
sedoheptulose, became labelled with tracer carbon as rapidly as hexose
during the first seconds of radiocarbon incorporation by photosynthesizing algae. Quite recently investigations by Bassham have revealed
that the distribution of this tracer carbon within the carbon chain of C_6,

C_7 and C_5 sugars is that shown in Table 2. The resultant distributions were explained on the basis of a series of condensation and transfer reactions. Leaving out the now familiar condensation of trioses to hexoses, and also disregarding the participation of free trioses, the scheme to give C_5 and C_7 sugars can be written in a condensed form as follows (the number of asterisks indicate the degree of labelling):

C C C* C*/C C → C C C* C C   Pentose

C C C* /C* C C → C C C* C* C* C C   Heptulose

↓ +C* C C   Triose

C C C* C C + C C C* C* C*

Average pentose, therefore, is labelled CCC***C*C* as required by the data in Table 2. Fixation of tagged carbon dioxide by means of the pentose molecule would give phosphoglyceric acid predominantly tagged in the carboxyl group. Still, we are not sure that the precursor molecules

Table 2. *Distribution of radioactivity in sugars from flow experiments*
(According to Bassham and co-workers)

Photosynthesizing system exposed to $^{14}CO_2$ for 5·4 sec. Radioactivity of each carbon of sugar expressed as percentage of total radioactivity of whole molecule.

Fructose	Sedoheptulose	Ribulose
—	2	—
3	2	—
3	28	11
43	24	10
42	* {27	69
3	{ 2	5
3	2	3

* Total radioactivity of carbon nos. 5 and 6 = 29 %. Radioactivities of carbon nos. 6 and 7 assumed equal.

must be formed exclusively via pentose. Pentose and heptose may arise merely because of the availability of active C_2, C_3 and C_4 fragments. The simplest but perhaps least likely way to account for the autocatalytic rise of tracer concentration in the α and β carbons of glyceric acid is to have the hexose molecule undergo a stepwise decomposition into three C_2 fragments. Bassham's newest experiments support Calvin's hypothesis that an undivided C_5 sugar is the direct precursor in the carboxylation reaction

$$C_5 + C_1 \rightarrow 2C_3.$$

Whenever the course of photosynthesis is suddenly altered, the sizes of the C_5 and C_3 reservoirs are seen to change in opposite directions. In one type of experiment, a gas mixture containing an ample percentage of carbon dioxide is quickly replaced by one without carbon dioxide while the illumination continues. As a consequence, the reservoir of labelled C_5 compound fills up while that of the C_3 acid decreases. The number of counts gained by the first compound are apparently lost by the second. On the other hand, if instead of removing the carbon dioxide, the light is turned off, the reverse occurs, the amount of the C_5 compound in the stationary state goes down and that of the C_3 compound goes up, with the result that practically only phosphoglyceric acid is found. This latter observation fully agrees with the studies of Fager reported at the Sheffield meeting (Symposium of the Society for Experimental Biology, 1951) on the exclusive fixation of carbon dioxide in phosphoglyceric acid during the first moments of darkness after a light period.

Our present knowledge about the cycle can, therefore, be summarized in shorthand by adding to the left-hand side of the scheme shown in Fig. 1 an arrow marked 'C_2 complex', leading back from carbohydrate to carbon dioxide fixation.

In the course of a discussion of quantum yields, it was stated by Warburg that the best yields could be obtained only with a carbon dioxide concentration of nearly 0·05 atm. The need for so much carbon dioxide even at the maximum rate has never been reported before, and we all know that the plants can do rather well with the very low carbon dioxide content of the air. How much more carbon dioxide than the 0·0003 atm. in the air is needed to realize the maximum efficiency or the highest rate of carbon assimilation in plants is an old and often debated question. The concentration present in air is definitely below optimum for photosynthesis. It was unexpectedly found during measurements of the carbon dioxide exchange of algae with the glass electrode that at light saturation the photosynthetic process in *Chlorella* or *Scenedesmus* required, indeed, a concentration of carbon dioxide corresponding to c. 0·01 atm. However, the dependence of rate of photosynthesis on carbon dioxide concentration is influenced by the light intensity, and for lower light intensities the requirement for carbon dioxide is definitely lower. Special studies by Rosenberg have shown that explanations based on the assumption of diffusion difficulties are not sufficient to remove the discrepancies between several of the published results. Since the quantum efficiency of photosynthesis must be low at high light intensities, it is only natural that efficiency measurements are made in a region where the rate of photosynthesis is either equal to or only a few times greater than that of respiration, and here 0·001 atm. carbon dioxide in the surrounding gas should be quite sufficient.

Another controversial point is the question of whether bicarbonate ions can be utilized directly by plants. If this is so, it would explain why algae have a nearly normal rate of photosynthesis in alkaline buffers in equilibrium with rather small carbon dioxide concentrations. If the plants are suspended in an acid medium, these same concentrations turn out to be too low for reaching normal saturation rates. It has been established beyond doubt that the plant cells show a certain variability or adaptability. We are thinking here mainly of the new work of Stiemann-Nielsen, Briggs and Wittingham. When, after a period of photosynthesis with a very high concentration of carbon dioxide, the algae are transferred to an alkaline buffer, the rate is definitely slower than it is an hour later. Incubation at the high pH improves the rate. Experiments with the glass electrode have shown the beneficial effect of higher bicarbonate concentrations. Since we are on the threshold of knowing exactly what the carboxylation reaction looks like in biochemical terms, we may soon be able to explain the effect of the concentration of either free carbon dioxide or of bicarbonate on the rate of carbon dioxide fixation.

V. REDUCTION OF PHOSPHOGLYCERIC ACID.
NUMBER OF LIGHT REACTIONS

As long as it seemed likely that several of the known reductive carboxylations could play a definite part in photosynthesis, one had to assume either an equal number of different photochemical reactions or the photochemical formation of a general hydrogen carrier, i.e. a reducing power capable of transferring hydrogen rather indiscriminately from the photochemical system to various fixation reactions going on in the cell. Some time ago Gaffron and his co-workers pointed out, however, that on the basis of experimental evidence it was far more likely that there were at the most two, perhaps only one, photochemical steps involved in the photosynthetic cycle. Bassham's analysis of the distribution of tracer carbon in the carbon chains of possible cycle intermediates definitely supports this view. At the present time we are, therefore, fortunately left with the choice between only two alternatives. Either light promotes directly the carboxylation of a C_2 fragment to the C_3 acid and, further, the reduction of the latter to triose and sugars; or only this one step leading from phosphoglyceric acid to triose, etc., constitutes the light reaction in photosynthesis.

We know that the carboxylation can happen in the dark. The task now consists in showing whether or not in the living cell there is normally a direct connexion with the photochemical mechanism. Since it may help further experimentation, we shall discuss one example for and one example against the proposal that light is normally involved.

Van der Veen found that leaves which have been momentarily heated still absorb a small amount of carbon dioxide immediately after exposure to light. Then nothing more happens until the leaves are darkened, whereupon the carbon dioxide previously taken up is released again. No corresponding effect with oxygen can be observed. This 'holding' effect of light for a restricted amount of carbon dioxide was confirmed by Butler. Franck explains the effect as due to a photochemically produced, unstable tautomeric form of an acceptor molecule. In the activated form it combines with carbon dioxide. Since the necessary enzymic reactions to complete the carboxylations are missing, on account of the heat treatment, the complex breaks down when the illumination ceases.

The observations of Fager, on the other hand, seem to speak for the existence of only one light reaction. He succeeded in reconstructing the natural carboxylation reaction leading to phosphoglyceric acid by combining a heat-stable plant extract, chloroplasts and carbon dioxide. The fixation of carbon dioxide proceeded here without light, catalysed by enzymes present in the chloroplast preparations. The effect of light in enhancing the yield of the Fager reaction suggests a possible direct coupling of this same carboxylation with the photochemical system, which would mean a second light reaction. More carbon dioxide is fixed in phosphoglyceric acid when the cell-free preparation is illuminated. The conclusion, however, that we are dealing here with one of the natural light reactions as they occur in the living cell is not necessarily correct. It is just as likely that the chloroplasts on illumination reactivate a part of the enzyme system which has become accidentally inactivated by oxidation during the preparation of the green juice, and this too would enhance the rate and the yield in the system described.

The mechanism one chooses to describe the photoreduction of phosphoglyceric acid is still a matter of preference. A biochemist would immediately favour the idea that it should be a literal reversion of the glycolytic dehydrogenation of carbohydrate, and he may point to the fact that Ochoa and co-workers succeeded, indeed, by skilfully assembling the necessary enzyme systems, in accomplishing such step-by-step reversal in an artificial system driven by illuminated chloroplasts. On the other hand, we can argue that the pathway of glycolysis leads to pyruvate and its decarboxylation, and not to a decarboxylation of glyceric acid. Enzymes decarboxylating the latter have so far not been found, and pyruvate is not included in the photosynthetic cycle but arises in a side reaction. If the photochemical reaction were really identical with the glycolytic pathway, its sensitivity to specific poisons should also be identical. Experiments with iodoacetic acid have shown that it is not so. Thus, if it should turn out that this is the only photochemical step in the entire cycle of carbon assimilation, it would be surprising if it were not specific and not different from the known enzymic reactions. Very little is known at the moment about what happens here. The obvious experiment of adding phosphoglyceric acid in the hope of having it reduced in the light by chloroplasts was not very successful. A certain increase in the rate of oxygen evolution resulting from addition of this compound under anaerobic conditions was observed by Tolmach, but the yield was very poor. We must admit that it would have been a piece of unusually good luck if enough phosphoglyceric acid had been 'activated' by the necessary enzyme, and thus metabolized in the proper order. The problem has to be attacked with cell-free preparations where membranes do not present too much of an obstacle. By working with large cells of *Chara*, Tolbert was able to separate chloroplasts and protoplasm from the sap of the vacuole. Under these conditions, he found that photosynthesis may continue normally for some time, but the least disturbance of structural arrangement in the protoplasm leads to a decay of the original faculty. To understand this, we may, without taking it too literally, consider the requirements for the nonphotochemical reduction of glyceric acid to triose. We know that in the living cells the carboxyl group of glyceric acid is never reduced as such. The carboxyl has to be phosphorylated, and this again requires energy. The question of how phosphoglyceric acid is reduced to triose is, therefore, intimately linked with the other problem concerning the phosphorylation and activation of the compounds taking part in the photosynthetic cycle. It seems inescapable that we may have some coupled reactions which are not reduction steps as such, and yet are driven by the photoreactions.

VI. THE WARBURG-BURK THEORY. BACK REACTIONS; CHEMILUMINESCENCE; PHOSPHORYLATIONS

The easiest way to explain how the cell provides the necessary energy-rich phosphate bonds taking part in one or more photosynthetic steps is to assume that they are provided by the normal respiration of the cell. At high light intensities photosynthesis is, however, often more than twenty times as fast as the normal respiration, as measured before or afterwards in the dark. It is, therefore, quite unlikely that normal respiration is the source of any phosphate donors for photosynthesis at saturation or near saturation rates. The usual answer to this is, of course, that we do not know the rate of respiration during photosynthesis. It might be very much enhanced. This is a very old argument with a long troublesome history. Luckily enough we can dispense with repetitions because the discussion has been simplified in two ways. Warburg and Burk have gone to the extreme by launching a revolutionary theory centring on respiratory back reactions, while at the same time new methods have been developed by which this hypothesis can be tested. The new theory requires that in the light, respiration with molecular oxygen goes along at twice the speed of the overall photosynthetic process. The substrates for the accelerated oxidation must, therefore, be produced even faster and this is accomplished in the following way: the substrates are produced by photochemical reduction while oxygen is evolved. Their re-oxidation provides energy which is stored quantitatively in a complex. Because the latter now contains so much energy the absorption of one more light quantum suffices to reduce a molecule of carbon dioxide with the release of one molecule of free oxygen. Warburg concludes from his observations that, under optimum conditions, this mechanism works with 97 % efficiency. In continuous photosynthesis, we observe only the balance of the two fundamental processes, and this is supposed to be the reason why the true mechanism has escaped detection for so long. The theory is the outcome of a prodigious amount of work on gas exchange during photosynthesis as measured manometrically under conditions where the plants were submitted to intermittent illumination. The length of the dark and light period varied between 1 and 3 min., and it was found that during the first 30 sec. or so of either a dark or a light period the pressure changes were considerably larger than they should have been if respiration or photosynthesis respectively had begun at their maximum stationary rates. Usually one finds just the contrary. Photosynthesis shows an induction period. The rate starts slowly and increases autocatalytically. As for respiration, abnormally high rates immediately after the end of the light period have been reported and discussed since van der Pauuw's work in 1932. On the other hand, it is known that carbon dioxide fixation overshoots the end-point of illumination and falls off in about 20 sec. The students in the field have been well aware of some irregularities occurring after each of the two transitions, dark to light and light to dark. And it is for this reason that in most of the published experimentation for measuring the quantum efficiency of photosynthesis, transition periods have been carefully excluded from the evaluation of the results.

Let us examine first some of the consequences of the Warburg-Burk theory:

(1) Since it claims universal validity, this mechanism of photosynthesis must be observable with any plant and not only in *Chlorella*.

(2) Whenever photosynthesis proceeds at all, a correspondingly enhanced oxygen exchange must occur. Thus, in order to test the theory, it is not necessary to make quantum yield measurements, but simply to ascertain whether or not the back reactions occur.

(3) The back reactions with oxygen should stop and in consequence photosynthesis also, if the oxygen developed in the light is immediately removed.

(4) If there were means by which one could follow respiration while the plants are illuminated, it should be possible to detect a respiration going on at twice the speed of the overall gas exchange in the light.

Beginning with the last point, it is urged that the reader, who wants to form an opinion on this matter, study carefully the results obtained by A. H. Brown with the mass-spectrometer. The plants were suspended in aqueous media containing the natural very small concentration of ^{18}O. The air above was, on the other hand, enriched with the oxygen isotope ^{18}O. During respiration in the dark, the plants took up ^{18}O as shown by the automatically recorded graphs of the instrument. In the light only ^{16}O was produced to any extent by photosynthesis. As for respiration, the uptake of ^{18}O continued as if light had no influence whatsoever. Exceptions to this behaviour were found to occur mainly in the direction of a slight respiratory inhibition. Several organisms were studied and with the alga, *Anabaena*, inhibition as well as some acceleration of respiration in the light could be demonstrated. But in all of the numerous experiments there was no indication that an accelerated uptake of free oxygen occurred during illumination; in other words, there were no results which would in any way support the theory.

Being a counterpart to the question of how respiration interferes with photosynthesis, the possible effects of removing oxygen have been discussed ever since Willstätter and Stoll's original observations. They showed that after an anaerobic period in the dark lasting several hours, photosynthesis in leaves was totally inhibited, but recovered completely under constant illumination. Since then experiments of this kind have been performed many times, always with the same results. A preceding anaerobic dark period produces an initial inhibition of photosynthesis, i.e. a lag or prolonged induction period, which disappears autocatalytically once the first small amount of oxygen is formed. Such results have been and still are sometimes taken as proof that photosynthesis requires the presence of free oxygen, which, according to the new theory, is required to feed the back reactions.

The argument as well as the experiments, however, is not conclusive. To subject plants to a period of anaerobiosis in the dark and then to find a prolonged induction period at the start of illumination does not prove that oxygen participates in the process of photosynthesis during the stationary phase. It shows merely that oxygen can remove an inhibition caused by the enforced fermentation metabolism which went on during the preceding dark period. This explanation of the Willstätter effect was given fifteen years ago and has so far not been contradicted by any new observations. Two types of experiments designed to test the Warburg-Burk theory and not complicated by the effect of anaerobic induction periods were recently performed. In manometers filled with hydrogen the oxygen developed by photosynthesizing algae was constantly removed through reduction to water by a large surface of a special palladium catalyst. The effectiveness of this device is such that it can cope with oxygen produced at a rate equivalent to 50 mm. Brodie fluid per min. No manometrically detectable partial pressure of oxygen builds up if the rate of photosynthesis is kept at a few mm. per min. The rates of photosynthesis were found to be nearly the same under these 'anaerobic' conditions and in air; if different, those in air were somewhat lower.

The same but much more rigorous proof was furnished by Allen working with *Scenedesmus* and *Chlorella*. He used an apparatus capable of determining exactly very small admixtures of oxygen in a stream of pure nitrogen. The gas stream passed over a suspension of a tiny amount of algae contained in 0·05 ml. of water. Stirring was done magnetically after control experiments had shown that this was very efficient. The rate of oxygen production at saturation light intensities was kept so low that the partial pressure of the oxygen escaping into the nitrogen was not higher than 10^{-5} mm. Hg. With *Scenedesmus* at pH 8·2 the saturation rate is as high or somewhat higher than under aerobic conditions. In acid suspensions the rate falls slowly with the duration of anaerobiosis, while the induction periods become longer. *Chlorella* is definitely more sensitive to lack of oxygen. The decline of the saturation rate and the much prolonged induction can be seen after 10 min. of anaerobic incubation.

Observations recently reported by Wittingham and Hill are in contradiction with the above results. These authors used their well-known haemoglobin complex method to measure the photosynthetic oxygen. They came to the conclusion that the photosynthetic rate starts to decline when the partial pressure of oxygen falls below 1·5 mm. Hg.

We have the impression that what they observed by their method was the prolongation of induction periods, a phenomenon which, as we have said, must be distinguished from steady-state conditions. In this respect it should be noted that Clendenning was able to eliminate almost completely the usual aerobic induction periods of normal algae by treating them with 100 % oxygen before their exposure to light.

Confronted with these experiments of Allen and of Gaffron, critics have said that perhaps in all of them the back reaction went too fast to be observed. The oxygen of photosynthesis may have been recaptured and used 'before it could leave the cell'. The answer to this is that Brown's recordings of photosynthesis with the mass spectrometer were aerobic experiments. The cells were in equilibrium with the surrounding oxygen containing ^{18}O. If the catalyst promoting the accelerated combustion in the light has such an affinity for oxygen, it is unimaginable that only ^{16}O and not ^{18}O also should have reacted with it. If, on the other hand, 'before it leaves the cell' is interpreted to mean back reactions with a precursor to free oxygen, e.g. the 'photoperoxides', there is no point left to argue about. We are agreed, then, that molecular oxygen is not necessary for photosynthesis.

We may now turn around and ask: are there no internal back reactions in photosynthesis and no effects of external oxygen? The answer is, of course, there are, as proved by observations.

There are no less than three ways in which atmospheric oxygen may influence photosynthesis, and since oxygen is always present these effects deserve to be called normal. Firstly, the indirect influence via respiration; secondly, the direct competition with the organic intermediates for the hydrogen in the photochemical reaction; and thirdly, rather unspecific, photosensitized auto-oxidations similar to the reactions which have been studied *in vitro*. Chlorophyll in organic solvents can promote oxidation with a quantum yield of one. To avoid or to minimize this third form of interference is actually one of the problems which the plant has successfully solved. As early as 1919 Warburg described an influence of oxygen on photosynthesis. The saturation rate at high light intensities was somewhat lower in air or in oxygen than in nitrogen. Since the effect varies with the organism and with experimental conditions, it has to be determined in each case whether it can be explained as a spontaneous photo-oxidation or a 'Hill' reaction, meaning that we interpose a hydrogen-donor XH or XH_2.

Oxygen as a 'Hill' reagent has been described by Mehler. Later experiments with the mass-spectrometer and oxygen ^{18}O have revealed an oxygen exchange by illuminated green algae when they are deprived of carbon dioxide. The photosynthetic production of oxygen is balanced by its reduction to water via hydrogen peroxide. In a way we may call this a back reaction, although it would be less confusing to call it a cycle.

The decisive achievement of the plant, however, is to avoid immediate back reactions subsequent to the photochemically enforced changes. While reactions with molecular oxygen have mostly to overcome a barrier posed by activation energies, these barriers are presumably low or missing between the radicals first formed after the conversion of light into chemical energy.

Despite the high efficiency of photosynthesis at low light intensities, it has been shown that back reactions do occur in cells and illuminated chloroplasts. There is a weak chemiluminescence of the active chlorophyll in the living cell which can be picked up by a photomultiplier immediately after a period of photosynthesis. This phenomenon, discovered by Strehler and Arnold, can hardly be interpreted without assuming an internal reaction between the photoperoxides and some reduced counterpart. While normal photosynthesis with carbon dioxide is going on, the intensity of the emitted light is low. It is strong when carbon dioxide is removed and more so when narcotics hinder the normal photosynthetic reactions. As Franck and Brugger have found, the intensity of the chemiluminescence fluctuates during induction periods in a manner closely paralleling those reported for the fluorescence of active chlorophyll in the living cell. In this connexion we must mention Linschitz's observation that not only porphyrin but also chlorophyll is able to catalyse the decomposition of organic peroxides while emitting light having the characteristic wavelength of its red fluorescence band.

Ever since Ruben expressed the likelihood of the need of phosphate bond energy for certain steps in the reduction mechanism, it has been generally agreed that the intermediates might be phosphorylated compounds. This has now been shown to be true by the work of Calvin and co-workers. There was also agreement that the photosynthetic cycle has to provide for its own phosphorylation reactions. Observations on the uptake of orthophosphate and the formation of adenosine triphosphate (ATP) during the first minute of illumination and the reverse reaction at the end of a light period can be regarded as evidence in favour of this assumption (Kandler, Strehler).

On the other hand, suggestions as to special mechanisms for the formation of the energy-rich phosphate esters must, for lack of information, be considered purely speculative at the present time. On account of the experiments described above, internal back reactions seem a more probable source of energy than reactions with free oxygen. To illustrate how much our position in respect to the question of photosynthetic phosphorylations has been improved, let us quote from the review of 1946. (The context there happens to be an unfavourable critique of the idea that the light energy may be utilized to produce nothing but energy-rich phosphate bonds.) 'The resulting phosphorylated compounds should be stable enough to survive the end of an illumination period and cause the reduction of carbon dioxide for some time afterward in the dark.... Nothing of that kind has ever been observed, although many an investigator looked for it.' The anaerobic pre-illumination treatment introduced by Calvin proved that some carbon dioxide could be fixed in the dark subsequent to the light period, and Fager showed that an extract made from cells so treated contains a phosphorylated compound which may add carbon dioxide to form phosphoglyceric acid.

If the arguments against the Warburg-Burk theory are valid, we also see no reason to believe that free oxygen rather than light energy is required to furnish the phosphate bonds in the reductive carbon cycle. Nevertheless, under special conditions, e.g. during induction periods or when carbon dioxide is limiting, the plant may suffer a lack of internal oxidants, and the photochemical coupling between the functional back reaction and the production of phosphorylation energy may have been broken. Under these circumstances, it may be found that formation of energy-rich phosphate bonds, via oxidation with atmospheric oxygen, is the only way to revive the photosynthetic activity. This would be a 'priming' effect alternative to the one introduced by Franck in his explanation of induction phenomena.

The present situation can be summarized by saying that we have found, or expect to find, back reactions on each successive level of the carbon assimilating process. Which one of them should be regarded not as an accidental loss but as an essential part of the photosynthetic mechanism has to be decided in each case on the basis of careful experimentation.

VII. REACTIONS IN CHLOROPLAST PREPARATIONS

The reactions which are of special interest in this section are usually entitled 'Hill reactions' and they are effects which appear more physiological and nearer to the natural activity of chloroplasts *in vivo* (see above, § III). Thus one is inclined to give more attention to those experiments where the oxidant or hydrogen- (electron-) acceptor is a familiar natural substance, such as oxygen, cytochrome, diphospho-pyridine nucleotide, triphosphopyridine nucleotide, methaemoglobin, glutathione, etc. But it stands to reason that these reactions might be just as artificial as those which we observe with other chemical compounds, for instance, quinone. For the study of the photochemical reactions of the chlorophyll complex, artificial systems containing natural acceptors may be no better than those made up with unphysiological compounds. The one exception is an assembly of chloroplasts and plant extracts which promotes an uptake of carbon dioxide to give phosphoglyceric acid. Here, there is a good chance that the test-tube reaction is more than just a vague model of the process in the intact cell. The reduction of quinone and of oxygen by illuminated chloroplasts, or by algal cells treated with quinone but seemingly intact, continues to be a useful research tool. The importance of several observations in different laboratories on the inhibition or stimulation of this reaction by added substances lies in the fact that they indicate the complexity of the system functioning in the extracted green particles, and thus invite further analysis.

In discussing the influence of oxygen on photosynthesis, it was stated that in living cells oxygen and the organic intermediates compete for the energy source. In the absence of carbon dioxide, oxygen is easily reduced; in the presence of carbon dioxide it becomes a harmless competitor. When illuminated chloroplasts are given quinone and oxygen, the quinone is always reduced first. The reduction of oxygen to hydrogen peroxide as shown by Mehler starts only when the quinone reduction has come to completion. Adding more quinone immediately stops the utilization of oxygen. Mehler observed that pretreatment of spinach chloroplasts with a little quinone in the light (hydroquinone has no effect) doubled the rate of alcohol oxidation in the presence of catalase, and enormously stimulated the rate of the photo-oxidation of ascorbic acid and of glutathione. Similar effects can be obtained with 10^{-3} M-manganous ions. Aerobic pretreatment of live *Scenedesmus* cells for 10 min. in the dark with 6-thioctic acid caused under certain conditions a doubling of the rate of a subsequent quinone reduction in the light. As Bradley and Calvin report, so far this effect is specific for *Scenedesmus* and was not found, for instance, in *Chlorella*.

Usually the rate of chloroplast reactions is slower than that of the corresponding photosynthesis in the live cell. There is room for improvement, and the mentioned stimulations may at present be summarily explained on the basis of introducing auxiliary hydrogen- or electron-transfer agents. Thioctic acid, because it happens to occur in the plant and has a known role as a coenzyme, offers, according to Calvin and Barltrop, interesting possibilities for speculation.

Vishniac treated chloroplasts with acetone, extracted the dry preparation with aqueous buffer, and thus obtained a nearly colourless extract which we may call E. This was mixed with oxidized glutathione, glutathione reductase, TPN, ferrocyanide and a fresh methanolic solution of chlorophyll. On exposure of this mixture to light, the chlorophyll sensitized the reduction of the oxidized glutathione with a concomitant oxidation of ferro- to ferri-cyanide. An overall gain in free energy is involved because ferric salts will spontaneously oxidize reduced glutathione. In experiments to be done, a successful replacement of chlorophyll by other dyestuffs may prove that this model is nearer to the known non-biological photoreductions than to the processes which occur in the living cell. Nevertheless, one wonders why extract E was necessary to promote the reaction.

The fixation of carbon dioxide by green cells in the dark immediately after a light period, has been the source of much speculation. Fager has shown that in all cases, whether after normal photosynthesis or after pre-illumination in nitrogen freed of carbon dioxide, the final product is phosphoglyceric acid. The question arises: what photo-product survives to cause this fixation? Fager's experiments with chloroplasts and plant extracts have now shown that it is very probably a phosphate ester, because such a substance can be extracted from pre-illuminated algae. When the extract is mixed with chloroplasts and carbon dioxide, a carboxylation occurs and the product is phosphoglyceric acid. Illumination increases the yield. At the time of writing, the identity of this 'acceptor' is still unknown. It has been established so far that it is a phosphate ester capable of surviving extraction with hot water. The same chloroplast preparations with and without ATP did not give carbon dioxide fixation when combined with either 2-phosphoglycolaldehyde or vinyl phosphate. Sugar phosphates were also inactive. Exchange reactions with preformed phosphoglyceric acid could be excluded. Thus the acceptor may be an as yet untried sugar ester, or an 'active' coenzyme-bound C_2 complex.

The carboxylating enzyme is present in the chloroplasts. The rate of the reaction is slow, but the reaction can easily go on for half an hour. The rate of fixation depends on the concentration of the extract containing the acceptor and of the chloroplasts. Cyanide and reagents blocking sulphydryl groups inhibit the enzymic reaction between acceptor and carbon dioxide. Light reactivates, perhaps by reducing blocked sulphydryl groups. In view of the reduction of oxidized glutathione by chloroplast preparations (reported by Conn & Hendley, and by Vishniac), the influence of light in the Fager reaction may be due to an increase in active enzymes and need not be identical with one of the photochemical steps in photosynthesis unless one likes to think of this light effect along the lines proposed by Calvin and Barltrop. There is no doubt that, of all the cell-free preparations so far investigated, this one promises to give the most important results. Neither the identity of the C_2 fragment nor the mechanism of the carboxylation reaction can now remain hidden for very long.

VIII. INDUCTION PERIODS. TRANSIENTS

Our present knowledge of photosynthesis, incomplete though it is, has made it impossible for us to imagine an abrupt transition from the stationary metabolism of the green cell in the dark to its stationary metabolism in the light at the very moment of exposure. Thus we postulate now what we have long known to exist. There must be inductions and after-effects with each change in light intensity. The light reaction sets the pace and all other partial reactions have to adjust themselves to it, one after the other. This takes time. It does not have any meaning, therefore, to say that under such and such conditions there is no induction period. We may only say that the transition takes less time by comparison with another case which we have taken as a control.

Our picture of the photosynthetic mechanism tells us also that induction periods recorded for oxygen need not be the same as those recorded for carbon dioxide, and unless a special regulatory mechanism exists to ensure that one side does not outrun the other, we should expect deviations of the assimilatory quotient $(-CO_2/+O_2)$ during each transition period.

As long as it was believed that the hydrogen from water could only be transferred to carbon dioxide, and the hydroxyl radicals disposed of only as oxygen, the self-regulating mechanism was obvious. Any interference with the normal course of events on either side of the mechanism could be expected to bring into play back reactions equalizing the rate of oxidations and reductions. Now we have learned that substances other than just carbon dioxide can be reduced photochemically, and that oxygen will fail to appear if the photoperoxides are snatched away in internal oxidations. The often rather intricate metabolism of the purple bacteria should be regarded as an example of what perhaps may also happen, although to a much lesser degree, in the green cells.

An orderly induction period with oxygen and carbon dioxide running nicely hand-in-hand might be the exception if we only look closer, and even if an assimilatory quotient of unity is observed during the entire period until a stationary state has been reached, it is no guarantee that there is no initial interference with respiration. These matters have been authoritatively discussed by Franck. They are suggested here merely as a reminder that the most challenging event in the field, the announcement of the 1-quantum process theory of Warburg & Burk, was developed entirely on the basis of observations which fall within the category of transient phenomena and induction periods, as well as after-effects.

Although interested in seeing for themselves the impressive 'negative' induction effects obtained by Warburg & Burk, other workers did not feel competent enough to handle manometers with such skill. Different methods were adopted; methods which in theory should be even more suitable for observing transient phenomena. Brackett & Olson used a platinum electrode to follow oxygen, Gaffron and Rosenberg a glass electrode to record carbon dioxide exchange, and A. H. Brown the mass-spectrometer.

In literally hundreds of automatically recorded experiments, special transient effects with rates high enough to indicate a 1-quantum process have not been found. But since the records show a rather bewildering variety of induction and after-effects, some of them with at least a qualitative resemblance to Warburg & Burk's transient rate anomalies, it is by no means excluded that under very special conditions even the ones observed by Warburg & Burk might occur. If so, they must be explained in the same way as all the other transient rate anomalies. While that is a difficult and time-consuming task, we know at least that the particular anomalies under consideration can be explained (Franck) without resorting to the hypothesis of the '1-quantum theory of photosynthesis'.

The simplest experimental pattern emerges if alternating dark and light periods of, say, 2 or 3 min. are repeated often enough to annul the influence of the previous life history of the plant which is often deducible from the irregularities seen during the first and second light periods. The main results of all these observations with plants which have not been treated purposely in an abnormal way can be summarized as follows. The rough overall pattern of transient phenomena agrees with that described most often in the literature; i.e. an acceleration of the gas exchange during the induction in the light, and a small but by no means large increase in respiration after darkening.

A comparison of the separate recordings for oxygen and for carbon dioxide, as obtained by the platinum and the glass electrode, reveals one major difference. The reversion of the oxygen gas exchange from negative to positive and vice versa occurs faster than the corresponding change in the metabolism of carbon dioxide. Lags in the response of the recording instruments were not responsible for this difference.

The fluctuations in the rate of oxygen metabolism, so clearly visible in the experiments of Brackett & Olson, occur after the rate has at least momentarily reached values which are within a small percentage of those found 3 min. later (see Fig. 4, constructed from their published data). Of great importance is the occurrence of a 'negative' induction during the first minute in the light. During this time the rate falls off. Quantitatively the effect is much smaller than the one observed by Warburg & Burk.

On the other hand, Fig. 5 shows a typical recording made with the glass electrode; such transient phenomena in the rate of carbon dioxide exchange are found with astonishing regularity both in *Chlorella* and in *Scenedesmus*. The upper part is a tracing of the continuously recorded

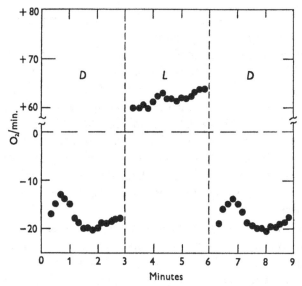

Fig. 4. Platinum electrode recording of oxygen exchange in *Chlorella*, according to Brackett, Olson & Crichard (*J. gen. Physiol.* **36**, 529, 563 (1953)). (*D* = darkness, *L* = light.)

changes of pH. The lower part gives the rates of carbon dioxide metabolism calculated for points chosen from the unbroken line above. A new phenomenon appears which earlier methods have not been sensitive enough to reveal. After each transition, there is a compensation period during which carbon dioxide is neither evolved nor absorbed. There is not the slightest doubt possible about the existence of this natural period, since it has been recorded innumerable times. Its length can vary from a few seconds to periods lasting more than a minute. What is surprising is the speed at which the change from respiration to compensation is accomplished: as far as the instrument can show, it is instantaneous. This agrees very well with observations by Blinks & Skow who, fifteen years ago, were the first to use both the glass and the platinum electrode for measurements of photosynthesis. The events at the end of a light period are more complicated. Several reactions overlap, and since it is not possible to go into details here, it must suffice to say that at least three reactions should be considered: the continued pick-up of carbon dioxide, respiration, and the reaction which causes the compensation. The most obvious explanation for these compensation periods is the reduction and reoxidation of intermediates without a concomitant fixation or release of carbon dioxide.

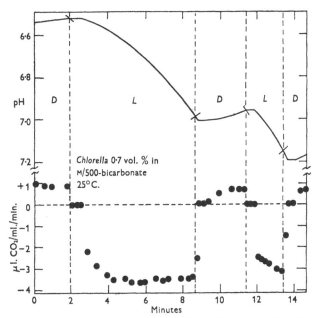

Fig. 5. Glass electrode recording of carbon dioxide exchange in *Chlorella pyrenoidosa*. One-day-old culture. Upper curve direct tracing of pH changes. Lower curve calculated rates. CO_2 not limiting. ($D=$darkness, $L=$light.)

It clearly follows that we are not forced to accept the Warburg-Burk theory until the authors have proved that their initial manometric pressure changes have no connexion with the irregularities so clearly seen in other laboratories. Let us now return to the problem of the need for, or the influence of, oxygen, discussed above in another context.

Figure 6 shows two curves of the pH changes under aerobic and anaerobic conditions for otherwise identical samples of algae. The initial pH is such that the carbon dioxide concentration in the buffer begins to limit the rate of the photosynthetic reaction as indicated by the decreasing numbers calculated for the pH values noted on the graph. The recordings show a conspicuous difference. During each of the one-minute light periods, the algae assimilate a little more carbon dioxide when oxygen is easily available than without oxygen, but this amount is promptly lost again in the subsequent one-minute dark periods. The overall amount of carbon dioxide permanently assimilated by means of photosynthesis is the same in nitrogen and in air. In air there is superimposed a light-driven reversible carboxylation. This example is given without further

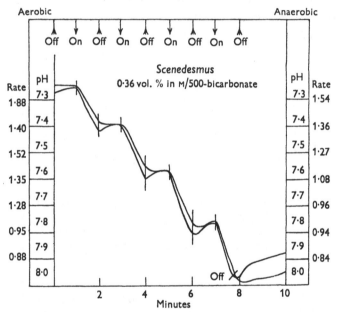

Fig. 6. Glass electrode recording of carbon dioxide exchange in *Scenedesmus*. 1 min. light and dark periods. Upper curve in nitrogen (photosynthetic oxygen accumulating). Lower curve in air. Increasing CO_2 limitation.

detailed comment merely to demonstrate that inductions are really as complex as pointed out at the beginning of this section.

Summing up, it can be said that the increase in our detailed knowledge of the mechanism of photosynthesis during the last few years has been greater than in any similar period before. Some of the discoveries have been rather surprising; yet not one of them forces us to give up the overall picture of the mechanism of photosynthesis as it already existed a few years ago. On the contrary, the new observations extend and strengthen it.

REFERENCES

Some recent reviews on photosynthesis and related subjects

BROWN, A. H. & FRENKEL, A. W. (1953). Photosynthesis. *Annu. Rev. Plant Physiol.* **4**, 23.

CALVIN, M., BASSHAM, I. A., BENSON, A. A. & MASSINI, D. (1952). Photosynthesis. *Annu. Rev. Phys. Chem.* **3**, 215.

HILL, R. (1951). Oxidoreduction in chloroplasts. *Advanc. Enzymol.* **12**, 1.

VAN NIEL, C. B. (1952). Bacterial photosyntheses. In *The Enzymes* (ed. J. B. Sumner & K. Myrbäck), **2**, part 2, p. 1074. New York: Academic Press.

PIRSON, A. (1953). Stoffwechsel organischer Verbindungen (Photosynthese). *Fortschr. Bot.* **14**, 289.

RABINOWITCH, E. (1951). *Photosynthesis and Related Processes*, **2**, part 1. New York: Interscience.

RABINOWITCH, E. (1952). Photosynthesis. *Annu. Rev. Plant Physiol.* **3**, 229.

UTTER, M. F. & WOOD, H. (1951). Mechanism of fixation of carbon dioxide by heterotrophs and autotrophs. *Advanc. Enzymol.* **12**, 41.

WITTINGHAM, C. P. (1952). The chemical mechanism of photosynthesis. *Bot. Rev.* **18**, 245.

THE PHOTOLITHO-AUTOTROPHIC BACTERIA AND THEIR ENERGY RELATIONS

HELGE LARSEN

Department of Chemistry, Norges Tekniske Högskole, Trondheim, Norway

The object of the present paper is to present a discussion on the two groups of micro-organisms known as the purple and the green sulphur bacteria. These organisms, which are able to live and thrive as true autotrophs in a strictly inorganic medium, require light energy for their growth and use carbon dioxide as carbon source. The type of photosynthesis carried out by these organisms is, however, somewhat different from that of algae and higher plants, as was shown by the masterly investigations of van Niel (1931). They do not liberate molecular oxygen during the photosynthetic process, and require the presence of special electron-donors for their assimilatory activity. Whenever these organisms are found in nature, hydrogen sulphide acts as electron-donor; its oxidation to elementary sulphur and sulphate is concomitant with the reduction of carbon dioxide to cell material. Under laboratory conditions a number of reduced sulphur compounds may be substituted for hydrogen sulphide.

The behaviour of the photo-autotrophic bacteria towards sulphur compounds resembles to a certain extent that of the colourless sulphur bacteria, which also make use of sulphur compounds as electron-donors. In the latter group of organisms the function of the electron-donor is that of an energy producer, and one is led to wonder if the electron-donor may not play a similar role in the metabolism of the purple and the green sulphur bacteria. This problem has from time to time been raised in discussions on the photo-autotrophic bacteria, but the various views expressed have for the most part been based upon rather meagre experimental material. It may be of some interest to devote part of the following discussion to this particular problem. For the rest, I shall attempt to present a picture of the present status of the general microbiology of these organisms. One particular aspect of their biochemistry, namely, their utilization and transformation of organic compounds, is treated in detail by Dr Elsden (p. 202).

The following discussion is based on the experience gained by the writer when working in Prof. van Niel's laboratory a few years ago.

I would like to avail myself of this opportunity to express my gratitude to Prof. van Niel for introducing me to this fascinating group of micro-organisms, and for arousing my interest in the problems of their microbiology and biochemistry through stimulating discussions.

GENERAL MICROBIOLOGY

The purple- and green-coloured bacteria have attracted the attention of microbiologists for more than a century. This is especially true of the purple bacteria, which seemed to lend themselves to experimentation with greater ease than did the green bacteria and which have been the object of several extensive studies. It soon became evident that these organisms are ubiquitously distributed in nature in mud and stagnant water containing hydrogen sulphide, and may, under proper conditions, give rise to quite spectacular mass developments in such localities. It may suffice at this point to refer to the interesting descriptions of the 'Lago di Sangue', whose name is derived from its blood-red colour which, according to the studies of Forti (1932), is due to the presence of enormous numbers of purple sulphur bacteria. Butlin (1953) has made similar observations on mass developments, especially of purple bacteria which give a distinct reddish colour to the water and shoreside of sulphur lakes in North Africa.

The purple sulphur bacteria

The first successful attempts to culture purple bacteria from waters containing hydrogen sulphide were made by Winogradsky (1888). He devised an ingenious though extraordinarily simple method by which enrichment cultures of these bacteria could be obtained from a variety of natural inocula. His arrangement, known as the 'Winogradsky column', merits a somewhat detailed discussion because even to-day it is one of the simplest methods for demonstrating the ecology of a number of different organisms belonging to the society which Lauterborn so aptly called 'die sapropelische Lebewelt'.

Winogradsky mixed mud samples with plaster of Paris and some organic material, e.g. straw, and placed the mixture in a tall glass cylinder which was subsequently filled completely with well water. When exposed to light, the significance of which Winogradsky did not realize at that time, these cylinder cultures gave rise to the development of colonies of purple sulphur bacteria.

The principle underlying the 'Winogradsky column' is as follows: First, heterotrophic organisms, e.g. cellulose decomposers, attack the organic material and produce, under the prevailing anaerobic conditions, carbon dioxide and simple organic compounds such as lower fatty acids.

The fatty acids in turn serve as carbon sources for a secondary flora of sulphate-reducing bacteria which produce hydrogen sulphide from the plaster of Paris. Thus, a continuous production of carbon dioxide and hydrogen sulphide ensues, and these substances, together with light, constitute the essential factors for the growth of the purple bacteria which develop as a tertiary flora.

Anyone who has made use of the 'Winogradsky column' will have noticed the characteristic changes gradually taking place in the microbial population in the cylinder. Colonies of sulphate-reducing bacteria first make their appearance: they reveal themselves as black spots in the mud mixture provided that traces of ionic iron are present. Once hydrogen sulphide production has started, colonies of purple and also of green sulphur bacteria appear as spots of the appropriate colour; these generally develop against the wall of the glass cylinder. In the upper part of the column, in the region where hydrogen sulphide comes into contact with air, an abundant development of colourless sulphur bacteria belonging to the genera *Beggiatoa*, *Thiothrix* and *Thiobacillus* may take place. Colonies of these organisms will also develop in the dark. In the light, various types of blue-green algae may flourish in the same region where the colourless sulphur bacteria are found. At the very top of the column, where a large amount of air is accessible and hydrogen sulphide is absent, diatoms, green algae and protozoa develop, although generally somewhat later than the aforementioned organisms. For closer examination or subculturing of the various types of organism, material is brought up from a chosen 'spot' by means of a Pasteur pipette.

The 'Winogradsky column' is certainly reliable as a means of obtaining visible growth of the photo-autotrophic bacteria from natural inocula. It is not, however, an ideal enrichment culture in the true sense of the word, on account of the many other types of micro-organism developing in the column. A good enrichment culture ought to provide a population in which at least 90 % of the organisms are of the type being sought. This is readily attained by the so-called 'bottle method', first introduced by Beijerinck (1893) and later improved and extensively used by van Niel (1931). The methodology is simple; a glass bottle is filled with a medium at the desired pH and containing sulphide, bicarbonate and nutrient salts, and is inoculated with a suitable amount of mud. The bottle is then tightly stoppered so that no air is left inside, and incubated in a light cabinet. After a few days the development of the photo-autotrophic bacteria is visible to the naked eye. Purple sulphur bacteria develop preferentially when the pH of the medium is originally adjusted to about 8·5, while green sulphur bacteria predominate in cultures with an initial pH of about 7·0. From the bottle stage it is a relatively simple matter to obtain pure cultures of the organisms by the shake-culture method devised by van Niel (1931).

The purple sulphur bacteria, or Thiorhodaceae, can utilize hydrogen sulphide in a photosynthetic process. The hydrogen sulphide is oxidized to sulphur and sulphate, the sulphur usually accumulating as droplets inside the cells. Within the group, one finds a large number of seemingly unrelated structural forms: cocci, sarcinas, rods, vibrios, spirilla, and a multitude of irregular, swollen and club-shaped structures. The cells may occur singly or combined into aggregates of different size and shape; they may be motile or non-motile. It was reported at an early date that transition forms between the extreme types can easily be detected, and this point soon became an issue in the rather heated discussions going on during the second half of the last century about the possible pleomorphism of bacteria. When Winogradsky started his work on the sulphur bacteria in de Bary's laboratory in 1885, de Bary was deeply engaged in this problem. He strongly defended the monomorphistic doctrine, and his view came to have a marked influence on Winogradsky. This point is of considerable interest to us even to-day, because the division of the group of purple sulphur bacteria into genera and species proposed by Winogradsky in 1888, on the basis of his morphological studies, is still in use and widely accepted.

Winogradsky (1888) used a quite unique procedure for his morphological studies of these bacteria. He transferred a small colony, containing only a few cells, to a drop of water containing hydrogen sulphide, placed between a slide and a cover-slip. By continually watching the cells through the microscope he observed that they accumulated sulphur globules and multiplied as long as hydrogen sulphide was present, and that no development took place in the absence of hydrogen sulphide. By repeated addition of water containing hydrogen sulphide he was able to 'grow' the bacteria in such slide cultures, and was thus able to follow, under the microscope, the development and multiplication of the cells. From his extensive studies, performed on material obtained from the 'enrichment column' and from natural sources, he concluded that several distinct and constant types could be recognized, and proposed that the group of purple sulphur bacteria be divided into twelve genera on the basis of their cell shape and their tendency to form characteristic aggregates during growth. The various species were differentiated mainly by the size and colour of the individual cells.

During the forty years that followed, most students of the purple sulphur bacteria did not seem to meet any great difficulties in adapting their observations to the taxonomic key which Winogradsky had proposed. It is true that this rather orderly picture was somewhat disturbed by Molisch (1907), who claimed that all purple bacteria use organic material for growth and do not require hydrogen sulphide, but Buder (1919) pointed out that one should distinguish between two fundamentally different groups of purple bacteria: (1) the purple sulphur bacteria (Thiorhodaceae) which use hydrogen sulphide for growth in an autotrophic manner, and (2) the purple non-sulphur bacteria (Athiorhodaceae) which cannot use hydrogen sulphide but require organic material for growth. The bacteria studied by Winogradsky naturally belong to the former group.

Our knowledge of the group of purple sulphur bacteria has in the meantime been greatly expanded by the investigations of van Niel (1931). His methods of culturing the organisms made it possible to perform morphological studies with pure cultures under conditions of much better control than those employed by Winogradsky. The results of these studies clearly show that the morphological invariability of the purple sulphur bacteria is not as pronounced as had been claimed by Winogradsky. The investigations made by van Niel on a strain of purple sulphur bacteria designated as a 'Chromatium-type' may be cited as an example. According to Winogradsky the genus *Chromatium* comprises the purple sulphur bacteria which occur as motile, ovoid to kidney-shaped single cells. Winogradsky recognized four or five species, mainly differing in size of the cells, the smallest measuring 1 by 2 μ., and the largest 6 by 7–15 μ. The particular strain studied by van Niel was cultured in liquid media with different sulphide and hydrogen-ion concentrations. Depending on the external conditions, cells greatly differing in size developed in the different cultures, including all the forms which Winogradsky had differentiated as five separate species. In old cultures the cells tended to settle and form aggregated masses. According to Winogradsky, most genera of the purple sulphur bacteria are defined by their characteristic aggregate formation. However, in his pure culture studies on the 'Chromatium-type' van Niel recognized, in various parts of the same sediment, aggregates which might be considered as members of Winogradsky's genera *Lamprocystis*, *Amoebobacter*, *Thiodictyon* and *Thiopolycoccus*. Other strains of purple sulphur bacteria, designated as 'Thiocystis-type', showed a similar behaviour. In old cultures various aggregates were found which corresponded to Winogradsky's definition of the genera *Thiocystis*, *Thiopedia*, *Amoebobacter*, *Thiocapsa*, *Thiothece* or *Thiopolycoccus*.

Thus van Niel clearly showed that the taxonomic key proposed by Winogradsky is of doubtful diagnostic value, but he hesitated to advocate its abandonment. He pointed out that a far more detailed investigation was needed before a better classification could be devised. Such an investigation has not been undertaken as yet, and the unsatisfactory state of the taxonomy still remains a major problem in the microbiology of these organisms.

The green sulphur bacteria

Many early investigators of the purple sulphur bacteria reported the occurrence of small green-coloured bacteria in their crude cultures. Winogradsky (1888) believed that these organisms were minute algae, which carried out a normal photosynthesis and furnished the molecular oxygen necessary for the oxidation of hydrogen sulphide by the purple sulphur bacteria. However, later investigations by Nadson (1912) indicated that these green-coloured bacteria do not produce oxygen, although they contain a pigment which is undoubtedly quite closely related to the plant chlorophylls.

When van Niel (1931) started his investigations on the purple sulphur bacteria, he corroborated the earlier reports about small green-coloured organisms being present in the crude cultures. Furthermore, he showed that at high sulphide concentrations and at neutral pH these green bacteria develop even faster than the purple bacteria. Thus, a rather selective medium could be devised for enrichment cultures of green bacteria from natural inocula. van Niel's findings indicated that the purple sulphur bacteria and the green bacteria were physiologically closely similar, and further studies on pure cultures, obtained by the same methods as those used for the purple sulphur bacteria, fully substantiated this viewpoint. The green bacteria oxidize hydrogen sulphide to elementary sulphur in a photosynthetic process during carbon dioxide assimilation. As in the case of the purple sulphur bacteria, no release of oxygen can be detected during the photosynthetic process, even by the most sensitive methods.

The green sulphur bacteria have been considered more difficult to culture than the purple sulphur bacteria. This belief stems partly from the observation that members of the group of purple sulphur bacteria quite frequently develop in liquid enrichment cultures started for growth of green sulphur bacteria. This phenomenon has recently been discussed in detail (Larsen, 1952, 1953), and it was concluded that it might be due to the presence of a vastly greater number of purple than green sulphur bacteria in the inoculum. In such cases it is advisable to use a 'Winogradsky column' as a first step in the enrichment. The 'Winogradsky column' has the advantage of supplying a semisolid substrate, isolating in space the colonies developing at the interface between the mud and wall of the glass cylinder. Even though a large number of purple colonies may appear, one can often observe discrete, localized mass development of green bacteria. Material from such green patches can easily be collected by means of a Pasteur pipette and used as inoculum for a liquid enrichment culture. This inoculum, containing a much greater number of green than purple sulphur bacteria, ensures a successful liquid enrichment culture.

Only two species of green sulphur bacteria have been studied in pure culture, both belonging to the genus *Chlorobium* (Larsen, 1952, 1953). Under optimum growth conditions these organisms appear as non-motile, small ovoids to short rods, measuring 0·7 by 0·9–1·5 μ., and excrete elementary sulphur outside the cells in media containing sulphide. Under less favourable conditions, e.g. at high sulphide concentrations and in slightly acid media, involution forms appear, revealing themselves as irregular and often considerably elongated rods. Old cultures, and also young cultures grown under unfavourable conditions, usually contain clumped masses at the bottom of the culture bottle.

The two species of the genus *Chlorobium*, viz. *C. limicola* and *C. thiosulphatophilum*, both oxidize hydrogen sulphide completely to sulphate through elementary sulphur as an intermediate stage. They are, however, distinguishable from each other in that only the latter species is able to utilize thiosulphate and tetrathionate in the photosynthetic process.

In the 6th edition of *Bergey's Manual* (1948) the group of green sulphur bacteria is treated as a separate family, the Chlorobacteriaceae, provisionally comprising six genera. The division into genera is based partly upon an alleged symbiosis of certain types of green bacteria with other organisms, and partly upon the tendency of the bacteria to form characteristic aggregates. The genus *Chlorobium* is the only one comprising bacteria not united into aggregates. Apart from the members of the genus *Chlorobium* none of the other green bacteria have been studied under controlled conditions, and their description and classification are based solely upon cursory observations made on material collected from natural habitats. Obviously, a classification based upon such fragmentary reports is very unsatisfactory, and can be no more than provisional. Furthermore, it should be borne in mind that Pringsheim (1949) has recently presented arguments to the effect that several of the organisms now classified among the true green sulphur bacteria may in fact be minute blue-green algae which have not yet been sufficiently studied to permit a decision as to their proper taxonomic position.

PHYSIOLOGY AND ENERGY RELATIONS

Winogradsky's studies of the sulphur bacteria (1887, 1888) led him to his well-known postulate of the existence of the chemosynthetic mode of life, i.e. the ability of an organism to build up its cell material using energy derived from the oxidation of inorganic substances. He suggested that the purple sulphur bacteria use energy derived from the oxidation of hydrogen sulphide, the oxidation being carried out in two steps: first to elementary sulphur and then to sulphate.

Several years earlier Engelmann (1882) had argued that the purple bacteria are photosynthetic organisms, because the presence of light seemed to be essential for their activity. Winogradsky was aware of Engelmann's hypothesis, but in view of his preoccupation with the essential role of hydrogen sulphide Winogradsky discarded the idea that light may be used by these organisms as an energy source.

Subsequently, Winogradsky's concept of a chemosynthetic mode of life was fully substantiated, first of all by his own brilliant investigations of the nitrifying bacteria, and later on by work on the hydrogen bacteria and the colourless sulphur bacteria. The metabolism of the purple bacteria was, however, not clearly understood. Some investigators believed that these micro-organisms carried out some sort of photo-synthetic metabolism, others that they lived by chemosynthesis. Buder (1919) even suggested that these organisms are equipped with both a photosynthetic and a chemosynthetic apparatus, operating independently of each other under different conditions.

This was the situation when van Niel (1931) started his work on the purple and green sulphur bacteria. He first proved beyond doubt that both light and reduced sulphur compounds are required for growth of these organisms, and that they can use carbon dioxide as the only carbon source; hence they must be regarded as true autotrophs. Measuring the transformations taking place in his bottle cultures by chemical analyses, he was able to show simple stoicheiometric relationships between the amounts of carbon dioxide assimilated and of hydrogen sulphide oxidized. Thus, the conversions taking place in cultures of the green sulphur bacteria could be expressed by the following equation:

$$CO_2 + 2H_2S \xrightarrow{\text{light}} (CH_2O) + 2S + H_2O; \qquad (1)$$

similarly, those in cultures of purple sulphur bacteria by the equation

$$2CO_2 + H_2S + 2H_2O \xrightarrow{\text{light}} 2(CH_2O) + H_2SO_4. \qquad (2)$$

Hydrogen sulphide acts as electron-donor and carbon dioxide as electron-acceptor. Molecular oxygen, claimed by Winogradsky to be used in the oxidation of hydrogen sulphide, and by Engelmann to be released during the photosynthetic process, is neither consumed nor evolved; the bacteria behave as obligate anaerobes.

The above equations present a striking resemblance to the classic equation for green plant photosynthesis:

$$CO_2 + 2H_2O \xrightarrow{\text{light}} (CH_2O) + O_2 + H_2O. \tag{3}$$

The main difference is that in the equations for the bacterial photosyntheses hydrogen sulphide takes the place of water in the equation for green plant photosynthesis; and instead of water being oxidized to molecular oxygen, hydrogen sulphide is oxidized to elementary sulphur or sulphate.

These findings induced van Niel (1931, 1935) to develop the general concept of photosynthesis, which is most simply expressed by the equation

$$CO_2 + 2H_2A \xrightarrow{\text{light}} (CH_2O) + 2A + H_2O, \tag{4}$$

where H_2A represents an electron-donor which is different in the different types of photosynthesis. In green plants water acts as electron-donor, whereas in the purple and green sulphur bacteria molecular hydrogen or any one of a number of reduced sulphur compounds may play this role. The purple sulphur bacteria can also make use of a number of organic compounds as electron-donor, as do the purple non-sulphur bacteria.

The introduction of a general formulation of the photosynthetic process, based upon studies of bacterial photosyntheses, also proved a very significant contribution with respect to an understanding of the mechanism of green plant photosynthesis. At the time when the above studies were presented, it was commonly believed that the oxygen evolved in green plant photosynthesis comes from carbon dioxide. It was pointed out by van Niel (1931, 1935), however, that a logical consequence of his comparative studies of bacterial photosyntheses would be that in green plant photosynthesis the oxygen evolved is derived exclusively from water by photolysis of the water molecule, and not from carbon dioxide. Subsequently, this postulate was amply confirmed by other approaches, and to-day it is one of the foundation stones upon which our knowledge of the photosynthetic mechanism is built.

It seems worth while to emphasize the line of thought which led van Niel to the above conclusion, because it furnishes one of the finest examples ever given of the use of 'comparative biochemistry' as a tool in the elucidation of metabolic processes. The concept of 'comparative biochemistry' implies the search for a common denominator among different though related metabolic processes. This denominator may provide the very key to an interpretation of a phenomenon which may not otherwise be considered. In the present instance the electron-donor expressed by the symbol H_2A appears to be such a denominator, and when it had been firmly established that a number of different photo-synthetic processes could be described by such a generalized equation it seemed highly probable that it could also be applied to green plant photosynthesis where a direct approach to the problem of the origin of the molecular oxygen was not available at that time. The correctness of the inference drawn from comparative biochemical considerations was not established until ten years later as a result of the use of isotopic tracers and the epoch-making discovery of Hill that isolated chloroplasts use absorbed light energy to split water and yield oxygen.

The above investigations of the purple and green sulphur bacteria led to the recognition of the true photosynthetic nature of these organisms, and explained the participation of hydrogen sulphide as electron-donor in the photosynthetic process. Naturally, several new problems arose from the insight thus gained, and one of these, namely, the problem of the energy relations in the bacterial photosyntheses will now be dis-cussed in some detail.

If the various types of photosynthesis are considered from a thermo-dynamic viewpoint, it is clear that the theoretical energy requirement may vary greatly, depending on the chemical nature of the electron-donor. For example, green plant photosynthesis (equation (3)) is a strongly endergonic process, proceeding with an increase in free energy of about 115,000 cal. per mole of carbon dioxide converted to cell material. The energy necessary to make the reaction proceed from left to right is furnished by the light absorbed by the plant chlorophylls. Photosynthesis according to equation (1), however, proceeds with an increase in free energy of only about 15,000 cal. per mole of carbon dioxide. This means that a bacterial photosynthesis, making use of hydrogen sulphide as electron-donor by converting it to elementary sulphur, theoretically requires for the assimilation of carbon dioxide only about one-eighth of the amount of light energy required by green plants for the same purpose.

It appears from these considerations that in the photosyntheses of the purple and green bacteria with hydrogen sulphide as electron-donor, an amount of chemical energy is involved which is not available for green plant photosynthesis. This amount of chemical energy varies with the nature of the electron-donor, as well as with the degree to which it is oxidized. It is true that this chemical energy is not sufficient to make the organisms theoretically independent of light energy, i.e. to make carbon dioxide assimilation proceed in the dark, but it certainly reduces the theoretical light requirement to a considerable extent as compared with the requirement for cell synthesis in green plants.

Such theoretical considerations bring to mind the suggestion of Buder (1919) that the photosynthetic bacteria may possess both a photosynthetic and a chemosynthetic apparatus, each contributing a certain amount of energy to the synthesis of cell material. The value of this reasoning is further strengthened by the fact that certain purple bacteria, which under anaerobic conditions are able to carry out photosynthesis according to the equation

$$CO_2 + 2H_2 \xrightarrow{\text{light}} (CH_2O) + H_2O, \tag{5}$$

can also grow in the dark at the expense of the oxidation of hydrogen, provided molecular oxygen is present (van Niel, 1943). In the latter case the bacteria behave as true chemosynthetic organisms, making use of the energy derived from the oxidation of hydrogen by oxygen for the assimilation of carbon dioxide. It thus appears that these organisms do in fact possess the ability to carry out chemosynthesis when *oxygen* is the electron-acceptor in the transformations producing energy, and it might seem reasonable to postulate that a similar process may also operate when carbon dioxide is the ultimate electron-acceptor, even if the latter process is insufficient to furnish *all* the energy needed for cell synthesis. The deficit in the necessary energy could then be derived from light.

This line of reasoning has been adopted by some investigators. Blum (1937), for example, developed an attractive hypothesis on the evolution of photosynthesis which was, indeed, based upon the assumption that in the bacterial photosyntheses, chemical energy is used to supplement the light energy involved in green plant photosynthesis. He postulated that the bacterial photosyntheses might represent intermediate evolutionary stages between chemosynthesis and green plant photosynthesis, the photosynthetic organisms having evolved from the chemosynthetic ones. The latter organisms carry out a synthesis of cell material from carbon dioxide in the dark, coupled with the oxidation of an inorganic substrate. The extent of the oxidative process is such that the decrease in free energy amply suffices to cover the requirements of the assimilatory reactions. In the photosynthetic bacteria, on the other hand, the amount of carbon dioxide assimilated along with the oxidation of a unit quantity of inorganic substrate is very much greater, and the chemical energy released by the oxidation process is inadequate to account for the extent of carbon dioxide assimilation. Hence an additional amount of energy has to be supplied in the form of radiant energy. Thermodynamically, a reaction corresponding to equation (1) requires an energy supply of 14,700 cal.; this could be furnished by a single mole quantum of red light, representing about 40,000 cal. per mole of carbon dioxide assimilated, and a photosynthetic reaction of this sort would therefore be possible as a 1-quantum process. Since green plant photosynthesis requires a theoretical minimum of 3 quanta per molecule of carbon dioxide assimilated, an evolution of green plant photosynthesis from chemosynthesis via 1- and 2-quanta photosyntheses could be envisaged. Thus Blum argued that the evolution of green plant photosynthesis might have developed from chemosynthetic bacteria, through green sulphur bacteria (1-quantum) and purple sulphur bacteria (2-quanta photosynthesis). It will be clear that this reasoning implies that the chemical energy released in the oxidation of the inorganic substrate must be fully utilized in the metabolism of the photosynthetic bacteria.

Meanwhile another line of thought has been developed by van Niel (1941, 1949 a, b), who reaches a different conclusion as to the function of the electron-donor in the bacterial photosyntheses. His arguments are based upon the experimentally established fact that the photosynthetic bacteria are also equipped with enzyme systems capable of oxidizing the electron-donor in darkness. Such an oxidation in the dark has been shown to occur in a number of cases, provided an electron-acceptor *other than carbon dioxide* is available to the bacteria, e.g. molecular oxygen, methylene blue, quinone, etc. This makes it unlikely that the electron-donor itself participates in the photochemical reaction, and rather suggests that it takes part in dark reactions with some electron-acceptor *furnished by* the photochemical reaction. Furthermore, there are ample reasons for postulating that the actual conversion of carbon dioxide to cell material is also accomplished exclusively by means of a series of dark reactions (van Niel, 1949 a). This leaves only one component of the chemical environment free to participate directly in a photochemical reaction, namely, water. If this be accepted, then one might envisage a photochemical reaction in the bacterial photosyntheses quite similar to that occurring in green plant photosynthesis, viz. a decomposition of water into a reducing and an oxidizing component. The reducing component furnishes the energy necessary to convert carbon dioxide into cell material in a series of dark reactions, while the oxidizing component is reduced by the electron-donor, also in a series of dark reactions. In the special case of green plant photosynthesis, in which an external electron-donor other than water is not necessary, the oxidizing component formed during the photochemical process undergoes a series of spontaneous transformations, resulting in its reduction with the simultaneous liberation of molecular oxygen. In the bacterial photosyntheses, however, the oxidizing component needs an external electron-donor for its reduction. van Niel (1949 b) has recently summarized his views in an instructive diagram, which is formulated as follows:

$$
H_2O \xrightarrow[\substack{\text{pigments}\\ \text{enzymes}}]{\text{light}}
\begin{cases}
\cdot H & \begin{pmatrix} E' \\ E'H \end{pmatrix} \begin{matrix} \diagdown CH_2O + H_2O \\ \diagup CO_2 \end{matrix} \begin{matrix} \diagdown CH_2O + H_2O \\ \diagup CO_2 \end{matrix} \begin{matrix} \diagdown CH_2O + H_2O \\ \diagup CO_2 \end{matrix} \\[2em]
\cdot OH & \begin{pmatrix} E''OH \\ E'' \end{pmatrix} \begin{matrix} \diagdown H_2A \\ \diagup A + H_2O \end{matrix} \begin{matrix} \diagdown \\ \diagup O_2 + H_2O \end{matrix} \begin{matrix} \diagdown H_2S \\ \diagup H_2SO_4 \end{matrix}
\end{cases}
$$

Dark reactions

Photochemical General | General | Example: H_2O as electron donor | Example: H_2S as electron donor

The consequence of the above line of thought is that the electron donor does not take part in the chain of events involved in the *direct* reduction of carbon dioxide, and as a result, the energy which might possibly be derived from its oxidation would be wasted for the formation of cell material.

A decision as to which of the above explanations for the function of the electron-donor in the bacterial photosyntheses is to be preferred can be reached by determining experimentally the minimum light-energy requirement of various bacterial photosyntheses. Such determinations have been carried out both with purple sulphur bacteria (Eymers & Wassink, 1938; Wassink, Katz & Dorrestein, 1942) and green sulphur bacteria (Larsen, Yocum & van Niel, 1952; Larsen, 1953). Bacteria in a state of maximum photosynthetic activity were used in these experiments. The bacteria were irradiated with monochromatic light of a wavelength corresponding to the absorption maximum of the chlorophyll of the bacteria, the light intensity being so adjusted that it limited the rate of carbon dioxide assimilation. In a set of otherwise identical experiments different electron-donors were applied, and the extent of carbon dioxide assimilation and the amount of light absorbed were measured. From the data thus obtained it was possible to calculate the *quantum yield*, i.e. the number of molecules of carbon dioxide assimilated per quantum absorbed, or the *quantum number*, i.e. the number of light quanta needed to convert one molecule of carbon dioxide to cell material. (For a detailed discussion of the early literature on the subject and of experimental techniques see Larsen, 1953.)

In the case of the purple sulphur bacteria the experiments were carried out with a strain of *Chromatium*. The following two photosyntheses were studied:

$$CO_2 + 2H_2 \rightarrow (CH_2O) + H_2O; \quad \Delta F^{\circ}_{298} = +1700 \text{ cal.}$$
$$CO_2 + 4S_2O_3^{2-} + 3H_2O \rightarrow (CH_2O) + 2S_4O_6^{2-} + 4OH^-; \Delta F^{\circ}_{298} = +98,500 \text{ cal.}$$

It was shown that in both cases an average minimum of about 9 light quanta were needed for the reduction of one molecule of carbon dioxide to cell material; individual determinations varied between 8 and 13 quanta (Wassink *et al.* 1942). In the experiments on green sulphur bacteria, *Chlorobium thiosulphatophilum* was used as test organism, and the following three types of photosynthesis were studied:

$$CO_2 + 2H_2 \rightarrow (CH_2O) + H_2O; \quad \Delta F^{\circ}_{298} = +1700 \text{ cal.}$$
$$CO_2 + \tfrac{1}{2}S_2O_3^{2-} + \tfrac{3}{2}H_2O \rightarrow (CH_2O) + SO_4^{2-} + H^+; \quad \Delta F^{\circ}_{298} = +29,200 \text{ cal.}$$
$$CO_2 + \tfrac{2}{7}S_4O_6^{2-} + \tfrac{13}{7}H_2O \rightarrow (CH_2O) + \tfrac{8}{7}SO_4^{2-} + \tfrac{12}{7}H^+;$$
$$\Delta F^{\circ}_{298} = +30,200 \text{ cal.}$$

Again it was found that in all three cases an average minimum of about 9 light quanta were needed to convert one molecule of carbon dioxide to cell material; in this case individual determinations varied between 8 and 12 quanta.

Table 1. *Theoretical and observed quantum requirements of various bacterial photosyntheses*

Organism	Electron-donor	Electron-donor oxidized to	ΔF°_{298}/mole CO_2 (cal.)	No. of quanta needed for assimilation of 1 mole CO_2 Theoretical minimum*	Observed minimum	Thermo-dynamic efficiency (%)
Chromatium	H_2	H_2O	1,700	0·05	9	0·6
	$S_2O_3^{2-}$	$S_4O_6^{2-}$	98,500	3·0	9	33
Chlorobium	H_2	H_2O	1,700	0·04	9	0·5
thiosulpha-	$S_2O_3^{2-}$	SO_4^{2-}	29,200	0·75	9	8·3
tophilum	$S_4O_6^{2-}$	SO_4^{2-}	30,200	0·77	9	8·5

* For *Chromatium*, calculated for light of wavelength 870 mμ.; for *Chlorobium*, for light of wavelength 730 mμ.

The experimental results of these studies have been summarized in Table 1. The table also contains data on the theoretical light requirements of the various types of photosynthesis, calculated from the free-energy changes of the equations given above. It should be noted that the free-energy values of the reactions express changes at the standard state, and thus signify first approximations only. In the same table are given values for the 'thermodynamic efficiency' of the various photosyntheses, i.e. the fraction of the absorbed radiant energy which is theoretically needed to permit cell synthesis from a thermodynamic viewpoint.

It appears from the data in Table 1 that the observed minimum light requirement of the bacterial photosyntheses is, in most cases, very much higher than the theoretical. Thus, in the case of molecular hydrogen as electron-donor, less than 1 % of the absorbed radiant energy is theoretically needed for photosynthesis. It is further to be noted that when thiosulphate is oxidized only to tetrathionate about one-third of the light energy absorbed fulfils the theoretical requirement, whereas in the case when this donor is oxidized completely to sulphate less than one-tenth of the absorbed light would be theoretically needed for cell synthesis. It thus appears that most of the bacterial photosyntheses are very wasteful from the point of view of energetics, and the more so the less positive the free-energy changes of the overall reactions, or, in other words, the more wasteful the greater the amount of chemical energy liberated by the oxidation of the electron-donor.

In this connexion it is rather striking that the actual light-energy requirement of the various types of bacterial photosynthesis is constant and independent of the chemical nature of the electron-donor. The constancy of the quantum number suggests that the amount of work the light has to carry out is the same in the various bacterial photosyntheses, and following this line of thought it might be postulated that the photochemical reaction is the same in these cases. Now it is a well-known fact that a quantum number of about 9 is also the value most frequently reported as the minimum for green plant photosynthesis. Applying the concept of 'comparative biochemistry', it thus seems logical to infer that in all types of photosynthesis the same photochemical reaction is involved. That this could be interpreted as a photolysis of water is convincingly shown by green plant photosynthesis.

Moreover, the constancy of the quantum number suggests that the chemical energy released by the oxidation of an electron-donor in a bacterial photosynthesis is not utilized for the assimilation of carbon dioxide, otherwise one should expect to find a lower quantum number for a photosynthesis involving a less positive change in free energy. (For example, a lower quantum number for photosynthesis with hydrogen than for photosynthesis with thiosulphate or water as electron-donor.) Since this is not the case it seems reasonable to infer that the extra chemical energy of the bacterial photosyntheses is not available for the formation of cell material.

Obviously the above conclusions support the general scheme for photosynthesis as outlined by van Niel (1949 a, b) on the basis of other lines of reasoning. On the other hand, the contention that the bacterial photosyntheses might involve a partial 'chemosynthetic' carbon dioxide assimilation must be rejected.

REFERENCES

BEIJERINCK, M. W. (1893). Ueber Atmungsfiguren beweglicher Bakterien. *Zbl. Bakt.* (1. Abt.), **14**, 827.

Bergey's Manual of Determinative Bacteriology, 6th ed. (1948). London: Baillière, Tindall and Cox.

BLUM, H. F. (1937). On the evolution of photosynthesis. *Amer. Nat.* **71**, 350.

BUDER, J. (1919). Zur Biologie des Bakteriopurpurins und der Purpurbakterien. *Jb. wiss. Bot.* **58**, 525.

BUTLIN, K. R. (1953). The bacterial sulphur cycle. *Research, Lond.* **6**, 184.

ENGELMANN, T. W. (1882). Zur Biologie der Schizomyceten. *Bot. Ztg*, **40**, 185.

EYMERS, J. G. & WASSINK, E. C. (1938). On the photochemical carbon dioxide assimilation in purple sulphur bacteria. *Enzymologia*, **2**, 258.

FORTI, A. (1932). Il 'Lago di sangue' a Pergusa in Sicilia e la prima piaga d'Egitto. *Nat. Sicil.* N.S. **8**, 63.

LARSEN, H. (1952). On the culture and general physiology of the green sulphur bacteria. *J. Bact.* **64**, 187.

LARSEN, H. (1953). On the microbiology and biochemistry of the photosynthetic green sulphur bacteria. *K. norske vidensk. Selsk. Skr.* **1**. Trondheim: Brun.

LARSEN, H., YOCUM, C. S. & VAN NIEL, C. B. (1952). On the energetics of the photosyntheses in green sulphur bacteria. *J. gen. Physiol.* **36**, 161.

MOLISCH, H. (1907). *Die Purpurbakterien nach neuen Untersuchungen.* Jena: Fischer.

NADSON, G. A. (1912). Mikrobiologische Studien. I. *Chlorobium limicola* Nads., ein grüner Mikroorganismus mit inaktiven Chlorophyll. *Bull. Imp. Bot. Gdn St Petersburg*, **12**, 55.

VAN NIEL, C. B. (1931). On the morphology and physiology of the purple and green sulphur bacteria. *Arch. Mikrobiol.* **3**, 1.

VAN NIEL, C. B. (1935). Photosynthesis of bacteria. *Cold Spr. Harb. Symp. quant. Biol.* **3**, 138.

VAN NIEL, C. B. (1941). The bacterial photosyntheses and their importance for the general problem of photosynthesis. *Advanc. Enzymol.* **1**, 263.

VAN NIEL, C. B. (1943). Biochemical problems of the chemo-autotrophic bacteria. *Physiol. Rev.* **23**, 338.

VAN NIEL, C. B. (1949a). The comparative biochemistry of photosynthesis. In *Photosynthesis in Plants* (ed. J. Franck and W. E. Loomis), p. 437. Ames: Iowa State College Press.

VAN NIEL, C. B. (1949b). The comparative biochemistry of photosynthesis. *Amer. Scient.* **37**, 371.

PRINGSHEIM, E. G. (1949). The relationship between bacteria and Myxophyceae. *Bact. Rev.* **13**, 47.

WASSINK, E. C., KATZ, E. & DORRESTEIN, R. (1942). On photosynthesis and fluorescence of bacteriochlorophyll in Thiorhodaceae. *Enzymologia*, **10**, 285.

WINOGRADSKY, S. (1887). Ueber Schwefelbakterien. *Bot. Ztg*, **45**, 489 (no. 31–37).

WINOGRADSKY, S. (1888). *Beiträge zur Morphologie und Physiologie der Bakterien*, Heft I. *Zur Morphologie und Physiologie der Schwefelbacterien.* Leipzig: Arthur Felix.

THE UTILIZATION OF ORGANIC COMPOUNDS BY PHOTO-SYNTHETIC BACTERIA

S. R. ELSDEN

Department of Microbiology and Agricultural Research Council Unit of Microbiology, University of Sheffield

Dr Larsen (p. 186, this volume) has given a most informative account of the biology and physiology of the green and purple sulphur bacteria and, by so doing, has spared me the task of providing an account of the fundamental aspects of bacterial photosynthesis. In addition, there are the reviews by van Niel (1941, 1944, 1949, 1952), and by Gest (1951).

It is my purpose to discuss certain aspects of the utilization of organic compounds by the photosynthetic bacteria and, since most of the work has been done with one species of the Athiorhodaceae, *Rhodospirillum rubrum*, this discussion will have serious limitations.

The Athiorhodaceae grow anaerobically on media containing a wide variety of organic compounds plus carbon dioxide and a source of growth factors, provided that the cultures are illuminated. These organisms contain a chlorophyll, spectroscopically identical with that found in the Thiorhodaceae, and which is closely related to chlorophyll *a* of green plants. The action spectrum for growth coincides precisely with the absorption spectrum of the bacteriochlorophyll (unpublished experiments of Arnold quoted by van Niel, 1944); and further, the action spectrum of photosynthesis was shown by French (1937) to be identical with the absorption spectrum of the pigment. No growth occurs if the cultures are incubated anaerobically in the dark, but, on the other hand, many strains will grow aerobically in the dark.

The ability to grow in the light in the presence of organic compounds is not restricted to the Athiorhodaceae. Muller (1933) showed that the strains of Thiorhodaceae he investigated would also multiply under these conditions. When these organisms were cultured in this way Muller observed that all the substrate disappeared and most of it could be accounted for as cell carbon. If the substrate were more oxidized than cell substance, then carbon dioxide was produced, whereas if it were more reduced carbon dioxide was used. van Niel (1944) in growth experiments with the Athiorhodaceae has made similar observations and has found that some 70 % of the substrate, in this case acetate, is converted into cell material, a figure which agrees well with those of Muller for the Thiorhodaceae.

Gaffron (1933, 1935) using washed suspensions of certain of the Athiorhodaceae obtained similar results. In a study of the photometabolism of the fatty acids (acetic to nonanoic) he observed that when acetate was the substrate carbon dioxide was produced, whereas with the higher acids carbon dioxide was fixed and the amount fixed was proportional to the chain length. van Niel (1944) confirmed this in growth studies; he also made a large number of estimations of the carbon dioxide produced or consumed under *in vitro* conditions with acetate, propionate and *n*-butyrate as the substrates (van Niel, 1941). In the case of acetate 0·2 mole carbon dioxide is produced per mole of acetate used, whereas in the presence of propionate and *n*-butyrate 0·31 and 0·65 mole carbon dioxide were consumed per mole of substrate respectively. It is clear from this brief review that, in the case of organic compounds it is impossible to establish a stoicheiometric relationship between carbon dioxide fixed and substrate oxidized such as was shown by van Niel (1931) for the Thiorhodaceae and green bacteria growing upon hydrogen sulphide. None the less, many of the Athiorhodaceae can carry out a reaction between hydrogen and carbon dioxide in the light which is depicted fairly accurately by the following equation:

$$2H_2 + CO_2 \xrightarrow{\text{light}} (CH_2O) + H_2O, \tag{1}$$

where (CH_2O) is a convenient but by no means precise representation of the product or products of the reaction. This reaction was first demonstrated in the Thiorhodaceae by Roelofsen (1934) and by Gaffron (1935) using the Athiorhodaceae: Larsen has now shown that it occurs in the green sulphur bacteria. To make the picture complete, van Niel (1944) has reported that certain of the Athiorhodaceae will grow if incubated under an atmosphere containing hydrogen and carbon dioxide provided that the cultures are illuminated and that the medium contains the necessary growth factors. Thus under certain conditions the Athiorhodaceae carry out a photosynthesis similar in all respects to that of the purple and green sulphur bacteria.

In addition, Foster (1940) isolated an organism which would grow in the light with *iso*propanol as the reducing agent. This organism is of particular interest in that it was unable to metabolize the oxidation product, acetone, which accumulated in the medium. The carbon balance sheet at the end of the growth period was in agreement with equation (2):

$$2\ \begin{matrix}CH_3\\ \\CH_3\end{matrix}\!\!\diagdown\!\!\diagup CHOH + CO_2 \xrightarrow{\text{light}} 2\ \begin{matrix}CH_3\\ \\CH_3\end{matrix}\!\!\diagdown\!\!\diagup CO + (CH_2O) + H_2O. \qquad (2)$$

The original culture died and an attempt was made by Siegel & Kamen (1950) to reisolate the organism, but this failed. They did isolate an organism, a strain of *Rhodopseudomonas gelatinosa*, which utilized *iso*propanol, but only small amounts of acetone accumulated. Consequently further work on this most interesting reaction has not been forthcoming.

These observations suggest that in the light, organic compounds are in part oxidized and in part assimilated, and that the amount of carbon dioxide fixed depends on the nature of the substrate, the more reduced it is the larger the amount of the gas fixed per mole of substrate utilized. van Niel (1941) has pointed out that heterotrophic organisms also assimilate a large part of their substrate, as was shown by Barker (1936) with *Prototheca zopffii* (see also Clifton, 1946). Barker found that *P. zopffii* consumed less oxygen and produced less carbon dioxide than was required for the complete oxidation of the substrate to carbon dioxide and water, and that the deficit could be explained on the assumption that part of the substrate was built up into cell substance.

The oxidation of the substrate by photosynthetic bacteria was studied by van Niel (1941). Using *Rhodospirillum rubrum*, he compared the rate of metabolism of both acetate and propionate in the light and in the dark. The light experiments were carried out anaerobically in the presence of carbon dioxide and the cells were suspended in bicarbonate buffer; the dark experiments were performed with an air-carbon dioxide gas phase. Under conditions of saturation illumination he observed that the rate of metabolism was the same as that in the dark. In the former case he measured carbon dioxide uptake in a system buffered with bicarbonate, and since the substrate was added as its sodium salt, he was in fact measuring the rate of disappearance of the carboxyl group. In the aerobic, dark experiment the rate of oxygen consumption was measured. In both cases, the gas exchange followed the same time course, indicating that the rates of acetate utilization were identical. This led van Niel to suggest that the same system of enzymes was concerned in the activation of the substrate under both light and dark conditions, and that metabolism of the substrate was a dark reaction. In other words, light and oxygen played analogous roles in the utilization of the substrate. It was of interest that in the dark, the oxygen uptake was considerably less than that required for complete oxidation of the substrate, and since the output of carbon dioxide was also low, it appeared that a considerable amount of the substrate was assimilated. This state of affairs is at least analogous to that found in the light.

It was next found that the oxidation of acetate in the dark by starved *Rsp. rubrum* occurred only in the presence of carbon dioxide, and later, van Niel (1949) reported that carbon dioxide could be replaced by catalytic amounts of succinate, fumarate, malate and oxaloacetate but not by α-ketoglutarate, citrate or *iso*citrate. It was tempting to think that, in the dark at any rate, acetate was oxidized by a cyclic mechanism which involved these compounds as carriers or catalysts. And to this end van Niel and Barker (van Niel, 1949) investigated the oxidation of labelled acetate in the presence of unlabelled succinate and found that the succinate which they recovered was labelled. This too suggested that acetate was being oxidized by a cyclic mechanism, possibly the tricarboxylic acid cycle. The question now arises, does this mechanism of oxidation operate in the light?

Mr J. G. Ormerod and I have attempted to obtain an answer to this question. The problem was to find a reliable test. We owe to Krebs the concept of the tricarboxylic acid cycle. This mechanism was postulated on the basis of four lines of evidence: (*a*) the respiration of preparations of animal tissues was stimulated by minute amounts of each of the constituent acids of the cycle; (*b*) malonate, which inhibits competitively the succinic dehydrogenase, inhibited also the respiration of these tissue preparations; (*c*) the tissue preparations contained all the necessary enzymes in adequate amounts; and (*d*) the malonate inhibition of respiration was overcome by the addition of fumarate or malate, and under these conditions succinate accumulated in amounts equivalent to the quantity of fumarate added and at a rate consistent with the functioning of the cycle. Since the succinic dehydrogenase was inhibited, it was considered that a direct reduction could not occur and that the succinate must have been formed by a mechanism which did not involve the succinic dehydrogenase. The tricarboxylic acid cycle explains this observation.

Malonate is thus a very useful compound with which to test for this system of enzymes, but unfortunately it has no effect on the oxidation reaction of *Rsp. rubrum*; and, indeed, certain strains will use it for growth!

The effects of monofluoroacetate have been much studied in recent years, and, thanks mainly to the work of Peters and his group, the mode of action of this compound has become clear. In fluoroacetate poisoning, citrate accumulates and Peters and his colleagues (Peters, 1952) have shown that this is due to the conversion of fluoroacetate into a tri-carboxylic acid (recently identified as fluorocitric acid (Peters, Wakelin, Rivett & Thomas, 1953) which inhibits competitively the enzyme aconitase thus causing the accumulation of citrate. The formation of fluorocitric acid suggests that fluoroacetate can be activated in much the same way as acetate, and that by itself it is not inhibitory. It is the compound produced from fluoroacetate which is toxic. This limits the usefulness of this compound in metabolic studies. As far as we know, fluoroacetate behaves precisely like acetate and consequently there is no *a priori* reason for assuming that only one inhibitory compound is produced from it. It is true that, to date, the inhibitory actions of fluoroacetate can all be interpreted in terms of an inhibition of the tricarboxylic acid cycle resulting from the formation of fluorocitrate. Despite this, I am firmly of the opinion that in the case of any reaction whose detailed mechanism is unknown, it is more prudent to regard an inhibition by fluoroacetate as an indication that a C_2 fragment is involved, rather than to assume that the tricarboxylic acid cycle is playing a part in the reaction.

I have digressed a little on the mode of action of fluoroacetate because we know from the work of both the Calvin and the Gaffron groups that one of the key reactions in photosynthesis involves the combination of a C_2 compound with carbon dioxide to give a C_3 compound, probably phosphoglyceric acid; and hence it is possible that an inhibition of photosynthesis by fluoroacetate is due not to the formation of fluorocitric acid but to the formation of a fluorinated C_3 compound which inhibits either the enzyme responsible for the fixation reaction, or the further metabolism of phosphoglyceric acid. The type of reaction I have in mind is expressed by the following scheme:

$$CH_2F \ COOH \rightarrow (CF\!\!-\!\!C) + CO_2 \rightarrow (CF\!\!-\!\!C\!\!-\!\!COOH). \qquad (3)$$
$$\underset{\text{acid}}{\underset{\text{monofluoroacetic}}{}} \qquad \underset{\text{C_2 compound}}{\text{fluorinated}} \qquad \underset{\text{C_3 compound}}{\text{fluorinated}}$$

We have used *Rsp. rubrum* almost exclusively for our experiments. The cultures were grown for 40 hr. in an illuminated incubator in a medium which contains malate as the carbon source plus Difco yeast extract as the growth-factor supplement. Washed suspensions of the organism grown in this way metabolize a variety of substrates anaerobically in the light in the presence of bicarbonate and under an atmosphere of nitrogen containing 5 % carbon dioxide. The substrates we employed were acids which were added in the form of their sodium salts, and thus the major part of the carbon dioxide uptake observed is due to bicarbonate formation resulting from the destruction of carboxyl groups, and not to fixation by the cells.

We usually used sodium monofluoroacetate at a final concentration of 8.3×10^{-4} M, and it quickly became apparent that there were in fact two classes of substrate: those whose utilization was inhibited by the drug and those whose metabolism was scarcely affected. In the former group were acetate, n-butyrate, pyruvate, oxaloacetate and hydrogen plus carbon dioxide. In the latter group were succinate, fumarate,

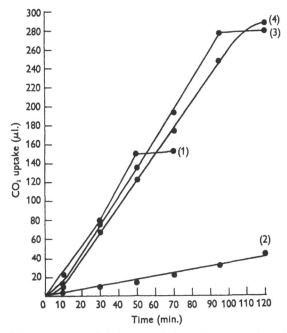

Fig. 1. Effect of fluoroacetate on the photometabolism of acetate and succinate by *Rhodospirillum rubrum*. (1) 10 μmoles acetate; (2) 10 μmoles acetate + 3.3×10^{-4} M-fluoroacetate; (3) 10 μmoles succinate; (4) 10 μmoles succinate + 3.3×10^{-4} M-fluoroacetate.

DL-malate, propionate and DL-lactate. Fig. 1 shows the progress curves for acetate and succinate in the presence and absence of the inhibitor. It will be noted that with the former compound there is a potent inhibition, whereas with the latter the inhibition is slight.

The cases of pyruvate and oxaloacetate are particularly instructive. When the metabolism of these compounds was followed manometrically the pressure changes were extremely small, even in the absence of the inhibitor, and at first sight it appeared as though these substrates were not attacked; yet, when the bicarbonate content of the medium was measured at the end of the experiment, it was found that it had increased

by an amount equivalent to the substrate added and that tests for the
presence of the substrate at the end of the experiment confirmed that it
had, indeed, been used. The explanation is simple: these substrates are
much more oxidized than the assimilation products, and the amount of
carbon dioxide produced counterbalances almost exactly the uptake of
carbon dioxide due to carboxyl group destruction. Consequently, the
effect of fluoroacetate on the photometabolism of these compounds had
to be followed by estimating directly the amount of substrate used.
Fig. 2 shows the effect of fluoroacetate on the utilization of pyruvate and

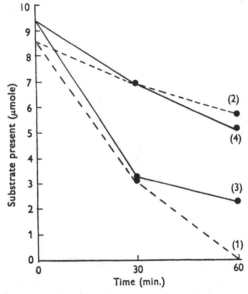

Fig. 2. Effect of fluoroacetate on the photometabolism of pyruvate and oxaloacetate by
Rhodospirillum rubrum. (1) 8·6 μmoles pyruvate; (2) 8·6 μmoles pyruvate + 3·3 × 10⁻⁴ M-
fluoroacetate; (3) 9·4 μmoles oxaloacetate; (4) 9·4 μmoles oxaloacetate + 3·3 × 10⁻⁴ M-
fluoroacetate.

oxaloacetate in the light, and it can be seen that there is a powerful
inhibition. Table 1 shows the effect of fluoroacetate on the light metabolism of a number of substrates.

It was unexpected to find that the organism reacted in different ways
to the drug depending on the nature of the substrate supplied, and in
our next group of experiments we looked for the accumulation of citrate
in suspensions poisoned with fluoroacetate. I must emphasize at this
point that our evidence that the material accumulating is citrate rests
entirely on the fact that it reacts as citrate in the method of Weil-
Malherbe & Bone (1949) as modified by Taylor (1953); on the other
hand, the specificity of the method leaves little to be desired. In this

case the results (Table 2) fell more or less into three categories. (*a*) Substrates whose metabolism was inhibited by the inhibitor but in the presence of which only small amounts of citrate accumulated; such substrates were *n*-butyrate and hydrogen plus carbon dioxide. (*b*) Substrates the utilization of which was inhibited and in the presence of

Table 1. Q_{CO_2} *(light) (expressed as μl. CO_2/mg. dry weight cells/hr.) on various substrates in presence and absence of monofluoroacetate* ($8\cdot3 \times 10^{-4}$M)

	Q_{CO_2}	
Substrate	+substrate	+substrate+fluoroacetate
Acetate	17·2	2·4
Propionate	14·8	13·3
n-Butyrate	9·7	4·3
Succinate	14·8	11·2
Fumarate	8·0	5·9
DL-Malate	8·8	7·3

Table 2. *Effect of monofluoroacetate* ($8\cdot3 \times 10^{-4}$M) *on citrate formation in the light*

	Citrate formed (μmoles)				Substrate used (μmoles)	
Substrate	No substrate	No substrate +fluoro-acetate	+substrate	+substrate +fluoro-acetate	No fluoro-acetate	+fluoro-acetate
Acetate	0·22	0·90	0·22	2·16	91	22
n-Butyrate	0·13	1·46	0·17	1·53	17	4
Pyruvate	0·28	1·44	0·45	3·70	43	13
Oxaloacetate	0·30	1·03	0·78	4·22	39	26
Succinate	0·06	0·67	0·30	0·86	46	43
Fumarate	0·16	1·74	0·15	1·89	29	21
DL-Malate	0·18	1·91	0·18	2·02	45	44
DL-Lactate	0·13	1·18	0·13	1·58	24	21
Hydrogen + carbon dioxide	0·14	0·37	0·14	0·37	272 mm. mano-metric fluid	78 mm. mano-metric fluid

which significant amounts of citrate were produced. Pyruvate, oxaloacetate and acetate fell into this category. (*c*) Substrates whose utilization was only slightly inhibited and from which little or no citrate was produced. In this group were succinate, propionate, DL-lactate, DL-malate and fumarate.

So far it would appear that the utilization of certain compounds, acetate, *n*-butyrate, pyruvate, oxaloacetate and hydrogen plus carbon dioxide by illuminated suspensions of *Rsp. rubrum* is inhibited by fluoroacetate and the characteristic feature of all of them, with the exception of hydrogen plus carbon dioxide, is that they can be readily converted to C_2 compounds. This set of experiments, considered in isolation, does not permit us to draw the conclusion that these compounds are oxidized by the tricarboxylic acid cycle. However, the experiments in which an inhibition by the drug is accompanied by the accumulation of citrate are very suggestive.

Further evidence that the photometabolism of acetate is associated with certain components of the tricarboxylic acid cycle was obtained

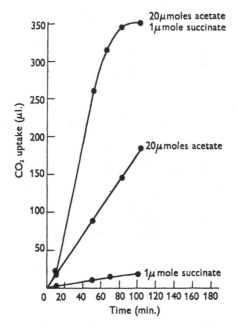

Fig. 3. Effect of succinate on the photometabolism of acetate by *Rhodospirillum rubrum*.

using cells grown on yeast autolysate. With washed suspensions of such cells it was possible to demonstrate that the rate of photosynthesis with acetate (20 μmoles) as the substrate was increased by small amounts (1 μmole) of succinate, fumarate and malate. The results of an experiment with succinate are given in Fig. 3; it is clear that the effect of the succinate is catalytic.

It will be recalled that van Niel found that the dark oxidation of acetate by *Rsp. rubrum* is stimulated by small amounts of oxaloacetate, malate, fumarate, succinate and carbon dioxide, and the similarity between the experiments quoted above and van Niel's is obvious. In view of this we examined the effect of fluoroacetate on the dark oxidations carried out by washed suspensions of *Rsp. rubrum*, and we found that the oxygen uptake in the presence of all the substrates tested (acetate, propionate, *n*-butyrate, DL-lactate, pyruvate, DL-malate, succinate, fumarate and oxaloacetate) was powerfully inhibited at a concentration of $8 \cdot 3 \times 10^{-4}$M. This was in contrast to the light experiments where the metabolism of only a few substrates was affected. The results expressed as Q_{O_2} (μl. O_2/mg. dry wt./hr.) are given in Table 3. In another series of experiments the cell suspensions were analysed for the presence of citrate at the end of the experimental period, and here we found that citrate accumulated in the presence of fluoroacetate with the following substrates: succinate, fumarate, DL-malate, pyruvate, oxaloacetate DL-lactate and propionate; but not with acetate and *n*-butyrate (Table 4). The relatively small yield of citrate from pyruvate may be due to the fact that the experiments were carried out in the presence of a carbon

Table 3. *Dark oxygen uptake of* Rhodospirillum rubrum *in presence and absence of monofluoroacetate* $(8 \cdot 3 \times 10^{-4}M)$

	Q_{O_2}			
Substrate	No substrate	No substrate + fluoroacetate	+ substrate	+ substrate + fluoroacetate
Acetate	4·0	1·0	14·0	1·0
Propionate	3·9	1·8	20·0	2·0
n-Butyrate	4·2	1·7	16·6	3·4
Pyruvate	4·8	1·5	17·8	3·0
Succinate	4·8	1·6	14·0	2·9
Fumarate	4·8	1·6	13·0	2·3
DL-Malate	3·4	1·7	15·0	3·4
Oxaloacetate	4·8	1·4	8·6	2·0

dioxide absorber. Finally, we carried out an experiment similar to those reported by van Niel in which he showed the catalytic effect of succinate and related compounds on the oxidation of acetate. In addition to showing the catalytic effect of succinate the experiment shows that the oxidation is inhibited by fluoroacetate (Fig. 4).

These experiments were all carried out with whole cells. Recently, Vernon & Kamen (1953) have reported that cell-free extracts of *Rsp. rubrum* contain malic dehydrogenase, succinic dehydrogenase and the malic enzyme of Ochoa, and Eisenberg (1953), working in van Niel's laboratory, has shown that extracts of dried *Rsp. rubrum* contain the following enzymes: aconitase, *iso*citric dehydrogenase, α-ketoglutaric oxidase, succinic oxidase, the condensing enzyme and pyruvic oxidase.

Thus, the cell contains the enzymic components of the tricarboxylic acid cycle; the dark oxidation of substrates is inhibited by fluoroacetate and under these conditions citrate accumulates; in the dark, the oxidation of acetate is catalysed by succinate, fumarate, malate and oxaloacetate. It would seem reasonable to conclude that, in the dark, all substrates are oxidized to the stage of a C_2 compound and that this is further oxidized by means of the tricarboxylic acid cycle. In the light, however, the situation is not so clear-cut; the selective action of fluoroacetate and the

Table 4. *Effect of monofluoroacetate* ($8\cdot3 \times 10^{-4}$M) *on citrate formation by* Rhodospirillum rubrum *under aerobic dark conditions*

	Citrate formed (μmoles)				Substrate used (μmoles)	
Substrate	No substrate	No substrate +fluoro-acetate	+substrate	+substrate +fluoro-acetate	No fluoro-acetate	+fluoro-acetate
Acetate	0·15	1·60	0·16	1·61	52	13
Propionate	0·10	1·13	0·16	2·60	34	10
n-Butyrate	0·13	1·66	0·13	1·78	37	2
Pyruvate	0·36	1·12	0·73	1·48	22	10
Succinate	0·07	2·80	0·07	4·90	48	20
Fumarate	0·16	1·82	0·13	2·99	49	9
DL-Malate	0·03	1·83	0·10	3·72	35	23
Oxaloacetate	0·32	2·30	0·77	4·45	21	5

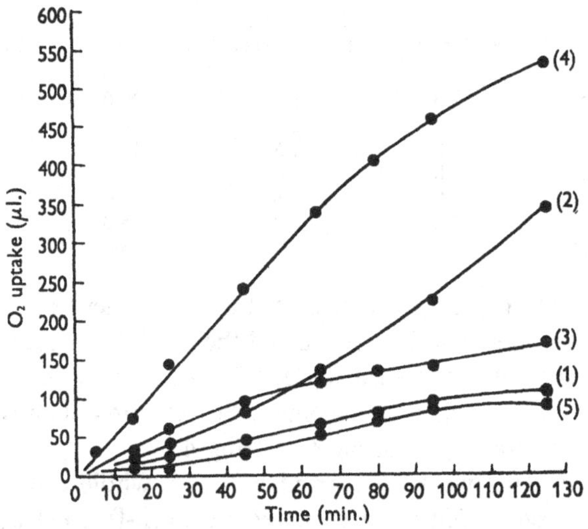

Fig. 4. Effect of succinate and fluoroacetate on the dark oxidation of acetate by *Rhodospirillum rubrum*. (1) no additions; (2) 20 μmoles acetate; (3) 1 μmole succinate; (4) 1 μmole succinate+20 μmoles acetate; (5) 1 μmole succinate+20 μmoles acetate+$3\cdot3 \times 10^{-4}$M-fluoroacetate.

fact that the hydrogen-carbon dioxide reaction is inhibited, all necessitate caution in interpretation.

This effect of fluoroacetate on the hydrogen-carbon dioxide reaction is difficult to interpret in terms of its established mode of action, i.e. by interference with the citric acid cycle. This may, therefore, be an indication that other reactions, such as that which I discussed earlier, may be inhibited. This effect obviously needs investigating further. The fact that the photometabolism of certain substrates is not affected by fluoro-acetate can be interpreted as follows. The cycle brings about the oxidation of the C_2 fragment, acetyl coenzyme A, and possibly is the mechanism by which α-ketoglutarate is synthesized. It seems to me to be unnecessary to postulate that all substrates have to be oxidized to completion in the light; indeed, all the evidence goes to show that a considerable portion of the substrate is converted into cellular material. Thus in the case of a substrate such as succinate, oxidation to the stage of pyruvate is all that is required to produce a substance which can, on paper at least, be readily converted to carbohydrate:

$$\left.\begin{array}{l} \begin{array}{l} CH_2COOH \\ | \\ CH_2COOH \end{array} \longrightarrow CH_3 . CO . COOH + 4(H) + CO_2, \\ CH_3 . CO . COOH \xrightarrow{2H} (C_3H_6O_3). \end{array}\right\} \qquad (4)$$

Additional information on the metabolism of acetate by *Rsp. rubrum* has been provided by Kamen and his colleagues and by Ehrensvärd and his group. Kamen, Ajl, Ranson & Siegel (1951) have used $^{14}CH_3COOH$ and $CH_3{}^{14}COOH$ to investigate the utilization of acetate. In the light, Kamen *et al.* found that almost all the small amount of carbon dioxide produced was derived from the carboxyl group of acetate, whereas all the methyl group and that part of the carboxyl group not converted to carbon dioxide were transformed into cellular material. On the other hand, aerobically in the dark, the carbon dioxide evolved was derived from both the carboxyl and the methyl groups of the substrate.

Glover, Kamen & van Genderen (1952) have adopted a different approach based on that used by Calvin *et al.* (1951) and Gaffron, Fager & Rosenberg (1951) for the study of carbon dioxide fixation by photosynthesizing algae. If, to a suspension of *Rsp. rubrum* using acetate at maximum rate in the light, labelled acetate is given and the reaction stopped a few seconds later, it is to be expected that the key intermediates in the oxidation of acetate will all be labelled; whereas, because of the shortness of the exposure, intermediates in secondary reactions will not. Glover *et al.* used $^{14}CH_3COOH$, and exposure times were for the most part in the range of 3–30 sec. The reaction was stopped with boiling ethanol and the soluble and insoluble fractions examined. Paper partition chromatography was used to separate the various constituents of the soluble fraction, and the labelled compounds, detected by radioautography, were eluted from the paper and their radioactivity measured.

In a light experiment in which the cells were exposed to methyl-labelled acetate and unlabelled carbon dioxide for 24 sec. the following labelled compounds were found (figures in parentheses are the counts/min.) succinate (715), α-ketoglutarate (405) and malate (170). No other labelled components of the tricarboxylic acid cycle were detected and the authors subsequently commented: 'It is of interest that in these short-term experiments on photo-assimilation of labelled acetate, no, or very little, labelling was ever observed in the citric (*iso*citric) region of the chromatogram, whereas regions corresponding to succinic acid and α-ketoglutaric acid were among the first to become labelled.'

In addition to these compounds two other areas on the chromatogram were found to be radioactive. One, adjacent to, but not identical with, the position occupied by α-ketoglutarate, and the other in the region occupied by the phosphate esters. The radioactive compound occupying the former area, called compound X, was not identified. It would seem to be an important intermediary, for it was the most active material on the paper having a count of 3030/min. It should be remembered, however, that all the figures given refer to total activity and not to specific activity.

When similar experiments were carried out with unlabelled acetate and labelled carbon dioxide it was found that, 3 sec. after adding the carbon dioxide, phosphoglyceric acid became radioactive, and was in fact the only labelled compound detected. This observation suggests strongly that the mechanism of carbon dioxide fixation is similar to that found in the algae.

The fact that the area of the chromatogram associated with citrate was not labelled when the suspensions were given labelled acetate is an argument against the view that the light oxidation of the acetate is brought about by the citric acid cycle. But, if this is so, then α-keto-glutarate must be produced from acetate by a mechanism involving neither citrate nor *iso*citrate. In contrast, when the exposure time was increased, all the catalytic acids of the citric acid cycle were labelled.

Cutinelli, Ehrensvärd & Reio (1950) cultured *Rsp. rubrum* anaerobically in the light on two media each containing acetate and bicarbonate. In one, the components were $^{13}CH_3^{14}COO^-$ and unlabelled $H^{12}CO_3^-$, and in the other $^{12}CH_3^{12}COO^-$ and $H^{14}CO_3^-$. By using these combinations it was possible to determine the fate of each of the carbon atoms. At the end of growth the cells were harvested and analysed. It was found that some 45 % of the protein carbon was derived from the methyl carbon of acetate, and about 25 % from the carboxyl carbon. The bicarbonate carbon accounted for 10 % of the protein carbon, and the remainder was thought to be derived from the yeast extract which the medium contained. Subsequently Cutinelli, Ehrensvärd, Reio, Saluste & Stjernholm (1951) and Cutinelli, Ehrensvärd, Högström, Reio, Saluste & Stjernholm (1951) examined the labelled protein in more detail. It was hydrolysed and twelve of the amino-acids were isolated. These were then degraded, and the isotopic nature and thus the origin of each carbon atom determined. The results for alanine, aspartic acid and glutamic acid are summarized in Table 5.

Table 5. *Distribution of acetate carbon and carbon dioxide carbon in amino-acids*

	C from methyl	C from carboxyl	C from carbon dioxide
Alanine	β-C	α-C	carboxyl-C
Aspartic acid	β-C	α-C	carboxyl-C
Glutamic acid	middle 3 C atoms	γ-carboxyl-C and at least 1 of middle 3 C atoms	α-carboxyl-C

The distribution of the three isotopes in alanine suggests that the carbon skeleton is produced by the carboxylation of a C_2 compound derived from acetate, and it is reasonable to suggest that pyruvate is the immediate precursor:

$$^{13}CH_3{}^{14}COOH \rightarrow (^{13}C{-}^{14}C) \xrightarrow{^{11}CO_2} {}^{13}CH_3 . {}^{14}CO . {}^{12}COOH. \qquad (5)$$

The isotopic composition of aspartic acid is consistent with its formation by the carboxylation of pyruvate (equation (6)):

$$^{13}CH_3 . {}^{14}CO . {}^{12}COOH + {}^{12}CO_2 \rightarrow \begin{array}{l} ^{13}CH_2 . {}^{12}COOH \\ | \\ ^{14}CO . {}^{12}COOH \end{array} . \qquad (6)$$

Equation (6) is a paraphrase of the Wood-Werkman reaction, but clearly the same labelling would result were the carboxylation brought about by the malic enzyme of Ochoa.

There are good reasons for believing that glutamic acid is synthesized by the amination of α-ketoglutarate. The labelling of the glutamic acid isolated is just that which one would expect were the α-ketoglutarate formed from oxaloacetate and acetyl coenzyme A with citrate and *iso*-citrate as intermediaries:

$$\begin{array}{l} ^{14}CO . {}^{12}COOH \\ | \\ ^{13}CH_2 . {}^{12}COOH \end{array} + (^{13}CH_3 . {}^{14}COOH) \rightarrow \begin{array}{l} ^{13}CH_2 . {}^{14}COOH \\ | \\ ^{14}C(OH) . {}^{12}COOH \\ | \\ ^{13}CH_2 . {}^{12}COOH \end{array} \rightarrow \begin{array}{l} ^{13}CH_2 . {}^{14}COOH \\ | \\ ^{14}CH_2 \\ | \\ ^{13}CO^{12}COOH \end{array} \qquad (7)$$

This observation is particularly important, since it implies that the α-ketoglutarate could not have been produced by a carboxylation of succinate, that is to say, by a reversal of the system of enzymes which oxidize α-ketoglutarate, for this would lead to the following types of labelling:

$$
\begin{array}{c}
{}^{14}\text{CH}_2{}^{12}\text{COOH} \\
| \\
{}^{14}\text{CH}_2{}^{12}\text{COOH} \qquad\qquad \diagup\, {}^{13}\text{CH}_2.{}^{12}\text{CO}.{}^{12}\text{COOH} \\
| \qquad\qquad\qquad +\,{}^{12}\text{CO}_2+(2\text{H})\big\langle \qquad\qquad\qquad\qquad\qquad\qquad ; \qquad (8)\\
{}^{13}\text{CH}_2{}^{12}\text{COOH} \qquad\qquad \diagdown\, {}^{14}\text{CH}_2.{}^{12}\text{CO}.{}^{12}\text{COOH} \\
| \\
{}^{13}\text{CH}_2.{}^{12}\text{COOH}
\end{array}
$$

this is quite different from that actually found. I make this point because the suggestion has frequently been made that carbon dioxide fixation in photosynthesis involves running the citric acid cycle in reverse.

The work of Ehrensvärd supports the view that, in the light, that part of the cycle leading to the synthesis of α-ketoglutarate is at work. Whether the cycle is functioning as an entity is, however, not clear. The fact that, during the photometabolism of acetate, most of the carbon dioxide produced is derived from the carboxyl group suggests that a different system is operating, but much more experimentation is required before this can be accepted as fact.

Table 6. *Origin of carbon dioxide when acetate is oxidized by one mole of oxaloacetate*

Moles acetate oxidized	Moles carbon dioxide from carbon of		
	Oxaloacetate	Acetate COOH	Acetate CH$_3$
1	2	0	0
2	3	1	0
3	3·5	2	0·5
4	3·75	3	1·25

Table 6 shows the origin of the carbon dioxide when one, two, three and four moles of acetate are oxidized in the presence of one mole of the catalytic acid (oxaloacetate) (these figures are based on the assumption that only succinate behaves as a symmetrical molecule in this system). It will be noted that even when four moles of acetate are oxidized only a small amount of the carbon dioxide evolved is derived from the methyl group of acetate. If, as would seem likely, there is a continuous removal of some of the α-ketoglutarate for synthetic purposes, then for the maintenance of a maximum rate of acetate removal by this mechanism it would be necessary to make up for the α-ketoglutarate lost by an equivalent amount of newly synthesized oxaloacetate. Such a process would accentuate the difference in behaviour of the methyl and carboxyl groups of acetate as can be readily seen from Table 6. Thus it is possible to picture a system in which a small fraction of the acetate is oxidized to completion by the cycle, and a larger amount is converted to α-keto-glutarate, to oxaloacetate and to pyruvate, all of which are required for synthetic purposes. In such a system little of the carbon dioxide produced would be derived from the methyl group of acetate. The major objection to this scheme comes from the observation of Glover *et al.* (1952) that in the early stages of the metabolism of methyl-labelled acetate no radioactivity was detected in the citrate or *iso*citrate, but α-ketoglutarate, succinate and malate were labelled. However, I do not think further progress on this problem will be made until we know more about the first steps in the metabolism of acetate. The unknown compound, described by Glover *et al.* (1952), merits much more attention, for it may well be identical with, or closely related to, the C_2 compound which is carboxylated to form phosphoglyceric acid.

Some five years ago a novel type of light reaction was discovered by Gest & Kamen (1949). They observed that if *Rsp. rubrum* was cultured in the light under certain well-defined conditions, large amounts of gas were formed. The gas was analysed and found to consist of hydrogen and carbon dioxide. The essential components of the medium were glutamate, malate and a supply of growth factors added as yeast autolysate plus biotin. The organism was grown in glass-stoppered bottles completely filled with the medium. The medium differed from all media previously used in that it contained neither ammonium salts nor materials, such as peptone, which on decomposition were likely to give rise to ammonia.

Washed suspensions of such cultures failed to produce hydrogen when illuminated no matter what substrate was examined. The gas phase used in these experiments was nitrogen. It was then found that if nitrogen was replaced by either hydrogen or helium a vigorous gas formation took place on illumination. Clearly nitrogen was in some way inhibiting the reaction. The substrates, in the presence of which hydrogen was produced, were malate, fumarate, oxaloacetate and pyruvate. Hydrogen formation was inhibited by nitrogen and also by ammonium ions (Kamen & Gest, 1949; Gest, Kamen & Bregoff, 1950).

Photoproduction of hydrogen is not restricted to *Rsp. rubrum*. Siegel & Kamen (1951) have found that *Rhodopseudomonas gelatinosa* will carry out the reaction if suitably cultured, and Bregoff & Kamen (1952) have observed a similar phenomenon in a strain of *Chromatium*. In the latter, hydrogen was formed in the presence of malate but not in the presence of thiosulphate, despite the fact that it would photosynthesize actively in the presence of the latter. The range of substrates in the presence of which hydrogen was produced was extended. Siegel & Kamen (1951) found that the presence of carbon dioxide was essential, and in the presence of this gas, hydrogen was evolved when acetate, propionate or *n*-butyrate were the substrates. Indeed, amongst the Athiorhodaceae it seems not improbable that all substrates which can be utilized for growth can also evoke hydrogen formation. The carbon dioxide effect probably explains the failure of Gest *et al.* (1950) to observe hydrogen formation in the presence of such substrates as succinate, since they carried out their experiments in the presence of carbon dioxide absorbers. The yield of hydrogen can be very large; thus Siegel & Kamen (1951) found as much as three moles of hydrogen per mole of malate.

The inhibitory action of nitrogen suggested that this gas was activated by the organisms, and Gest *et al.* (1950) followed up this clue and showed that *Rhodospirillum rubrum* could indeed fix gaseous nitrogen. Subsequently the group at the University of Wisconsin found that many, if not all, photosynthetic bacteria, including the Thiorhodaceae and the Chlorobacteriaceae, were able to utilize nitrogen gas as a source of nitrogen (Lindstrom, Burris & Wilson, 1949; Lindstrom, Tove & Wilson, 1950).

The mechanism of hydrogen formation, whether by heterotrophic bacteria such as *Bacterium coli*, or by the photosynthetic bacteria, is unknown. Further, in the case of photosynthetic bacteria we do not know the source of the hydrogen; whether it is produced from the substrate or whether from water remains to be tested. Nor can we say at the moment how this type of light reaction is related to the others I have described. All that we can be certain of is that the photoproduction of hydrogen represents an oxidation of the substrate.

The work I have so far described has been concerned with the changes which the substrates undergo both in the light and in the dark. I propose to conclude with a brief discussion of the part played by light in the metabolism of these organisms. Larsen (p. 197, this volume) gives an account of the theory advanced by van Niel (see also van Niel, 1952). The essence of this theory is that the primary reaction consists of the photolysis of water with the production of an oxidized and a reduced component. The oxidized component is the electron-acceptor (hydrogen-acceptor) in the oxidation of the substrate and the reduced component is the electron-donor (hydrogen-donor) in the reduction of carbon dioxide. This theory explains all the facts with the possible exception of the photoproduction of hydrogen; but in this case, until we know whether the hydrogen originates from water or from the substrate, there is not much advantage in speculating. The fact that photoproduction of hydrogen from an inorganic substrate has not so far been observed may indicate that hydrogen is produced from the organic substrate.

Roelofsen (1935) and Wassink (1947) have shown that when suspensions of purple sulphur bacteria are illuminated in the absence of substrate the redox potential is increased, and this is taken to indicate the formation of an oxidized component. van Niel (1949) attempted to demonstrate by chemical means the formation of an oxidized component by illuminated cells but without success. If the amount of oxidized component in the cells is small, and this may well be the case if it is continuously regenerated in the light, then perhaps the tests employed were too insensitive.

A possible clue to the nature of the oxidized component may be provided by the work of Davenport & Hill (1952) who have discovered a new cytochrome, cytochrome f, which is peculiar to chloroplasts. Its function is not known. Recently, Vernon (1953) has described the isolation, from *Rhodospirillum rubrum*, of a cytochrome whose absorption spectrum is identical with that of cytochrome c. Vernon found that *Rsp. rubrum* contains very large amounts of this compound. I have also extracted a similar pigment from *Rsp. rubrum*, though by a different method from that used by Vernon, and Dr Lascelles had told me that acetone powders of *Rhodopseudomonas spheroides* show absorption bands corresponding to those of reduced cytochrome c. The large amounts present suggest that these cytochromes play a part in the photosynthetic process and, indeed, a cytochrome would have just the properties of the oxidized component. Were this the case we could then picture the light reaction as the oxidation of a cytochrome and the reduction of an, as yet unknown, electron-acceptor brought about by the photolysis of water as shown in equation (9) (where Fe^{2+}cyt. and Fe^{3+}cyt. represent reduced and oxidized cytochrome respectively).

$$H_2O + Fe^{2+}cyt. + X \xrightarrow{\text{light}} Fe^{3+}cyt. + OH^- + XH. \qquad (9)$$

The oxidation of the substrate would then be brought about in a manner similar to that found in aerobic organisms, i.e. by the transfer of electrons to the cytochrome system. Such an oxidation could well result in the phosphorylation of adenosine diphosphate to the triphosphate by reactions analogous to those known to occur in mitochondria prepared from the tissues of higher plants and animals. If such a mechanism does occur then it would explain in a most satisfactory way the conversion of radiant energy into chemical energy. It is of considerable interest that Strehler (1953) has found that illumination of suspensions of *Chlorella* leads to the formation of adenosine triphosphate. No comparable experiments have been carried out with the photosynthetic bacteria, but Gest & Kamen (1948) have observed that illumination of suspensions of *Rhodospirillum rubrum* causes a disappearance of inorganic phosphate. It is also relevant that Vishniac & Ochoa (1952) have found that when chloroplast preparations are incubated with animal tissue mitochondria, diphosphopyridine nucleotide, adenosine triphosphate and radioactive phosphate, the adenosine triphosphate becomes labelled. But in this case the phosphorylation results from the oxidation of the reduced diphosphopyridine nucleotide (the reduced component in this particular chloroplast reaction) by the mitochondria.

Vernon & Kamen (1953) have begun a study of the light-dependent reactions of cell-free extracts of photosynthetic bacteria. They have discovered that such extracts, on illumination, will oxidize reduced cytochrome c. The reaction is dependent on the presence of oxygen, but the cytochrome oxidase is not involved because the experiments were carried out in the presence of sufficient cyanide to inhibit this enzyme. By using a large amount of reduced cytochrome c it was possible to demonstrate an approximate equivalence between the amount of cytochrome oxidized and the amount of oxygen consumed. The mechanism of this most interesting reaction has still to be explained. It was thought that perhaps hydrogen peroxide was involved, but attempts to demonstrate the presence of this substance failed. These latter experiments were carried out in the presence of ethanol and catalase which, had hydrogen peroxide been produced, would have resulted in the formation of acetaldehyde by the coupled oxidation reaction described by Keilin & Hartree (1945).

If it is assumed that XH, the reduced component, is either autoxidizable or can reduce an autoxidizable compound, and that the oxidized component, YOH, has a potential such that it can oxidize reduced cytochrome c, then the reaction can be represented as follows (equation (10)):

$$2H_2O + 2X + 2Y \xrightarrow{\text{light}} 2XH + 2YOH$$
$$2YOH + 2Fe^{2+}cyt.c \longrightarrow 2Fe^{3+}cyt.c + 2OH^- + 2Y$$
$$\underline{2XH + O \longrightarrow 2X + H_2O}$$
$$H_2O + O + 2Fe^{2+}cyt.c \xrightarrow{\text{light}} 2Fe^{3+}cyt.c + 2OH^- \qquad (10)$$

The chloroplast cytochrome, cytochrome f, has a potential more positive than that of cytochrome c (Davenport & Hill, 1952). Should

Vernon's cytochrome from *Rsp. rubrum* prove to have a potential similar to that of cytochrome *f* it would oxidize reduced cytochrome *c*. Clearly it is essential to repeat the experiments of Vernon & Kamen (1953) using the bacterial cytochrome in place of cytochrome *c*. The earlier observation made by French (1940) that cell-free extracts of photosynthetic bacteria will, on illumination, oxidize ascorbic acid, would also be explicable were a cytochrome to participate in the light reaction, for, as is well known, ascorbic acid will reduce oxidized cytochrome *c*.

The Athiorhodaceae thus seem to occupy a position somewhere between the autotrophs and the heterotrophs. The evidence I have discussed suggests strongly that in the light, organic substrates are oxidized by mechanisms similar to those found in heterotrophic organisms, radiant energy being used to make the electron-acceptor for these reactions. It is also clear that the fixation of carbon dioxide proceeds by a mechanism similar to that found in green plants. Unfortunately, most of the work has been done using one organism, *Rsp. rubrum*, and an investigation of the other Athiorhodaceae and the Thiorhodaceae is obviously needed before we can generalize with safety. The nature of the light reaction is still far from clear, but a start has been made by Vernon & Kamen (1953), and there is the hope that, in time, we shall understand how the light reaction is coupled both with the oxidation of organic compounds and with the assimilation of carbon dioxide and part of the substrate.

REFERENCES

BARKER, H. A. (1936). The oxidative metabolism of the colourless alga, *Prototheca zopffii*. *J. cell. comp. Physiol.* **8**, 231.

BREGOFF, H. M. & KAMEN, M. D. (1952). Photohydrogen production in *Chromatium*. *J. Bact.* **63**, 147.

CALVIN, M., BASSHAM, J. A., BENSON, A. A., LYNCH, V. H., OUELLET, C., SCHOU, L., STEPKA, W. & TOLBERT, N. E. (1951). Carbon dioxide fixation in plants. In *Carbon Dioxide Fixation and Photosynthesis. Sym. Soc. exp. Biol.* Cambridge University Press.

CLIFTON, C. E. (1946). Microbial assimilation. *Advanc. Enzymol.* **6**, 269.

CUTINELLI, C., EHRENSVÄRD, G. & REIO, L. (1950). Acetic acid metabolism in *Rhodospirillum rubrum* under anaerobic condition. I. *Ark. Kemi*, **2**, 357.

CUTINELLI, C., EHRENSVÄRD, G., REIO, L., SALUSTE, E. & STJERNHOLM, R. (1951). Acetic acid metabolism in *Rhodospirillum rubrum* under anaerobic condition. II. *Ark. Kemi*, **3**, 315.

CUTINELLI, C., EHRENSVÄRD, G., HÖGSTRÖM, G., REIO, L., SALUSTE, E. & STJERNHOLM, R. (1951). Acetic acid metabolism in *Rhodospirillum rubrum* under anaerobic condition. III. *Ark. Kemi*, **3**, 501.

DAVENPORT, H. E. & HILL, R. (1952). The preparation and some properties of cytochrome *f*. *Proc. Roy. Soc.* B, **139**, 327.

EISENBERG, M. A. (1953). The tricarboxylic acid cycle in *Rhodospirillum rubrum*. *J. biol. Chem.* **203**, 815.

FOSTER, J. W. (1940). The role of organic substrates in photosynthesis of purple bacteria. *J. gen. Physiol.* **24**, 123.

FRENCH, C. S. (1937). The rate of CO_2 assimilation by purple bacteria at various wavelengths of light. *J. gen. Physiol.* **21**, 71.

FRENCH, C. S. (1940). The pigment-protein compound in photosynthetic bacteria. I. The extraction and properties of photosynthin. *J. gen. Physiol.* **23**, 469.

GAFFRON, H. (1933). Über den Stoffwechsel der schwefelfrein Purpurbakterien. *Biochem. Z.* **260**, 1.

GAFFRON, H. (1935). Über den Stoffwechsel der Purpurbakterien. II. *Biochem. Z.* **275**, 301.

GAFFRON, H., FAGER, E. W. & ROSENBERG, J. L. (1951). Intermediates in photosynthesis. Formation and transformation of phosphoglyceric acid. In *Carbon Dioxide Fixation and Photosynthesis*. *Symp. Soc. exp. Biol.* Cambridge University Press.

GEST, H. (1951). Metabolic patterns in photosynthetic bacteria. *Bact. Rev.* **15**, 183.

GEST, H. & KAMEN, M. D. (1948). Studies on the phosphorus metabolism of green algae and purple bacteria in relation to photosynthesis. *J. biol. Chem.* **176**, 299.

GEST, H. & KAMEN, M. D. (1949). Photoproduction of molecular hydrogen by *Rhodospirillum rubrum*. *Science*, **109**, 558.

GEST, H., KAMEN, M. D. & BREGOFF, H. M. (1950). Studies on the metabolism of photosynthetic bacteria. V. Photoproduction of hydrogen and nitrogen fixation by *Rhodospirillum rubrum*. *J. biol. Chem.* **182**, 153.

GLOVER, J., KAMEN, M. D. & VAN GENDEREN, H. (1952). Studies on the metabolism of photosynthetic bacteria. XII. Comparative light and dark metabolism of acetate and carbonate by *Rhodospirillum rubrum*. *Arch. Biochem. Biophys.* **35**, 384.

KAMEN, M. D., AJL, S. J., RANSON, S. L. & SIEGEL, J. M. (1951). Non-equivalence of methyl and carboxyl groups in photometabolism of acetate by *Rhodospirillum rubrum*. *Science*, **113**, 302.

KAMEN, M. D. & GEST, H. (1949). Evidence for a nitrogenase system in the photosynthetic bacterium, *Rhodospirillum rubrum*. *Science*, **109**, 560.

KEILIN, D. & HARTREE, E. H. (1945). Properties of catalase. Catalysis of coupled oxidation of alcohols. *Biochem. J.* **39**, 293.

LINDSTROM, E. S., BURRIS, R. H. & WILSON, P. W. (1949). Nitrogen fixation by photosynthetic bacteria. *J. Bact.* **58**, 313.

LINDSTROM, E. S., TOVE, S. R. & WILSON, P. W. (1950). Nitrogen fixation by green and purple sulphur bacteria. *Science*, **112**, 197.

MULLER, F. M. (1933). On the metabolism of the purple sulphur bacteria in organic media. *Arch. Mikrobiol.* **4**, 131.

VAN NIEL, C. B. (1931). On the morphology and physiology of the purple and green sulphur bacteria. *Arch. Mikrobiol.* **3**, 1.

VAN NIEL, C. B. (1941). The bacterial photosyntheses and their importance for the general problems of photosynthesis. *Advanc. Enzymol.* **1**, 263.

VAN NIEL, C. B. (1944). The culture, general physiology, morphology and classification of the non-sulphur purple and brown bacteria. *Bact. Rev.* **8**, 1.

VAN NIEL, C. B. (1949). The comparative biochemistry of photosynthesis. In *Photosynthesis in Plants*. Ames: Iowa State College Press.

VAN NIEL, C. B. (1952). Bacterial photosyntheses. In *The Enzymes* (ed. J. B. Sumner and K. Myrbäck), **2**, part 2. New York: Academic Press.

PETERS, R. A. (1952). Croonian Lecture. Lethal synthesis. *Proc. Roy. Soc. B*, **139**, 143.

PETERS, R. A., WAKELIN, R. W., RIVETT, D. E. A. & THOMAS, L. C. (1953). Fluoro-acetate poisoning: Comparison of synthetic fluorocitric acid with the enzymically synthesized fluorotricarboxylic acid. *Nature, Lond.* **171**, 1111.

ROELOFSEN, P. A. (1934). On the metabolism of the purple sulphur bacteria. *Proc. Acad. Sci. Amst.* **37**, 660.

ROELOFSEN, P. A. (1935). On photosynthesis of the Thiorhodaceae. Dissertation, Utrecht.

SIEGEL, J. M. & KAMEN, M. D. (1950). Studies on the metabolism of photosynthetic bacteria. VI. Metabolism of *iso*propanòl by a new strain of *Rhodopseudomonas gelatinosa*. *J. Bact.* **59**, 693.

SIEGEL, J. M. & KAMEN, M. D. (1951). Studies on the metabolism of photosynthetic bacteria. VII. Comparative studies on photoproduction of H₂ by *Rhodopseudomonas gelatinosa*, and *Rhodospirillum rubrum*. *J. Bact.* **61**, 215.

STREHLER, B. L. (1953). Firefly luminescence in the study of energy transfer mechanisms. II. Adenosine triphosphate and photosynthesis. *Arch. Biochem. Biophys.* **43**, 67.

TAYLOR, T. G. (1953). A modified procedure for the microdetermination of citric acid. *Biochem. J.* **54**, 48.

VERNON, L. P. (1953). Cytochrome *c* content of *Rhodospirillum rubrum*. *Arch. Biochem. Biophys.* **43**, 492.

VERNON, L. P. & KAMEN, M. D. (1953). Studies on the metabolism of photosynthetic bacteria. XV. Photo-autoxidation of ferrocytochrome *c* in extracts of *Rhodospirillum rubrum*. *Arch. Biochem. Biophys.* **44**, 298.

VISHNIAC, W. & OCHOA, S. (1952). Phosphorylation coupled to photochemical reduction of pyridine nucleotide by chloroplast preparation. *J. biol. Chem.* **198**, 501.

WASSINK, E. C. (1947). Photosynthesis as a light-sensitized transfer of hydrogen. *Leeuwenhoek ned. Tijdschr.* **12**, 281.

WEIL-MALHERBE, H. & BONE, A. D. (1949). The micro-estimation of citric acid. *Biochem. J.* **45**, 377.

THE ROLE OF ACCESSORY PIGMENTS IN PHOTOSYNTHESIS

L. R. BLINKS

Hopkins Marine Station of Stanford University,
Pacific Grove, California

I. OCCURRENCE, STRUCTURE AND PROPERTIES OF ACCESSORY PIGMENTS

Since all photosynthetic organisms contain at least one type of chlorophyll, this must probably be regarded as the essential photosynthetic pigment. In most cases the type present is chlorophyll *a*, but in bacteria other chlorophylls take its place. In all plants above the blue-green algae, one other chlorophyll (*b*, *c*, or *d*) is also present, but whether or not these other chlorophylls should be considered 'accessory' is still not clear. There is also at least one yellow pigment, generally *β*-carotene, and though this pigment is widely distributed, surprisingly little is known of its function. There is almost invariably one other yellow, orange or red pigment, one of the carotenoids; sometimes there are several of these.

Boresch (1932), Cook (1945) and Strain (1951) have all given good accounts of the occurrence and chemistry of plant pigments, with special reference to the algae.

It is hardly necessary to point out the chemical nature of chlorophyll, since we are here concerned with accessory pigments. Carotene is perhaps even more familiar to biochemists, the empirical formula $C_{40}H_{56}$ being generally accepted. The carotenoids differ from carotene in having in addition one or more oxygen atoms, and a different number of hydrogen atoms depending on whether hydroxyl, carboxyl or other groups are present (Strain, 1938; Karrer & Jucker, 1950). The commonest carotenoids found in algae are: fucoxanthin, neofucoxanthin, lutein, zeaxanthin, flavoxanthin, neoxanthin, diatoxanthin, diadinoxanthin and myxoxanthin (see Strain & Manning, 1942, 1943; Strain, Manning & Hardin, 1944). The names of many of these indicate their source and have been tabulated in a convenient fashion by Strain (1951). Among the bacteria, spirilloxanthin (rhodoviolascin) of *Rhodospirillum rubrum* is perhaps best known, but there are others such as rhodopin.

Except for the isolated case of floridorubin, reported from a single red alga (Feldman & Tixier, 1947), all the other known accessory pigments belong to the group of *phycobilins*. These impart characteristic colours to the organisms; for example, phycoerythrin in the red algae and phycocyanin in the blue-green, though both are usually found in each group (Kylin, 1910, 1912). The carotenoids appear to be *loosely* bound to proteins in the cell; they are easily broken free by heat, and extracted by acetone and methyl alcohol. On the other hand, the phycobilins are easily extracted by water and remain as chromoproteins in solution. They have been favourite objects for physico-chemical study because of their stability and colour; phycoerythrin was one of the first proteins to be studied by ultracentrifugation (Svedberg & Lewis, 1928; Svedberg & Katsurai, 1929; Svedberg & Eriksson, 1932), by electrophoresis (Tiselius, 1930, 1937), and very recently by chromatographic adsorption (Swingle & Tiselius, 1951). The molecular weight is quite high, 270,000, but the molecule can be split into fractions half this size. The constitution of the phycocyanin protein has been investigated by Kitasato (1925) and Wassink & Ragetli (1952); the latter found thirteen known amino-acids and three unknown substances; no arginine was present, and the composition was not markedly different from haemoglobin or the bulk proteins of *Chlorella*.

The chromophore is split from the protein only with difficulty (e.g. with boiling hydrochloric acid in the absence of oxygen); it has been studied chiefly by Lemberg (1928, 1929, 1930). This is in marked contrast to the carotenoids which have been studied by many biochemists. Lemberg concluded that the chromophore belongs to the general class of bile pigments, with an open chain of four pyrrole rings linked by CH and CH_2 groups. Thus phycoerythrobilin (derived from phycoerythrin) is apparently identical with mesobilierythrin, which is:

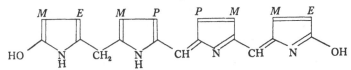

where M, E and P stand respectively for methyl, ethyl and propyl groups. It yields mesobiliverdin and mesobilirubin on heating with alkali in methanol. Phycocyanobilin is regarded as an oxidation product of phycoerythrobilin and it yields only mesobiliverdin. (For a brief review of bilin chemistry, see Lemberg & Legge (1949).)

Phycoerythrin absorbs strongly in the green part of the spectrum, giving it a lavender-red appearance to the eye. This colour is almost entirely complementary to that of chlorophyll, the two absorption spectra overlapping only slightly. There are at least three, possibly several more, phycoerythrins. The commonest, *R*-phycoerythrin, is characteristic of the higher red algae, the Rhodophyta, and it has three major absorption peaks at 495, 545 and 565 mμ. *C*-phycoerythrin, found in the blue-green algae (Cyanophyta), has but a single absorption peak at 550 mμ. A new phycoerythrin, which may be designated as B-phycoerythrin since it is found in one of the Bangiales, a primitive red algal group, has two absorption peaks, a major one at 545 mμ. and a very small one at 565 mμ. on the shoulder of the main one (cf. Koch, 1953). All of these phycoerythrins have been crystallized, and appear to be definite entities electrophoretically and chromatographically.

Two phycocyanins have to date been described. Phycocyanin C (from blue-green algae) absorbs largely in the orange and red, with a maximum at 615 mμ. A similar absorption is shown by purified phycocyanin from *Porphyra naiadum*. The higher red algae contain R-phycocyanin (sometimes very dilute), with absorption peaks at 615 and 550 mμ. There may now be some question as to whether the latter is a pure substance, since it may be a 'complex' between phycocyanin and phycoerythrin. There is increasing evidence for the latter in extracts of *P. naiadum*; an apparently homogeneous entity moves as an anion in the electrophoresis cell, even at pH 5, where phycocyanin and phycoerythrin are nearly isoelectric. At higher pH values this 'complex' breaks down (almost 70 % to phycoerythrin and phycocyanin at pH 7). A somewhat analogous behaviour of this extract is also displayed on the calcium phosphate adsorption column of Swingle & Tiselius (1951). At pH 5 and with low concentrations of acetate buffer, pure phycocyanin and phycoerythrin move successively through the column, but a purple fraction which remains on the column must presumably be highly ionized to be so strongly adsorbed. It is eluted only at high buffer concentrations and its absorption spectrum includes the peaks of both pigments.

A greater understanding of these chromoproteins is desirable. Apart from their consitituent amino-acids, molecular weight, ionic mobility, isoelectric point and solubility properties, little is known about such proteins and even less about how the proteins are attached to the bilin groups. That the linkage is firm is indicated by the difficulty with which the chromophore is detached by hydrolysis (Lemberg, 1930), but its nature has not been demonstrated. We are now investigating the possibility of phosphate groups or even of nucleic acid being involved, since the degree of ionization of some of the components is too great to be accounted for by the carboxyl groups of amino-acids.

If phycocyanin and phycoerythrin are linked in one complex or molecule in the plant, then it is feasible that energy transfer (see below) would be facilitated. Of critical importance in deciding this is the fluorescence of the 'complex' compared with its components. Work is now in progress on this.

The fluorescence of the phycobilins is one of their most striking properties. While the carotenoids (with the exception of phytofluene) are only slightly fluorescent, and the chlorophylls mildly so, phycoerythrin solutions 'glow' like yellow lamps and phycocyanin fluoresces a rich red. Strain (1951) mentions the quenching of phycoerythrin fluorescence by iodine. We have found this to be reversible, and on reduction of the iodine, fluorescence reappears. According to the work of J. H. McClendon (unpublished) a number of other substances also do this. Oxidants are not the only effective agents, and some oxidants, e.g. ferricyanide, are without effect. It is conceivable that their influence is on the protein, since tanning and precipitating agents are often effective. In many cases the quenching of fluorescence is reversible if reduction or dialysis is promptly carried out. Acidification decreases but does not by any means abolish fluorescence.

The case of a 'naturally non-fluorescent phycoerythrin' seems to be explicable in terms of both acidity and an oxidant. The red alga *Polysiphonia* yields extracts which absorb like other phycoerythrins but have feeble fluorescence (Kylin, 1931). We have been able to quench the fluorescence of other phycoerythrins with the colourless extract of this alga. A disulphonate of dibromohydroxybenzoic acid, found in *Polysiphonia* (Mastagli & Augier, 1949; Augier & Henry, 1950), may be responsible, together with a pH of about 3 in the extract.

Closely related to fluorescence is the photodynamic action of many dyes, and it is important to know whether phycoerythrin displays any such activity. A variety of enzymes were exposed to light in its presence and found to be scarcely affected. Nor was the photo-oxidation of thiourea increased by its presence. Coupled oxido-reductions such as that of riboflavin plus ascorbate (Krasnovskii, Evstigneef, Brin & Gavrilova, 1952) are not photosensitized by phycoerythrin, although they are by chlorophyll. Easily reduced dyes, such as indophenol, were likewise not photo-reduced in the presence of phycoerythrin. It will not alone carry out the Hill reaction. (In so far as they can be tested, carotenoids also seem to be inactive in most of these respects.)

These and a number of other pieces of evidence indicate that the accessory pigments are probably not direct photochemical mediators, but rather light absorbers which pass on their energy to other systems, especially to chlorophyll.

Before taking up the evidence favouring this interpretation, we should discuss the geometry and cytology of accessory pigmentation, i.e. whether the carotenoids or phycobilins are suitably located for energy transfer to other systems. In some cases they clearly are not; e.g. in the filamentous, epiphytic green algae *Trentepohlia* (which is brick-red due to β-carotene), the orange pigment is located mostly in oil droplets, readily visible under the microscope. The same is probably true in the flagellated, green algae *Dunaliella*, which is also orange when grown in saturated brine. Although the cell contents are very densely packed, much of the carotenes appears to be in oil droplets, as in *Trentepohlia*.

It would not be surprising if no energy transfer could occur from the carotene in such structures to the photosynthetic system in the plastids. This is borne out by the experimental evidence given later. In other cases, however, the location of the carotenoids is not so obvious: in bacteria and blue-green algae there is no clear separation of yellow and green (or blue and red) pigments in the ultimate 'grana', and certainly in the higher algae most of the pigments are all packed into the plastids, where no very great spatial separation is possible. Since proteins occupy a large percentage of the plastid volume, it is possible that the carotenoids and chlorophylls are crowded into mono-layers on the surfaces of protein platelets. There has been considerable speculation concerning the exact arrangement, but chemical affinities suggest that the hydrocarbon chains of the carotenoids are in close juxtaposition with the phytyl chains of the chlorophyll, or with other lipids (cf. Hubert, 1935). Arnold & Oppenheimer (1950) have stressed the necessity of small intermolecular distances between absorbing and receiving molecules. Transfer of course would be optimal with molecules in direct contact, or in chemical combination.

A carotenoid-chlorophyll-protein does occur in extracts, and it migrates homogeneously in the electrophoretic cell. But we have not yet been able to isolate a water-soluble protein complex containing chlorophyll and the phycobilins. The complex may, perhaps, break down during the extraction of pigments; if so, the process is rapid, for it is possible to injure red algae (e.g. by heat, anoxia, ammonia, or sodium hydrosulphite) so that phycoerythrin emerges from the plastids into the cytoplasm, and later into the vacuole, while the chlorophyll and carotenoids remain in the plastids. If the plastids themselves are isolated, and made to swell osmotically without bursting, the phycoerythrin can be seen in a swollen 'vacuole' or bleb at the side of the plastid, quite separate from the green grana. We have described several suspending agents for red algal plastids, notably polyethylene glycol (McClendon & Blinks, 1952), which prevent the loss of red pigment due to osmotic bursting. When so protected, these plastids display the Hill reaction. As soon as they have lost their phycoerythrin, they cease to reduce ferricyanide; but of course other substances, including enzymes, are probably lost when they burst.

II. PHOTOSYNTHETIC FUNCTION OF
ACCESSORY PIGMENTS

The presence of pigments in cells, or even in plastids, constitutes no assurance that they are photosynthetically effective. Carotenoids, for example, often play a tropistic role, as in some purple bacteria and in many motile cells or gametes. The only definitive method so far available

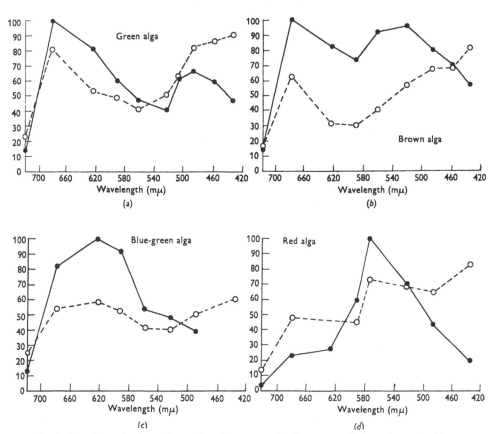

Fig. 1. Engelmann's absorption and action spectra for four types of algae, determined by his motile bacteria method: O– – –O absorption values; ●——● numbers of bacteria accumulated at the indicated parts of the spectra (redrawn from Engelmann, 1884).

for testing the light effects of pigments in cells is to determine the action spectrum of the given function. Noteworthy results have of course attended such analysis: visual purple (in vision), carotenoids or flavins (in tropisms) and haem compounds (in respiration) have all been implicated by this technique, and the same method suggests new light absorbers in photoperiodism.

With regard to chlorophyll, the situation in photosynthesis is reasonably clear. Many organisms photosynthesize well in spectral regions absorbed by chlorophyll, and little or not at all in regions transmitted by it. Before taking up the exceptions to this rule, we may indicate the organisms which adhere to it.

(a) Role of carotenoids

(i) Green algae

Seventy years ago, Engelmann (1883, 1884) exposed *Spirogyra* and other green algae to a microspectrum and found that oxygen-sensitive motile bacteria congregated chiefly in the red and blue light; photo-

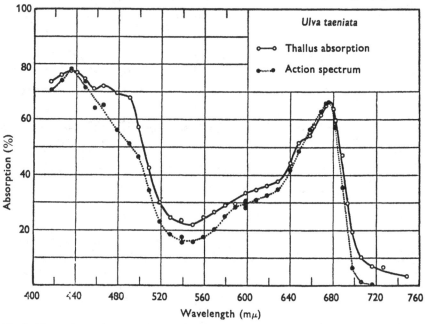

Fig. 2. Absorption and action spectra for the green alga *Ulva taeniata* (Haxo & Blinks, 1950).

synthesis was most active in the regions of chlorophyll absorption (Fig. 1a). The green alga *Ulva* has recently been studied by polaro-graphic methods with results shown in Fig. 2. Since its thallus is only two cells thick, the absorption maximum and minimum are emphasized especially well; in thicker tissues (leaves) and dense suspensions, the peaks and valleys tend to become levelled out due to repeated absorption of light by lower cells. On the whole the correspondence between absorption and action spectra is good, except for the regions around 490 mμ. and above 700 mμ., where photosynthesis is relatively lower than absorption. (A correction for a small error in the method used for measuring absorption makes the correspondence still better in the green part of the spectrum.) The ineffectiveness of light around 700 mμ. on *Chlorella* has been discussed by Emerson & Lewis (1943), who suggest that the energy per quantum at this wavelength may not be adequate for the photochemical work. This is in contrast to the effectiveness of considerably longer wavelengths of infrared light in bacterial photosynthesis (Gaffron, 1934, 1935; French, 1937; Thomas, 1950; Duysens, 1952; Larsen, Yocum & van Niel, 1952). The deviation around 480 mμ. is more relevant in discussing accessory pigments, since this is a region where carotenoids are the chief light absorbers, contributing (*in extracts*) nearly 80 % of the absorption. Obviously there is some ineffective absorption, but photosynthesis in this region does not fall by any means to 20 % of its peak value. Throughout the region 420–460 mμ. the carotenoids contribute almost 50 % of the absorption (again in *extracts*, let us emphasize; there is still no way to estimate it exactly in cells), yet photosynthesis is generally as great as in the red region, where chlorophyll is the only absorber. This action spectrum, as well as many others, forces one to conclude that most of the pigments, *including carotenoids*, are photosynthetically effective in green plants. The particular carotenoid implicated around 480 mμ. has not yet been determined. This need not be a truly ineffective pigment, but might be one stimulating photo-oxidation: the respiration rates of some tissues appear to be more enhanced immediately after exposure to this region of the spectrum than after exposures to other wavelengths.

In green plants, then, the accessory pigments contribute, though not completely, to photosynthesis. There are, however, some marked exceptions. These are the algae mentioned earlier, *Dunaliella* and *Trentepohlia*, which have high concentrations of β-carotene, mostly in oil droplets. The action spectra of these algae indicate practically no photosynthesis in the blue end of the spectrum. Sometimes, indeed, there is a slightly enhanced respiration. Clearly the β-carotene in this cellular locus (possibly also when it is in the plastid) acts as a filter, removing blue light by absorption before it can reach the chlorophyll. The ecological significance of this is not clear; both of these algae live in extremely 'xerophytic' situations (in saturated brine, or epiphytically), often exposed to very bright light and drying. Why blue light should be excluded from photosynthetic activity in such situations is difficult to understand; it is probably an accident of carotenoid storage. We have not been able to set conditions so that *Dunaliella* displays only 'trace' amounts of chlorophyll, as described by Fox & Sargent (1938), and in view of the inactivity of carotenoids in normal *Dunaliella*, it may be that such a stage is a resting one, with little or no photosynthesis. It would, however, be desirable to investigate the situation further; if photo-synthesis should be found to occur in such cells, it would be the first case in which it is possible with carotenoids as the chief substances absorbing light.

(ii) *Bacteria*

A similar ineffectiveness of certain carotenoids occurs in photo-synthetic bacteria. French (1937) showed that the photosynthesis of *Rhodospirillum rubrum* occurred only in the spectral regions absorbed by bacteriochlorophyll, but not in those absorbed by the red carotenoid, spirilloxanthin (rhodoviolascin). These spectral regions do not greatly overlap (as they do in green plants), so that the absorption of bacterio-chlorophyll in the shorter wavelengths is not interfered with, and there is active photosynthesis around 588 mμ. (the lower absorption band of this chlorophyll) as well as in the infrared. Thomas (1950), however, found that other bacterial carotenoids are photosynthetically effective. Bacteria also accumulate in spectral regions absorbed by carotenoids; but, as French has already pointed out, this may be a tropistic response.

(iii) *Diatoms (Bacillariophyceae) and brown algae (Phaeophyceae)*

The diatoms and brown algae are algae in which the function of an accessory pigment was first clearly worked out, and in which fucoxanthin can be confidently stated to aid photosynthesis. Engelmann's technique indicated its participation in the photosynthesis of diatoms (Fig. 1*b*): more oxygen was evolved in the green part of the spectrum than was the case with green algae, and this corresponds to an increased light absorp-tion in that region. Such absorption is itself remarkable, because the absorption spectra of solutions of fucoxanthin in ether and other organic solvents show no striking differences from those of other carotenoids. In the living cell the enhanced absorption between 500 and 560 mμ. is probably due to a complex between fucoxanthin and protein; the complex is broken down by heating, and this produces the well-known 'greening' of kelps and diatoms treated in this manner. It is also split by methyl alcohol.

Dutton & Manning (1941) showed by polarographic methods that the monochromatic mercury lines, violet (at 404–7 mμ.), blue (at 435 and at 496 mμ.) and green (at 546 mμ.), produced photosynthesis in diatoms almost equally in relation to the light absorbed, although the proportion absorbed by carotenoids varied from 38 to 93 %. More striking still, the efficiency was the same as for a band of red light absorbed entirely by chlorophyll. There was again a slight decrease of effectiveness in the region of the blue-green (496 mμ.) just as was found in green algae. But in general the conclusion was inescapable that light absorbed by carotenoids (especially fucoxanthin) was used about as well as that absorbed by chlorophyll. This conclusion was confirmed with diatoms by Wassink & Kersten (1946) at five spectral lines or zones, and by Tanada (1951) at some thirty points on the spectrum.

Haxo & Blinks (1950) also reported on a thalloid brown alga, *Coilodesme*, at some thirty-three wavelengths throughout the visible region of the spectrum, and the action spectrum they obtained is shown in Fig. 3. The overall correspondence between absorption and photo-

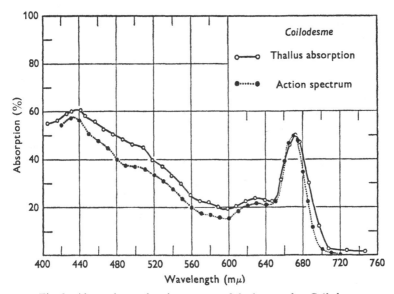

Fig. 3. Absorption and action spectra of the brown alga *Coilodesme*
(Haxo & Blinks, 1950).

synthesis is very good. It was later found that a small systematic correction is necessary in the better transmitted wavelengths of the absorption curve, and if account is taken of this, the absorption curve is brought even closer to the action curve. As with the green and blue-green algae the major deviation is again in the region of 480–500 mμ. Fucoxanthin is largely responsible for the enhanced absorption and action at wavelengths just above this region (500–560 mμ.). While Montfort (1941) has claimed an even greater activity for fucoxanthin than for chlorophyll, we can at least conclude that it is *equally* effective, and that light absorbed by it is very efficiently used in photosynthesis. The brown algae have, better than any other group, broadened the effective absorption and utilization of the solar spectrum, and it is perhaps not surprising that diatoms are the dominant algae in plankton and that the giant kelps, the largest of all algae, are the most valuable commercially.

We may turn now from the carotenoids to the phycobilins.

(b) Role of phycobilins

(i) Blue-green algae (Myxophyceae)

Engelmann's pioneering researches again pointed the direction; his motile bacteria accumulated along the orange part of a spectrum illuminating a filament of blue-green alga (Fig. 1c) and their distribution corresponded to light absorption by phyocyanin. Sixty years later Emerson & Lewis (1942) showed that in Chroococcus the light absorbed by this pigment was as effective in photosynthesis as that absorbed by chlorophyll. Both phycocyanin and chlorophyll were more effective than the carotenoids; efficiency dropped in the shorter wavelengths absorbed chiefly by these pigments. In this case we can regard phyco-cyanin as an accessory pigment, while the carotenoids apparently are not.

A few measurements made with filamentous blue-green algae (Anabaena and Oscillatoria) in my laboratory also indicate the high effectiveness of phycocyanin. However, chlorophyll and carotenoids were both much less efficient, and Duysens (1952) has reported a similar result in Oscillatoria. Why these genera differ from Chroococcus has not yet been explained; it may be concerned with methods of culture, especially light conditions. Since this situation becomes even more striking in red algae, we may now turn to this group.

(ii) Red algae (Rhodophyceae)

Engelmann again showed high oxygen production by filamentous red algae in the green part of the spectrum (Fig. 1d). This has been repeatedly confirmed, for broad spectral regions passed by glass filters or coloured solutions, by a number of later workers who used the Winkler titration procedure to determine dissolved oxygen (Ehrke, 1932; Montfort, 1934, 1936; Schmidt, 1937; Levring, 1947). Levring used a series of thirteen Schott glass filters with different 'cut-off' values, giving bands of 100–200 mμ. in width, some six or seven of which were fairly well separated. He found that photosynthesis by various red algae was sometimes two or three times as great in the green as in the blue or red regions of the spectrum. This was in agreement with some preliminary determinations made by Haxo & Blinks (1946) with monochromatic mercury blue and green light, and a narrow band of red light (the oxygen evolution being measured polarographically). The results found by all these workers with oxygen evolution were paralleled by Wurmser

(1921) and Klugh (1930) using the Osterhout-Haas indicator method for carbon dioxide.

Detailed action spectra for red algae were determined by Haxo & Blinks (1950), using the same monochromator and polarographic oxygen method as employed for *Ulva* and *Coilodesme*. The results were strikingly different from those obtained with the latter algae. In every

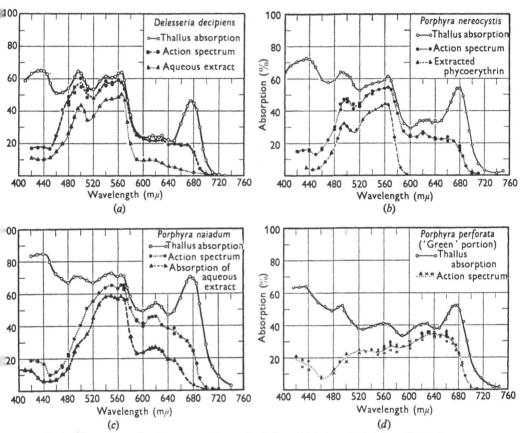

Fig. 4. Absorption and action spectra of several red algae: (*a*) *Delesseria decipiens*, (*b*) *Porphyra nereocystis*, (*c*) *Porphyra naiadum*, (*d*) *Porphyra perforata*. These have increasing amounts of phycocyanin, in the order indicated, and accordingly show increasing photosynthetic activity in the region of 620 mμ. Three absorption spectra of the aqueous extracts (phyco-bilins) are also shown (Haxo & Blinks, 1950).

red alga tested, the photosynthetic action spectrum closely paralleled the absorption of the phycobilins (Fig. 4*a–d*). When phycoerythrin was the chief phycobilin, photosynthetic peaks occurred at 495, 545 and 565 mμ. With larger amounts of phycocyanin, photosynthesis increased at 615–620 mμ.; *Porphyra perforata* showed its maximum in this region. *P. naiadum*, which has the anomalous phycoerythrin with only two absorption peaks (vide supra), reflects this situation with a two-peaked action spectrum (Fig. 4c). The extremely consistent correspondence between absorption by phycobilins and the algal action spectrum can leave little doubt of the real activity of these accessory pigments. In this respect the detailed action spectra provide confirmation of Engelmann's earliest observations.

The most unexpected results of these studies was the remarkable *inactivity* of the chlorophyll and especially of the carotenoids. Figs. 4a–d all show that the photosynthetic rate is almost at its lowest, or at least plunging rapidly toward its lowest point, at those spectral regions where chlorophyll absorbs most strongly, namely, at 435 and 675 mμ. There is sometimes a hint of an action peak at these wavelengths, though usually it is only a shoulder or 'foothill'. It should be emphasized that there is a high content of chlorophyll *a* in these algae; it compares well with that in *Ulva* (Fig. 2) and *Coilodesme*. There is also ample carotenoid. Yet in the blue end of the spectrum there is very little photosynthesis. We are forced to conclude that carotenoids are practically inactive, and surprisingly enough, that this also applies to perhaps 60–80 % (on an absorption basis) of the chlorophyll.

III. QUANTUM EFFICIENCIES

With several of the algae mentioned above, Dr C. S. Yocum (1951) has carried such action spectra studies (which indicate *relative* effectiveness only) on to an absolute basis. These are perhaps the first quantum efficiencies determined with naturally occurring 'wild' algae collected from their position of growth. While no attempt was made to culture these under optimum conditions and thus obtain the highest possible quantum yield, there is perhaps all the greater ecological interest in the values as indicating what 'natural' efficiencies, or 'productivity', might be. The method (manometric and volumetric gas determination) was also completely different from the polarographic one used for the action spectra, and lends confirmation to that rather unorthodox procedure. Yocum's results are summarized in Table 1.

The efficiencies which we found are somewhat less than those reported by many workers (Emerson & Lewis, 1942, 1943; Tanada, 1951) for *Chlorella* and diatoms (where 8–12 quanta were required for the release of one molecule of oxygen), but for *Ulva* and *Ilea* the values are relatively consistent throughout the spectrum. This confirms the essentially equal efficiency of all absorbing pigments in the green and brown algae.

The situation is quite different, however, in the red algae. Comparable efficiencies are indeed found in the regions absorbed by phycoerythrin and phycocyanin (14–15 quanta), but much lower values (25 quanta) prevail in the chlorophyll (red-absorbing) region, with extremely low efficiencies (35–50 quanta) in the carotenoid-chlorophyll (blue-absorbing) zone. There is still no clear explanation for this remarkable inefficiency of the chlorophyll of the red algae. (If an 'inaction spectrum' is plotted, it closely resembles the absorption of chlorophyll!) Perhaps chlorophyll d is not an adequate substitute for b or c as a 'co-pigment' for a (though *Chroococcus* seems to function with only chlorophyll a). Apparently a large fraction of the chlorophyll a is inactive, both in natural and in cultured red algae. Duysens (1952) gives an action spectrum for *Porphyridium* (Fig. 5) not very different from those of the red algae. He estimates chlorophyll to have only 20% of the effectiveness of phycoerythrin.

Table 1. *Number of absorbed quanta per molecule of oxygen released* (Yocum, 1951)

Wavelength (mμ.)	436	500	560	620	675
Green algae:					
Ulva lobata	14	18	23	16	12
Ulva lactuca	16	14	13	15	15
Brown alga:					
Ilea fascia	10	11	11	11	10
Red algae:					
Porphyra nereocystis	37	16	14	15	25 (high int.)
	50	17	14	15	23 (low int.)
Porphyra naiadum	35	24	13–17	18	25
Delesseria decipiens	26	16	16	20	32

The ineffective fraction of the chlorophyll might be:

(a) A slightly different chlorophyll (precursor, breakdown product, isomer). Yet such is not suggested either by absorption spectrum or by chromatography.

(b) A normal chlorophyll, but not bound to the proper protein (enzyme).

(c) Topographically inaccessible to the remaining systems of photosynthesis, e.g. too deep in the plastid, or with a 'non-conducting' barrier ('anti-sensitizers') between it and other chlorophyll molecules.

We can at present only speculate about these and other possibilities. Many red algae yield their chlorophyll to acetone very slowly, and to extract the chlorophyll it is often necessary to bring them into a series of dilute, then more concentrated solutions of methanol. It seems more tightly bound than in the green or brown algae. Whether this is a clue to its inactivity remains to be seen. Electron microscopy of normal plastids of red algae and of those extracted with methanol may yield important information about this subject.

Yet there is no evidence that in the red algae there are separate
systems for photosyntheses mediated by chlorophyll and by phycoerythrin. Yocum has shown that temperature, light intensity, partial pressure of carbon dioxide, anoxia, poisons, etc., affect red and green algae,
or photosynthesis due to red or green light in red algae, about equally.
When photosynthesis has been just saturated for red light (which occurs
at a much higher intensity than for green light), the addition of green

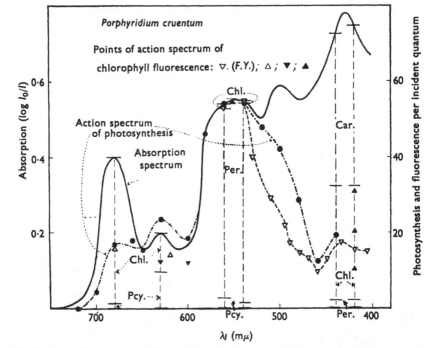

Fig. 5. Absorption spectrum ($\bullet$——$\bullet$), action spectrum of photosynthesis ($\bullet$-·-·-$\bullet$),
and action spectrum of chlorophyll fluorescence ($\triangledown$- - -$\triangledown$) of the red alga *Porphyridium
cruentum* (Duysens, 1952). Note that wavelength sequence is reverse of Figs. 2–4. Fluorescence action spectrum from French & Young (1952).

light to the saturating red does not increase the rate, i.e. no separate
pathway seems indicated, and the effects are not additive. Both pigments seem to be coupled to the same system. Yet there are a few
indications that the rates are *momentarily* different. These appear best
in red algae 'adapted' to red light, which we will now discuss.

IV. CHROMATIC ADAPTATION AND DE-ADAPTATION

Can red algae be forced or 'trained' to use their chlorophyll better?
Certain red algae live at high tide levels without any change in their
pigment ratios (an argument against the ecological significance of
phycoerythrin, which has formerly been stressed). Other species, how-

ever, may bleach out, and become almost green, so much so as to be mistaken for green algae by inexperienced botanical collectors! In California this occurs especially in a species of *Iridophycus* (*Iridaea*) and with *Halosaccion*. We have made action spectra of both of these 'greened' thalli, and they do indeed show increased activity in the region of chlorophyll absorption in the red end of the spectrum (but not much change in the blue end) indicating some increased activity of the chlorophyll. But the phycoerythrin still remains more effective, despite its decreased absorption. However, these plants are rather thick and not too well suited to action spectra determinations.

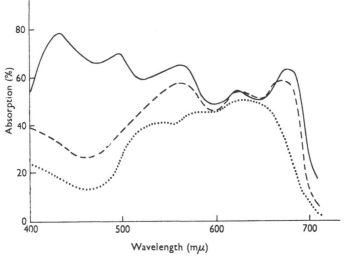

Fig. 6. Absorption (——) and action spectra (– – –) of *Porphyra perforata*, kept in mono-chromatic blue light for 8 days (Yocum, 1951). The action spectrum indicates the activation of chlorophyll by this blue light (red has similar effect). The absorption spectrum is not much altered (cf. Fig. 4*d* for normal spectrum). A normal action spectrum (......) is included for comparison.

More convincing experimental adaptation was found by Yocum (1951) using *Porphyra nereocystis* and *P. perforata*. The normal action spectrum of the latter is shown in Fig. 4*d*. We wanted to see whether its growth rate corresponded to its short-term photosynthetic action spectrum. Disks of the thallus were therefore kept in running sea water while exposed respectively to mercury blue and green light, and a band of red light, all of equal intensity. The growth (measured by increase in diameter) was at first fast in green light, and slow in both red and blue. After several days, growth increased in the red light but it never became very great in the blue. When the thalli adapted to red or blue light were tested for photosynthesis, it was found that the activity of the chloro-phyll was relatively enhanced; instead of being one-third as efficient as phycoerythrin, it became almost as effective (Fig. 6), and the action spectrum showed a distinct 'hump' at the region where chlorophyll has maximum absorption (675 mμ.). The quantum efficiencies together with growth and colour data are shown in Table 2. Clearly, although the total efficiencies have fallen, that due to chlorophyll has increased from about one-third that of phycoerythrin, to nearly as high. Carotenoids, on the other hand, have not been activated, even by blue light. Nor has absorption greatly altered during this time, although to the eye the thalli adapted to blue light looked purplish and the ones adapted to green light look tan. This may indicate a change of position of the pigments on the plastids, but apparently the adapted algae display chiefly a *functional* chromatic adaptation, rather than any great alteration of pigment ratios. Longer term experiments involving the growth of *Porphyridium* for several months in red and blue light have indicated that it too can display

Table 2. *Number of quanta required to release one molecule of oxygen in thalli of* Porphyra nereocystis *before and after adaptation periods in coloured lights and darkness* (after Yocum, 1951)

		Quanta/molecule			
Wavelength at which measured (mμ.)	Before exposure	After exposure for 1 week to			
		436 mμ.	546 mμ.	660 mμ.	Darkness
436	33	33	33	33	62
500	18	30	25	22	52
560	15	23	22	24	25
620	16	25	25	21	34
675	42	30	52	27	70
Thallus colour	Red	Dark red	Tan	Red	Red
Growth (%)		27	160	40	12

such functional chromatic adaptation, activating its chlorophyll, but not its carotenoids.

One of the most striking effects is chromatic *de-adaptation*. In *Porphyra nereocystis*, even a few hours of green light serves to abolish the response to red light which required a week or more of growth in red light to produce (Yocum, 1951). This might be called a 'competitive photo-inhibition'; not only does phycoerythrin utilize green light, but by its very absorption seems to inactivate, block, or disconnect chlorophyll from the photosynthetic system. If this is not carried too far the chlorophyll can be reactivated. In *P. nereocystis* adapted to red light brief exposure to intense mercury light lowers all photosynthetic response immediately thereafter ('solarization'), but the plants soon recover. However, the response to green light recovers more rapidly than that to red (Fig. 7). Some preferential affinity for the enzyme systems may be thereby displayed (there is no indication of a differential bleaching). Some interesting 'induction' or 'transitional' phenomena have also been observed in the thalli adapted to red light. On passing directly from red to green light of equal (and equally effective) intensity, the phenomena shown in Fig. 7 emerged. The 'cusps' indicate that green light is initially *more* effective, but photosynthesis quickly decreases; red light, on the other hand, is initially *less* effective, but becomes *more* so, although the *steady* photosynthetic rate is the same. This suggests that adaptation and de-adaptation may be occurring to some extent in a very brief time. This effect is under further study.

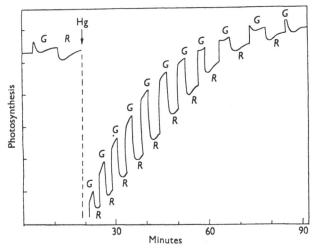

Fig. 7. Time course of photosynthesis in *Porphyra nereocystis* adapted to red light. Exposures to green light (*G*) and red light (*R*) are given alternately, with essentially equal photosynthetic response for equal incident energy. A very intense illumination (Hg) with the mercury green line (546 mμ.) followed, which 'solarized' the alga, so that the response was then greatly reduced. That to green was, however, considerably greater than that to red and remained so for about an hour. By this time the red response had become equal to the green, and both had nearly recovered to the initial rate. Note the upward and downward cusps as wavelengths are changed. Photosynthesis is measured in arbitrary units, recorded automatically from the polarographic circuit which measures oxygen production.

V. ENERGY TRANSFER AND FLUORESCENCE

We may now consider the mechanism of collaboration by the accessory pigments. In no organism have these light absorbers completely supplanted chlorophyll. Always in *content*, and usually in *function*, chlorophyll *a* remains involved. Attempts to find separate pathways of photosynthesis for the accessory pigments have so far failed. Does the energy absorbed by carotenoids and phycobilins have to be transferred therefore to chlorophyll before it can be effective in photosynthesis? In most speculations it is agreed that it does.

Dutton, Manning & Duggar (1943) were the first to show that light absorbed by *carotenoids* (largely fucoxanthin) could excite the red fluorescence of *chlorophyll* in living photosynthesizing diatoms. (Little or no transfer was observed in solutions of the pigments.) They concluded that 'carotenoid-sensitized photosynthesis in *Nitzschia* takes place principally through the transfer of absorbed energy from carotenoid to chlorophyll molecules, with subsequent reactions the same as though chlorophyll molecules were primary absorbers'.

They suggested three mechanisms for such a transfer: (1) collision; (2) intramolecular transfer (implying that carotenoid and chlorophyll were bound together in large molecules); (3) radiation by carotenoid and immediate re-absorption by chlorophyll, in a manner analogous to the internal conversion of nuclear gamma rays.

These results were essentially and independently confirmed by Wassink & Kersten (1946), who suggested the possibility that energy transfer might occur via the protein to which the pigments were attached (Wassink, 1948). The theory of transfer has also been discussed by Arnold & Oppenheimer (1950). The latter pointed out that the emission of fluorescent light (with 1 or 2 % yield) by the phycobilins, and its re-absorption by chlorophyll, is hardly efficient enough to be considered, while the carotenoids scarcely fluoresce at all. A direct 'resonance' or 'internal conversion' transfer seems possible, if the pigments are within a certain critical distance which must be less than a few wavelengths of light. Actually, values of 7–40 Å. units were calculated by Arnold & Oppenheimer for the distances between phycocyanin and chlorophyll in *Chroococcus*. In the living cells they estimate that 1–2 % of the energy absorbed by phycocyanin is re-emitted as fluorescent light (of phycocyanin), 4–8 % is degraded to heat, leaving 90–95 % transferred to chlorophyll and available for photosynthesis, a figure consistent with the findings of Emerson & Lewis in *Chroococcus*.

The most complete description of chlorophyll fluorescence as excited by phycobilins is that of French & Young (1952), who supplemented the earlier studies of van Norman, French & MacDowall (1948). Suspensions of *Porphyridium* were employed, and both the spectral distribution of fluorescent light and the action spectra of the exciting light were determined. The remarkable result (though quite in conformity with the photosynthetic action spectra) was that chlorophyll fluorescence was more excited by green light (absorbed by phycoerythrin) than by blue light absorbed by chlorophyll itself (and by carotenoids) (cf. Fig. 5). In other words, and paradoxically enough, the action 'spectrum' of *chlorophyll* fluorescence resembles the *absorption* spectrum of phyco-

erythrin. There can be no doubt in this case that energy has been transferred from phycobilin to chlorophyll, at least to the extent of causing
fluorescence. In this respect as well, the relative inactivity of light
absorbed by chlorophyll (and the carotenoids) in the shorter wavelengths can be ascribed to a large 'inactive' fraction of chlorophyll (and
to inactive carotenoids) comparable to that shown in the photosynthesis
of red algae (Fig. 4a–d). This implies that the energy absorbed by inactive pigments is not re-emitted even partially as fluorescent light, but
is degraded to heat. (Calorimetric studies would be desirable with the
red algae.) Further evidence that only the 'active' chlorophyll fluoresces
(although only emitting a small amount of its absorbed energy in this
manner) comes from some recent measurements kindly made for us by
Dr C. S. French on normal and on 'red-adapted' *Porphyra nereocystis*
(cf. action spectra, Fig. 6). The chief difference in the thalli adapted to
red light was that the 'self-excited' fluorescence of chlorophyll (due to
absorption of blue light) was enhanced, whereas that excited by green
light (absorbed by phycoerythrin) was not much changed. This is in
agreement with the photosynthetic measurements, which indicated a
greater relative content of 'active' chlorophyll in the plants adapted to
red light. It would appear that in the normal plants, the 'active'
chlorophyll is more effectively linked to the phycobilins than it is to its
optically identical, but photosynthetically ineffective congener, the
'inactive' chlorophyll.

Duysens (1951, 1952) has given reason for believing that the transfer
of energy from phycoerythrin is not direct to chlorophyll, but goes via
phycocyanin; the deep red fluorescence of the latter can be excited by
green light (absorbed largely by phycoerythrin), with an efficiency of
80 % of that due to orange light (absorbed by phycocyanin itself). Most
red algae have some phycocyanin, although in those found in deep
water or caves this is very much less than in surface species. One may
question whether it is always concentrated enough to form the *only* pathway of energy between phycoerythrin and chlorophyll.

An apparently new phycocyanin, probably to be designated *P*-phyco-
cyanin, has recently been isolated from several red algae and a blue-green
alga by Haxo, O'h Eocha & Strout (1954). (These authors also find several
slightly different phycoerythrins in various algae.) The new phycocyanin
has an absorption peak at 645–650 mμ., half-way between those of *C*- or
R-phycocyanin and chlorophyll *a*; it might therefore serve as a bridge for
energy transfer between these two. Since its fluorescence maximum is at
680 mμ., it is probably not the pigment postulated by Duysens to account
for the fluorescence of living *P. laciniata* farther in the infrared. Duysens

suggested that this might be due to chlorophyll *d*, which, in draining off energy from chlorophyll *a*, might render some of the latter inactive. Whether chlorophyll *d* is sufficiently concentrated in many red algae to accomplish this is open to some question.

VI. SUMMARY

The participation of accessory pigments in photosynthesis is well established, especially for fucoxanthin and the phycobilins. These serve to extend the range of light absorption into a greater part of the spectrum than is absorbed by chlorophyll alone; they probably operate by passing on the absorbed energy to chlorophyll. There is no clear evidence of an independent photosynthetic pathway through the accessory pigments which by-passes or parallels that through chlorophyll. In many cells some or all of the carotenoids appear inactive, either because they are topographically badly situated for energy transfer, or for some other reason. *β*-Carotene in some green algae and the carotenoids of purple bacteria are in this category. In the red algae, most of the carotenoids and a large proportion of the chlorophyll itself are inactive, leaving the major effective absorption to the phycobilins and the small amount of 'active' chlorophyll. Growth of the algae in red or blue light (free from green) activates a greater proportion of chlorophyll, but not the carotenoids. This functional adaptation is rapidly lost on exposure to green light, and presumably is the reason for the low efficiency of chlorophyll in the natural red algae.

REFERENCES

ARNOLD, W. & OPPENHEIMER, J. R. (1950). Internal conversion in the photosynthesis of blue-green algae. *J. gen. Physiol.* 33, 423.

AUGIER, J. & HENRY, M. H. (1950). Brome dans les Rhodophycées. *Bull. Soc. bot. Fr.* 97, 29.

BORESCH, K. (1932). Algenfarbstoffe. In *Handbuch der Pflanzenanalyse* (ed. G. Kleine). Vienna: Springer.

COOK, A. H. (1945). Algal pigments and their significance. *Biol. Rev.* 20, 115.

DUTTON, H. J. & MANNING, W. M. (1941). Evidence for carotenoid-sensitized photosynthesis in the diatom *Nitzchia closterium*. *Amer. J. Bot.* 28, 516.

DUTTON, H. J., MANNING, W. M. & DUGGAR, B. M. (1943). Chlorophyll fluorescence and energy transfer in the diatom *Nitzschia closterium*. *J. phys. Chem.* 47, 308.

DUYSENS, L. N. M. (1951). Transfer of light energy within the pigment systems present in photosynthesizing cells. *Nature, Lond.* 168, 548.

DUYSENS, L. N. M. (1952). Transfer of excitation energy in photosynthesis (Dissertation). Utrecht: Kemink.

EHRKE, G. (1932). Über die assimilation-komplementär gefärbter Meeresalgen im Lichte von verschiedenen Wellenlängen. *Planta*, 17, 650.

EMERSON, R. & LEWIS, C. M. (1942). The photosynthetic efficiency of phycocyanin in *Chroococcus*, and the problem of carotenoid participation in photosynthesis. *J. gen. Physiol.* 25, 579.

EMERSON, R. & LEWIS, C. M. (1943). The dependence of the quantum yield of *Chlorella* photosynthesis on wavelength of light. *Amer. J. Bot.* **30**, 165.

ENGELMANN, T. W. (1883). Farbe und Assimilation. *Bot. Ztg*, **41**, 1, 17.

ENGELMANN, T. W. (1884). Untersuchungen über die quantitativen Beziehungen zwischen Absorption des Lichtes und Assimilation in Pflanzenzellen. *Bot. Ztg*, **42**, 81; 97.

FELDMAN, J. & TIXIER, R. (1947). Sur la floridorubine, pigment rouge des plastes d'une Rhodophycée. *Rev. gén. bot.* **54**, 1.

FOX, D. L. & SARGENT, M. C. (1938). Variations in chlorophyll and carotenoid pigments of the brine flagellate, *Dunaliella salina*, induced by environmental concentrations of sodium chloride. *Chem. & Industr.* **57**, 1111.

FRENCH, C. S. (1937). The rate of CO_2 assimilation by purple bacteria at various wavelengths of light. *J. gen. Physiol.* **21**, 71.

FRENCH, C. S. & YOUNG, V. K. (1952). The fluorescence spectra of red algae and the transfer of energy from phycoerythrin to phycocyanin and chlorophyll. *J. gen. Physiol.* **35**, 873.

GAFFRON, H. (1934). Über die Kohlensäure-assimilation der roten Schwefelbakterien (I). *Biochem. Z.* **269**, 447.

GAFFRON, H. (1935). Über die Kohlensäure-assimilation der roten Schwefelbakterien (II). *Biochem. Z.* **279**, 1.

HAXO, F. & BLINKS, L. R. (1946). Photosynthetic action spectra in red algae. *Amer. J. Bot.* **33**, 836.

HAXO, F. & BLINKS, L. R. (1950). Photosynthetic action spectra of marine algae. *J. gen. Physiol.* **33**, 389.

HAXO, F., O'H EOCHA, C. & STROUT, P. M. (1954). Comparative studies of chromatographically separated phycochromoproteins (in the press).

HUBERT, B. (1935). The physical state of chlorophyll in the living plastid (Dissertation). Amsterdam: Mulder.

KARRER, P. & JUCKER, E. (1950). *Carotenoids.* Amsterdam: Elsevier.

KITASATO, Z. (1925). Biochemische Studien über Phykoerythrin und Phykocyan. *Acta Phytochim, Tokyo,* II, **2**, 75.

KLUGH, A. B. (1930). Studies on the photosynthesis of marine algae. I. *Contr. Canad. Biol.* **6**, 41.

KOCH, W. (1953). Untersuchungen an bakterienfreien Massenkulturen der einzelligen Rotalge *Porphyridium cruentum* Nägeli. *Arch. Mikrobiol.* **18**, 133.

KRASNOVSKII, A. A., EVSTIGNEEF, V. B., BRIN, G. P. & GAVRILOVA, V. A. (1952). Isolation of phycoerythrin from red algae, and its spectral and photochemical properties. *Dokl. obsch. Sobr. Ak. Nauk, SSSR,* **82**, 947; Abstr. in *Chem. Abst.* **46**, 7623.

KYLIN, H. (1910). Über Phykoerythrin und Phykoerythrin und Phykocyan bei *Ceramium rubrum. Hoppe-Seyl. Z.* **69**, 169.

KYLIN, H. (1912). Über die roten und blauen Farbstoffe der Algen. *Hoppe-Seyl. Z.* **76**, 397.

KYLIN, H. (1931). Einige Bemerkungen über Phykoerythrin und Phykocyan. *Hoppe-Seyl. Z.* **197**, 1.

LARSEN, H., YOCUM, C. S. & VAN NIEL, C. B. (1952). On the energetics of the photosyntheses in green sulphur bacteria. *J. gen. Physiol.* **36**, 161.

LEMBERG, R. (1928). Die Chromoproteide der Rotalgen. *Liebigs Ann.* **461**, 46.

LEMBERG, R. (1929). Pigmente der Rotalgen. *Naturwissenschaften,* **17**, 541.

LEMBERG, R. (1930). Die Chromoproteide der Rotalgen. II. Spaltung mit Pepsin und Säuren. Isolierung eines Pyrrolfarbstoffes. *Liebigs Ann.* **477**. 195.

LEMBERG, R. & LEGGE, J. W. (1949). *Hematin Compounds and Bile Pigments.* New York and London: Interscience.

246   L. R. BLINKS

LEVRING, T. (1947). Submarine daylight and the photosynthesis of marine algae. *Göteborgs VetenskSamh. Handl.* **5** (no. 6), 1.

McCLENDON, J. H. & BLINKS, L. R. (1952). Use of high molecular weight solutes in the study of isolated intracellular structures. *Nature, Lond.* **170**, 577.

MASTAGLI, P. & AUGIER, J. (1949). La structure de la substance phénolique de *Polysiphonia fastigiata. C.R. Acad. Sci., Paris,* **229**, 775.

MONTFORT, C. (1934). Farbe und Stoffgewinn im Meer. *Jb. wiss. Bot.* **79**, 493.

MONTFORT, C. (1936). Carotinoide, Photosynthese und Quantentheorie. *Jb. wiss. Bot.* **83**, 725.

MONTFORT, C. (1941). Die Photosynthese brauner Zellen im Zusammenwirkung von Chlorophyll und Carotinoiden. *Z. phys. Chem.* A, **186**, 57.

VAN NORMAN, R. W., FRENCH, C. S. & MACDOWALL, F. D. H. (1948). The absorption and fluorescence spectra of two red marine algae. *Plant Physiol.* **4**, 455.

SCHMIDT, G. (1937). Die Wirkung der Lichtqualität auf den Assimilationsapparat verschieden gefärbte Gewebe. *Jb. wiss. Bot.* **85**, 554.

STRAIN, H. H. (1938). *Leaf Xanthophylls. Publ. Carneg. Instn*, **490**.

STRAIN, H. H. (1951). The pigments of the algae. In *Manual of Phycology* (ed. G. M. Smith). Waltham: Chronica Botanica.

STRAIN, H. H. & MANNING, W. M. (1942). The occurrence and interconversion of various fucoxanthins. *J. Amer. chem. Soc.* **64**, 1235.

STRAIN, H. H. & MANNING, W. M. (1943). A unique polyene pigment of the marine diatom *Navicula torquatum. J. Amer. chem. Soc.* **65**, 2258.

STRAIN, H. H., MANNING, W. M. & HARDIN, G. (1944). Xanthophylls and carotenes of diatoms, brown algae, dinoflagellates, and sea-anemones. *Biol. Bull., Woods Hole,* **86**, 169.

SVEDBERG, T. & ERIKSSON, I. B. (1932). Molecular weights of phycocyan and phycoerythrin. III. *J. Amer. chem. Soc.* **54**, 3998.

SVEDBERG, T. & KATSURAI, T. (1929). The molecular weight of phycocyan and phycoerythrin from *Porphyra tenera* and of phycocyan from *Aphanozomenon flos-aquae. J. Amer. chem. Soc.* **51**, 3573.

SVEDBERG, T. & LEWIS, N. B. (1928). The molecular weights of phycoerythrin and phycocyan. *J. Amer. chem. Soc.* **50**, 525.

SWINGLE, S. M. & TISELIUS, A. (1951). Tricalcium phosphate as an adsorbent in the chromatography of proteins. *Biochem. J.* **48**, 171.

TANADA, T. (1951). The photosynthetic efficiency of carotenoid pigments in *Navicula minima. Amer. J. Bot.* **38**, 276.

THOMAS, J. B. (1950). On the role of the carotenoids in photosynthesis in *Rhodospirillum rubrum. Biochim. Biophys. Acta,* **5**, 186.

TISELIUS, A. (1930). Moving boundary method of studying the electrophoresis of proteins. *Nova Acta Soc. Sci. upsal.* **7** (4), 1.

TISELIUS, A. (1937). A new apparatus for electrophoretic analysis of colloidal mixtures. *Trans. Faraday Soc.* **33**, 524.

WASSINK, E. C. (1948). Some remarks on chromophyllins and on the paths of energy-transfer in photosynthesis. *Enzymologia,* **12**, 362.

WASSINK, E. C. & KERSTEN, J. A. H. (1946). Observations sur le spectre d'absorption et sur le rôle des carotenoids dans la photosynthèse des diatomées. *Enzymologia,* **12**, 3.

WASSINK, E. C. & RAGETLI, H. W. J. (1952). Chromatography of phycocyanin. *Proc. Acad. Sci. Amst.* (C), LV, **4**, 462. Abstr. in *Nature, Lond.* **171**, 376.

WURMSER, R. (1921). *Recherches sur l'Assimilation Chlorophyllienne.* Paris: Vigot.

YOCUM, C. S. (1951). Some experiments on photosynthesis in marine algae. (Dissertation). Abstr. in *Abstracts of Dissertations*, Stanford University, 1951.

PROBLEMS IN THE MASS CULTIVATION OF PHOTO-AUTOTROPHIC MICRO-ORGANISMS

E. C. WASSINK

Laboratory of Plant Physiological Research, Agricultural University, Wageningen, Holland

I. MASS CULTURING OF ALGAE IN RELATION TO THE WORLD ENERGY SITUATION

Interest in the mass culture of photo-autotrophic micro-organisms has largely arisen since the Second World War; various reasons may be responsible for this. The closer interrelation between people in remote parts of the world has led to a reconsideration of their very different material standards of living, and to a critical survey of available and potential sources of food. In addition, the availability of nuclear energy on a technical scale has made a reconsideration of the world-energy sources feasible. It should be noticed that, as yet, nuclear energy plants do not result in considerable overall energy gain. Furthermore, it cannot escape notice that some parts of the world consume natural resources, e.g. wood, on a scale larger than they produce them. It is also obvious that at the present day the greater part of the energy we use is derived from solar energy stored in coal and oils, which have arisen from photosynthesis carried on in the past, and estimations all point to the conclusion that these resources, especially their more accessible parts, are not extremely extensive.

It seems quite logical that not only the energy 'capital' but also the 'income', i.e. the solar radiation incident on the earth's surface, should be discussed. Figures about the estimations made will not be repeated in detail here. They point to the fact that the total fossil fuel energy, as far as is known, only equals at most a few years of solar radiation on earth, so that, considering the long time for which photosynthesis has been going on, only a very small fraction of the solar energy has been preserved. This, of course, is not surprising, since it is only under very specialized conditions that fossilization is possible.

In any case, it is clear that the daily irradiation by sunlight constitutes a powerful source of energy, provided that it can be trapped without being immediately dissipated to heat and other low-grade types of energy, such as easily occurs by accumulation at low temperature and in a diffuse form. The chief forms in which solar energy is effectively available on earth are:

(1) as plant material containing chemical energy produced by photosynthesis (this material is used either for food or as an energy source, via heat production, e.g. the use of wood as a fuel),

(2) as water power,

(3) as wind power,

(4) as food of animal nature, ultimately derived from plants,

(5) as the physical strength of men and animals.

De Vries (1948) mentions that nowadays 75 % of mankind still live by using practically only the effects of contemporary solar radiation. On the other hand, modern industry and traffic rely almost entirely on fossil fuel energy, and in highly industrialized countries with a modern type of civilization (e.g. the U.S.A.) nearly 90 % of the energy used per citizen is derived from energy deposits in coal and oil (de Vries, 1948).

In connexion with what has been put forward above, it is clear that a sound world-energy household should attempt to rely on renewable sources of energy as much as possible. Such sources are still the only ones available for the synthesis of foodstuffs, which are necessarily products of recent photosynthesis because of the limitations of.animal physiology. It should be kept in mind that in primitive civilizations the main energy consumption per person is along this line, whereas in a complicated industrial civilization the main part of the very much increased energy consumption per person is required to keep the industrial system going. The more the latter type of civilization spreads over the world (and in view of the figure quoted above from de Vries it is still likely to do so to a great extent), the greater the demands for industrial energy, whereas the increased world population and the increasing personal demands connected with a more complicated sort of civilization increase the demand for food. Industrial energy, from the view point of the human being, is 'lower grade' energy than that required for food. (Food can, eventually, be used as fuel, but not the reverse.)

The ways of achieving the purposes outlined seem to be the following:

(1) An increased yield from solar energy,

    (A) by increased yield of practical photosynthesis,

        (i) by a better utilization of known agricultural techniques and principles,

        (ii) by introducing new techniques and following new principles (including the production of foodstuffs in great demand);

(B) by trapping solar energy and converting it into a form of energy which can be used:

(i) for industrial purposes,

(a) as electrical energy ('photocell principle'),

(b) as chemical energy,

(ii) as food, by developing new principles and large-scale techniques to synthesize compounds utilizable as food (e.g. sugars, fats);

(C) by increased utilization of wind and water power.

(2) A development of the technical utilization of nuclear energy on a large scale so that this energy can be derived from materials that are available in sufficient amounts, and in such a way that a large gain in overall available energy is achieved. If so, this energy can be used along the same lines as indicated under (1)(B).

It is clear that basic work along these various lines will be carried out simultaneously, and it cannot be said at present whether a particular one of these possibilities (or others) will play the predominant role in the future, or whether these various possibilities will provide useful applications.

This brief general background survey can be concluded with the remark that the application of known agricultural principles to a level such as that found in many countries of Western Europe would yield very great increases in food production without extending the area used. So far, mental and technical difficulties largely prevent this.

It is clear that from the point of view of the present discussion, mass cultivation of algae belongs under heading (1)(A) (ii) above, and that the first question to consider is whether algae in mass culture are likely to convert solar energy into energy of cellular material with a higher yield than well-known agricultural crops, and/or whether they have other special advantages. It was from the point of view of this efficiency of energy conversion, that the work of our group was conducted. Some considerations and results will now be presented.

II. YIELD OF SOLAR ENERGY IN NATURAL PHOTOSYNTHESIS

Daniels (1949) has computed that the U.S.A. receive a daily amount of sunshine of $2 \cdot 8 \times 10^8$ kcal./person. The amount of energy used daily per person is $1 \cdot 5 \times 10^5$ kcal., i.e. $0 \cdot 5 \times 10^{-3}$ of the incident amount. However, of this, only about 2% is derived from contemporary photosynthesis, while the rest is from fossil fuel! It thus follows that, in America, only about 10^{-5} of the daily incident energy is used by man. This figure (2%) is somewhat arbitrary, representing the 3000 kcal. used daily as food. Part of the rest is also derived from contemporary photosynthesis, e.g. in so far as wood is used for building and paper. This probably covers another 3000 kcal.

In the Netherlands, the population per unit area is about ten times as high as in the U.S.A. The daily food would thus account for something of the order of 10^{-4} of the daily radiation. We will see that photosynthesis in crop plants traps on the average less than 1 % of the total solar radiation (less than 2 % of the photosynthetically utilizable radiation). This may at first sight appear consoling, since a high potential gain in energy still seems conceivable (viz. from 10^{-4} to 10^{-2}), but two factors offset this comforting thought. First, by no means the whole area irradiated by the sun can be used for food production, and secondly, with many crop plants only a small part of the material built up in photosynthesis is of direct nutritional value.

One may even say that in the Netherlands neither the agricultural area nor current agricultural practice are likely to be subject to considerable improvement in the future, whereas the population is still increasing. It thus seems that in a civilization like ours, not much more than 1 % of the potentially available solar radiation per person can be used for producing food (and this 1 % is only utilized with an efficiency of 1 % by the crop plants). Thus, a situation obtains in which a search for a more effective production of food, either for men or domestic animals, seems very reasonable.

The present author has made estimations of the yield of solar energy conversion in some current crop plants. Some data are collected in Table 1 (from Wassink, 1948). The yield varies from 0·45 % to slightly over 2 % of the photosynthetically utilizable radiation falling on the cultivated area during the growing season. Since some of the solar radiation is in the infrared, less than half this yield is obtained when all incident solar radiation during the growing season is taken into account, and probably again half this figure, when the radiation during the whole year is considered. (Many crop plants grow less than half of the year, but the greater part of solar energy is incident in the growth season, covering the longest days with the highest position of the sun. In Holland, for example, the total incident solar energy is about 1350 cal./cm.² in December, 11,000 in July (Wassink, 1948; Reesinck, 1947).)

Rabinowitch (1945) has assumed that the yield of energy conversion in green plants under natural conditions is about 2 % of the absorbed, utilizable radiation. He allows about 10–20 % for reflexion losses which brings this figure down to about 1·6 % for incident radiation. This estimation seems very high in view of the figures presented in Table 1, which are derived from high figures of crop yield (and based on agricultural data, including estimations for waste products such as roots, stems, etc.) in a highly developed agricultural country. For the world as a whole the agricultural returns are very much lower.

Calculations by the present author for the production of wood, yield figures of the same order of magnitude, and will be discussed elsewhere. Rabinowitch mentions that light losses by reflexion are less in this case than with field crops.

Table 1. *Efficiency of solar energy conversion during the growing season of some agricultural crop plants* (Wassink, 1948)

(1) Crop plant	(2) Production of dry matter* (0·01 g./cm.²)	(3) Chemical energy of dry matter evaluated as CH₂O (10¹⁰ergs/ cm.²)	(4) Growing† period	(5) Total radiation received (cal./cm.²)	(6) (cal./cm.²)	(7) (10¹⁰ergs/ cm.²)	(8) Efficiency column 3 column 7 (%)
					Solar radiation‡		
					Usable in photosynthesis (excluding infrared)		
Onions	3·5	0·55	April–Sept.	58,000	29,000	1·22	0·45
Carrots	6·86	1·07	May–Oct.	54,400	27,200	1·14	0·94
Potatoes	9·6	1·5	April–Sept.	58,000	29,000	1·22	1·23
Rapes	3·6	0·56	Aug.–Nov.	21,700	10,850	0·45	1·24
Wheat	10·45	1·62	Nov.–Aug.	61,000	30,500	1·28	1·26
Rye grass (Lolium)	10·2	1·60	March–Oct.	67,500	33,800	1·42	1·13
Beets, mangels	16·0	2·5	May–Oct.	54,400	27,200	1·14	2·20
Maize	12·8	2·0	10 May–10 Sept.	43,600	21,800	0·92	2·18
Sugar cane	33·0	5·2	April–March	129,000§	64,500	2·70	1·92

* From agricultural data.
† The months named are included.
‡ Calculated after Reesinck (1947), measurements made at Wageningen (except those for sugar cane).
§ Recalculated from recent measurements by Dee & Reesinck (1951) at Djakarta. This value is not far from the one used previously (Wassink, 1948), derived from data reported by Boerema (1920), viz. 120,000 cal./cm.², yielding an efficiency of 2·05 %.

More difficult to estimate is the production of the oceans. The carbon fixation of 3·75 tons carbon per hectare, indicated by Rabinowitch (1945, p. 7) amounts to an energy conversion of somewhat over 1 % of the photosynthetically utilizable radiation. According to de Vries (1948), about 1 kg. of fish is caught annually per hectare of sea surface. This figure might provide an independent estimation, provided some additional data were known. Since these are not known, however, they will be estimated below.

We assume that 1 kg. fish has grown from 5 kg. 'zooplankton' which may have grown from 25 kg. phytoplankton. We assume arbitrarily that

1 % of the total fish is caught, and that fish is 1 % of the organic matter in the sea. This would bring the yield to 250,000 kg. phytoplankton/year/ hectare or about 25,000 kg. dry weight. This would be of the order of very good agricultural crops (cf. Table 1), and would imply an energy conversion of about 2 % of the radiation capable of being used for photosynthesis. (It should be mentioned that fishing is often done in considerable concentrations of fish so that local catches would be likely to give erroneous results if taken as an average. For example, according to Tesch (1920) the North Sea yields about 7700 kg. fish per sq. mile per annum, which is about 35 kg. per hectare. It is clear that such figures could never represent the actual achievement of photosynthesis in this area.)

III. ENERGY CONVERSION IN PHOTOSYNTHESIS UNDER LABORATORY CONDITIONS

We have seen in the preceding section that photosynthesis under natural conditions appears to occur with an energy conversion of 1–2 %, as far as reliable data exist. On the other hand, it is well known that higher conversion levels are easily obtained in photosynthesis under laboratory conditions. Notwithstanding the fact that considerable confusion still obtains in this field, the assimilation of one molecule of carbon dioxide per 8–10 quanta of light absorbed is quite feasible. This corresponds to an energy conversion of 25–30 %.

In our opinion an analysis of the possibilities of the successful mass cultivation of algae has to start with a consideration of the possible reasons for this discrepancy. It should be kept in mind that the high laboratory efficiencies have mostly been obtained under conditions of light limitation with unicellular algae, in experiments of short duration with so-called 'resting cells', i.e. investigated in a medium that is very incomplete as a culture solution.

Various reasons may be accepted why field crops yield lower energy conversions. Firstly, light intensity may exceed the saturation value for photosynthesis. Under these circumstances, much light is wasted by ineffective absorption in the upper layer of leaves which receive too much light energy per unit time and are unable to convey this energy to leaves which are less adequately illuminated.

We have now assumed that the internal dark reaction system enters as a limiting factor at light intensities of the order of magnitude of natural daylight. But, in addition, some other external factor, e.g. carbon dioxide content of the atmosphere or water supply, may be responsible for inadequate light efficiency; unfavourable temperatures will also lower the efficiency. Fig. 1 is a scheme in which various cases are represented. Curve OAB represents the rate of production (photosynthesis) with a sufficient supply of all external factors. The yield is maximal for light intensities between O and E (slope OA), it decreases from E to F (slope OB for intensity F). With low temperature, or in presence of inhibitors (both these conditions reduce the capacity of the internal enzymes), or with a limited supply of an external factor supplied by the environment (e.g. carbon dioxide, water), curves of the type OGD usually result. At the intensity E, the efficiency is already much below maximum (cf. slopes OC and OA), and decreases further between E and F in much the same way as above.

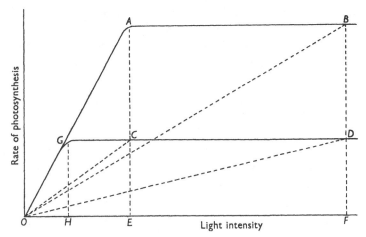

Fig. 1. Schematic representation of the relation between light intensity and the rate of photosynthesis with special reference to the efficiency of energy conversion.

It can be seen quite easily that in many of the mentioned respects, algal cultures are likely to be superior to field crops, e.g. the cultures can be stirred so that various cells can be alternately exposed to light and darkness. It has been known for a long time that the intercalation of short dark periods between brief light flashes improves the efficiency at high light intensity. For example, Warburg (1919) reported an increase of nearly 100 % (per unit of light) when light of high intensity was alternated with darkness in equal periods of $c.$ $1 \cdot 25 \times 10^{-4}$ min. More will be said about this subject later. Thus, by stirring a relatively thick suspension and so rapidly changing their position to light and darkness the yield of algae is likely to increase whereas this possibility does not arise in the case of higher plants. Moreover, water and carbon dioxide are more easily supplied in suitable concentrations, and temperature is easily regulated in tanks of water, whereas such controls are impossible in fields.

Thus, in principle, algal cultures seem to offer better prospects for effective utilization of high intensity radiation than do higher plants. In considering whether it is possible to realize an energy conversion of as much as 20–25 % (see above) two problems remain to be discussed, viz.

(1) Is the efficiency of the photosynthetic apparatus in algae higher than in higher plants?

(2) Is the efficiency found in short term experiments with algae also obtainable during active growth?

In connexion with the first point, some experiments of the present author may be mentioned in which the photosynthesis of leaf discs of various higher plants was investigated by the Warburg manometric technique (Wassink, 1946). When light is the limiting factor, quantum efficiencies of 0·07–0·1 (in sodium light) can be easily observed, indicating an energy conversion of about 20 %.

IV. EFFICIENCIES OBTAINED IN PRELIMINARY GROWTH EXPERIMENTS IN THE LABORATORY

The present author started preliminary experiments upon the second question raised in the preceding section with senior students in 1948, and early in 1949. *Chlorella vulgaris* was cultivated in 0·5 l. inverted, glass-stoppered bottles, each having two holes with cotton plugs in the bottom and an aeration filter mounted in the neck. Each bottle was placed in a small light cabinet with an opaline glass, and a BG 17, 3 mm. filter (Schott and Gen., Jena) in front. Four cabinets were arranged around a water-cooled 300 W. incandescent lamp. The temperature was about 22° C. At various places inside these light cabinets the intensities of incident and transmitted light were measured with a photocell, whereas special measurements were carried out in order to convert the results into absolute energy values. Aeration of the cultures was provided by a stream of air, containing 5 % carbon dioxide. The growth curves were of the normal S-shape. The efficiency of production of dry matter (which was taken to be energetically equivalent to 'CH$_2$O') was calculated over the period of active growth, which lasted for about 10 days. The efficiency was 12–15 % of the incident light at light intensities of 5000–8000 lux, and 20–24 % at 1500–3000 lux. These experiments, chiefly carried out by J. F. Bierhuizen, were the first to suggest that the energy conversion in a growing *Chlorella* culture is not much lower than that found in a photosynthesis experiment with resting cells.

It is relevant to mention here similar experiments, carried out with higher plants. Beet seedlings were raised under 'daylight' fluorescent tubes at 20° C., being illuminated with about 5000 lux for 24 hr./day (Pl. 1). Successive harvests were taken, and the yield compared with the amount of light energy that had fallen on the leaf surface. For this purpose, photographs were taken from above, and the leaf surface was estimated with a planimeter. As soon as the full soil surface was covered by the plants, this was no longer necessary. Taking all the light falling on the leaves to be absorbed, and the composition of photosynthetic products to be 'CH_2O', D. J. Glas and P. Gaastra, in 1949, found efficiencies of 12–19 %; in some later experiments by Kok and van Oorschot, the yield was 11–15 %. This suggests that during growth, higher plants are able to convert light of 'low' intensity with a considerable efficiency, much higher than the one recorded under field conditions. This observation thus constitutes a parallel to the findings in assimilation experiments of short duration with leaf disks (p. 254).

V. MORE DETAILED OBSERVATIONS CONCERNING THE ENERGY CONVERSION IN GROWING ALGAL CULTURES IN THE LABORATORY

A much more detailed examination of the efficiency of cultures of C. vulgaris growing under carefully established conditions was made in this laboratory by Kok (1952). The experiments were performed with a Warburg apparatus, and some of the results may be briefly mentioned. Vessels of a special type were used (Fig. 2), with a capacity of about 250 ml.; 100 ml. of nutrient medium (Knop's solution) were introduced, inoculated, and incubated in the Warburg thermostat in sodium light for 4–7 days. The maximum light intensity was 2×10^4 ergs/cm.^{2}/sec., which is within the range in which light is limiting. Six manometers were available, which could receive different light intensities. As may be seen from Fig. 2, the vessels were provided with a special long-tailed tap which permitted aeration of the vessels when attached to the manometers. During the night and when no measurements were taken, the suspension was aerated with air + 3 % CO_2. By turning the tap, the system was converted into an ordinary Warburg manometer and the gas exchange could be measured in light and in the dark; Pardee (1949) solution was in the central well, in equilibrium with the gas phase containing 3 % CO_2.

Besides daily measurements of the rates of photosynthesis and respiration (by temporary darkening of the vessels), daily measurements were made of the light absorption by the culture, and of the incident light intensities. The latter were determined with a standardized thermo-pile, and the absorption measurements with an Ulbricht white sphere (Vermeulen, Wassink & Reman, 1937; Kok, 1948). At the end of the culture period the amount of dry matter produced, its chemical com-position (elementary analysis), and in some cases the heat of combustion, were determined.

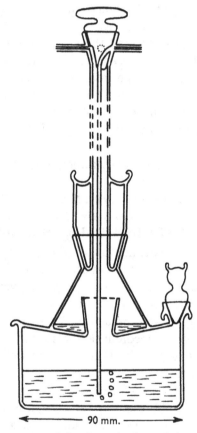

Fig. 2. Warburg manometer vessel for determinations of the efficiency of light energy conversion in growing cultures of *Chlorella* (Kok, 1952).

It was found that as the cultures grow older and denser, the rate of photosynthesis decreases, the rate of respiration increases, and the quantum number per molecule of oxygen increases from about 8 at the start to over 20 after 5 days of cultivation. The 'formula' of the algal material in one experiment (taking N as unity) was: $C_{6.85}H_{11.8}O_{2.65}N$. The overall equation of photosynthesis was:

$$6.85CO_2 + 6.4H_2O + NO_3^- \rightarrow C_{6.85}H_{11.8}O_{2.65}N + 9.7O_2 + OH^-,$$

and the efficiency, calculated as $100 \times$ cal. fixed/cal. absorbed, was 23.5%.

In a vessel with 'low' nitrogen, the elementary composition of the material was $C_{24.5}H_{43.3}O_{12.5}N$, indicating a much lower nitrogen content. The overall equation for growth was given as:

$$24.5CO_2 + 20.2H_2O + NH_4^+ \rightarrow C_{24.5}H_{43.3}O_{12.5}N + 28.6O_2 + H^+,$$

and the efficiency of energy conversion was 15·1 %.

The efficiencies of energy conversion obtained in Kok's experiments are collected in Fig. 3. The variation shown is not due to experimental errors but to differences in the environmental conditions applied. The efficiency values usually decrease with prolonged exposures or low nitrogen content. These values were obtained in sodium light. Theoretically, the efficiency values might increase by about 15 %, using red light,

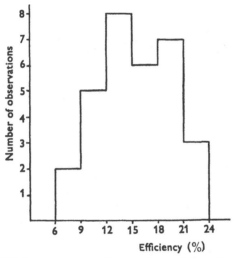

Fig. 3. Efficiency of light energy conversion as observed in various experiments with growing cultures of *Chlorella* (Kok, 1952).

the quanta of which contain about 15 % less energy. From Fig. 3 it is evident that the average yield is 12–18 %, which confirms the earlier experiments by Wassink and co-workers, described above (see also Wassink, Kok & van Oorschot, 1953).

The following additional observations may be mentioned which are in good agreement with related observations by Spoehr & Milner (1949). Increase in light intensity generally yields a decrease in protein content in favour of carbohydrate and, in high light intensities, of fat. The same holds for prolonged continuous illumination, and deprival of nitrogen. On the other hand, interruption of illumination by dark periods of considerable duration (e.g. 5 hr. or more per daily cycle) causes an increase in the protein content. If the rate of photosynthesis is P and that of respiration R, $P-R$ is the total action of light, and the fraction

$P/(P-R)$ results in growth (R taken as negative). Optimum values of $P/(P-R)$ were as high as 0·95; in dense cultures after long exposures the value fell and even reached 0·3.

As already mentioned prolonged exposure at high light intensity causes a relative decrease in protein content. Intercalation of dark periods leads to a 'rebalancing' of the chemical composition, and Kok assumes that this exerts a favourable influence on growth, which over-compensates the losses due to respiration during the dark periods (which, under normal day and night conditions, are of the order of 10 %).

VI. FURTHER OBSERVATIONS CONCERNING THE EFFECT OF THE COMPOSITION OF THE MEDIUM ON THE YIELD AND COMPOSITION OF THE CELL MATERIAL

More extensive experiments along the same lines as in the previous section were made in our group by van Oorschot (unpublished). He used an apparatus, similar to the one used by Kok; the light intensities applied were $1-5 \times 10^4$ ergs/cm.²/sec., in sodium light. The dry weight of cell material was determined at the beginning and at the end of the culture period which was usually 3–4 days. The elementary composition of the cell material and the amount of light absorbed were determined, as explained in § V. During the culture period, gas exchange was also measured. Protein, fat and carbohydrate contents were computed from the elementary composition, as described by Spoehr & Milner (1949).

The most extensive variations were in nitrogen content. Nitrogen was supplied as potassium nitrate. Decreasing concentrations of potassium nitrate in the medium yielded cell material with decreasing protein content, both at high ($5 \cdot 0 \times 10^4$ ergs/cm.²/sec.) and at low light intensity ($1 \cdot 5 \times 10^4$ ergs/cm.²/sec.). With sufficient amounts of potassium nitrate (e.g. 0·025 M at the start) the protein content remains high even at high light intensities (duration of the experiment 69 hr.). At the high light intensity, a decrease in the nitrate concentrations causes a decrease in the efficiency of energy conversion, whereas at low light intensities even much lower nitrate concentrations did not result in a decrease in efficiency. Obviously, decreasing the concentrations of potassium nitrate causes light saturation at a lower intensity (Table 2). Increasing the light intensity in a medium without nitrogen results in a decrease in protein content, increases in carbohydrate and fat content, and a decreased yield of energy conversion (e.g. from 21 to 5·5 %), whereas in the presence of nitrogen, in the same series of light intensities, the yield only drops from 21 to 16 %. Addition of nitrogen to nitrogen-deficient cultures causes recovery of the energy yield.

Cultures, precultivated for various times, were illuminated for 95 hr. in a nitrogen-free medium. Table 3 shows the efficiencies obtained and the compositions computed. It is seen that, with this treatment, young cultures still grow and rapidly shift to a composition which resembles that of old cultures. Old cultures themselves, under these conditions, do not assimilate and hardly change their composition.

The effect of renewal of the medium or repeated addition of nutrient salts on growth and composition of the cells was also investigated. It is seen that without renewal of the medium or addition of salts, the growth rate of the culture decreases relatively soon (Fig. 4), whereas renewal of

Table 2. *Relation between the light intensity and the influence of nitrogen deficiency upon the efficiency of energy conversion in* Chlorella

Incident intensity (10^4 ergs/cm.^{2}/sec.)

1·2–1·5		4·5–5·0	
KNO$_3$ (M)	Efficiency (%)	KNO$_3$ (M)	Efficiency (%)
0·005	17·8	0·02	16·8
0·0025	14·7	0·015	17·0
0·00125	17·7	0·010	14·7
0·000625	16·8	0·005	8·1

Table 3. *Efficiency of energy conversion and composition of cultures of different ages, which have been grown in a nitrogen-free medium for 95 hr.*

Age of culture...	29 days			8 days			3 days		
Incident light intensity (10^4 ergs/cm.^{2}/sec.) ...	Start	2·2	4·0	Start	2·1	4·1	Start	4·2	
% efficiency	—	0	0	—	7·4	3·2	—	7·2	
% proteins	14·5	12·4	14·1	21·5	12·2	12·6	56·2	13·4	
% lipids	27·9	30·4	30·8	18·4	30·3	26·3	15·1	32·6	
% carbohydrates	57·4	57·2	55·1	60·1	57·4	61·1	28·7	59·4	

the medium or addition of salts (potassium nitrate and potassium dihydrogen phosphate) results in unimpeded growth for a long time. The resulting changes in the composition of cell material during the culture period are given in Fig. 5.

Bongers (unpublished) investigated the changes in the nitrogen content of the cell material of *C. vulgaris* growing in the presence of limited amounts of nitrogen. The light intensity was about 400 ft.c. from fluorescent tubes and the temperature was 25° C. The conditions were such that the nitrate in the medium had disappeared after about 30 hr., but the production of dry weight still increased markedly up to about 75 hr. After 100 hr. the increase in dry weight is negligible. After

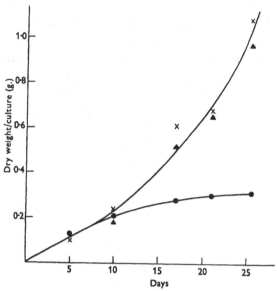

Fig. 4. Dry weight of *Chlorella* at increasing ages of the cultures. with renewal of medium (▲) or addition of salts every 2–3 days (×) and without renewal or addition (•). Incident intensity *c.* 4000 lux.

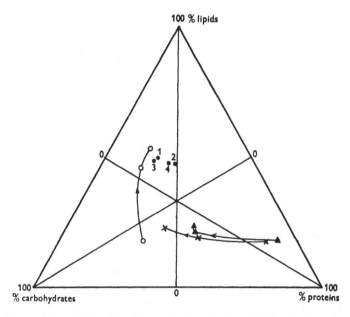

Fig. 5. Composition of *Chlorella* cell material after 10, 17 and 26 days (indicated by arrows) with renewal of medium (▲), addition of salts (×) and without additions (O) Incident intensity *c.* 4000 lux. Points 1, 2, 3, 4: observations with cultures, kept during the whole period in the same medium for 77, 74, 67, 30 days respectively.

exhaustion of the external nitrogen, the relative nitrogen content of the dry matter shows a sudden drop, and becomes constant at a low level. In this region the R-value (see Spoehr & Milner, 1949) shows a minimum, indicating a maximum carbohydrate content of the cells. Beyond this point the R-value increases again, indicating an increase in the fat content of the cells (Fig. 6).

VII. INTERMITTENCY OF ILLUMINATION IN RELATION TO MASS CULTURING

It has been mentioned above that Warburg (1919) was the first to observe that for a constant amount of light absorbed, the yield of photosynthesis at high light intensities is increased when short dark periods are alter-

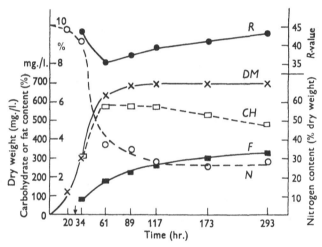

Fig. 6. Dry matter production (*DM*), nitrogen content (*N*) of cell material, *R*-value, and percentage of carbohydrate (*CH*) and fat (*F*) in dry matter as a function of age of a *Chlorella* culture with limited supply of nitrate. At ↓ nitrate in medium exhausted. (Bongers, L. H. J., 1953, unpublished.)

nated with short light periods. Emerson & Arnold (1932) extended this work and found that large amounts of light energy, administered in a very short flash, could be used efficiently provided dark periods of at least 0·02 sec. were intercalated. The obvious explanation is that during the short flash a quantity of some primary photoproduct is formed, the further utilization of which by the subsequent dark reaction is completed in a time of the order of 0·02 sec., under the conditions prevailing in Emerson & Arnold's experiments. In connexion with more recent views on photosynthesis it seems likely that some energy-acceptor is loaded up or activated or primarily transformed in the 'photic' link, yielding a product in a concentration which is large with respect to the capacity of some step in the subsequent dark enzyme systems.

It is obvious that this situation demonstrates one respect in which mass algal cultures may be superior to higher plants as regards utilization of high sunlight intensities (cf. §III). By stirring the suspensions, it would seem feasible to utilize sunlight more effectively, since this results in alternate exposure of the cells to higher and lower light intensities. In connexion with curve OAB (Fig. 1), it can be said that any sort of a cycle in which a cell is alternately exposed to light intensities higher and lower than E with sufficient rapidity of change from higher to lower will improve the yield from light of an intensity higher than E. In order to establish that light of an intensity F, for example (Fig. 1), is used with the same efficiency as that of intensity E, the primary requirement to be fulfilled is that sufficient dark time is intercalated so that the total light period is only E/F of the total time of observation. It is of course clear that this procedure would not yield any conspicuous improvement in the yield of conversion of light of intensity F, if an observation time of f min. were divided into e min. of light at intensity F, and $f-e$ min. darkness, such that $e/f = E/F$. Warburg (1919) has already established that a splitting of the exposure time into smaller periods of light and darkness up to a very high frequency of intermittency still improves the efficiency of energy conversion.

Kok (1953) has made an extensive investigation of the most suitable duration of light and dark periods in cultures of *Chlorella* when the incident intensity approximates to that of sunlight. He varied both duration of flash and relation between light and dark periods. He found that flash times of the order of 0·1 sec. still show very little 'flash effect' on the yield of energy conversion (combined with dark periods of 3, 10 or 30 times the flash duration). Only when the duration of the flash is decreased to about 0·01 sec. is a considerable rise in the yield found, and the yield is still further improved by further shortening of the flash. Kok observed that full efficiency (i.e. that corresponding to the slope OA in Fig. 1) is reached only when the flash time is around 0·003 sec. or less, and the dark period is 10–30 times the flash duration. It is clear that any condition which brings E nearer to F (Fig. 1) lowers the requirements for flash utilization. For example, increase of temperature can be mentioned, the influence of which was investigated by Kok. It is seen from Fig. 7 that, other conditions being equal, the flash experiments at the highest temperature show the highest efficiency.

It is clear that the considerations given do not establish a maximum for the duration of the dark period. A certain amount of light, given in a short flash, can only be utilized photosynthetically with maximum efficiency if the adjacent dark period is sufficiently long. Increasing the length of this dark period further does not decrease the efficient use of the light as such, consequently the only relevant factor is the shortness of the flash. In terms of algal mass cultures, this would mean that each cell should be at the surface, in the full sunshine, for a short moment only, after which it might dive into the dark lower layers of a dense culture for a considerable time. This could probably be achieved by adequate stirring. However, another problem arises in growing cultures (namely, the losses due to respiration), and this has also been discussed by Kok.

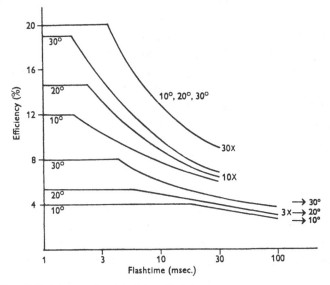

Fig. 7. Effect of flashtime, ratio of darktime/flashtime (3, 10, 30), and temperature on the efficiency of energy conversion in photosynthesis by *Chlorella*; → indicates efficiency reached in continuous light (after Kok, 1953).

If the dark periods are too long, the losses by respiration will become of the same order of magnitude as photosynthesis during the light periods, and although the absorbed light is used with maximum efficiency, no further increase in dry matter will result. This puts a limit to the length of the dark periods, and to the useful 'density' (concentration and thickness of layer) of the algal culture. Kok has estimated that the optimal density of a culture as regards effective intermittency would be some 100–400 μl. of cell material per cm.^{2} of irradiated surface. Kok (1953) finally remarks that algae with a higher AB level (see Fig. 1) would meet the requirements of intermittency more easily. At the moment he is now working in our group with that purpose in view.

It has been remarked (Burlew, 1953, p. 19) that a fundamentally different approach to the problem of overcoming light saturation at the surface may be provided by 'light dilutors', i.e. some sort of light transmitting system protruding from the surface into the depth of the culture. The same idea has arisen in Holland in connexion with the work of our group, and has been worked out theoretically in some detail (Tilstra, 1952), although experiments along this line have not yet been made.

A discussion on the effects of light saturation and intermittency should not be concluded without mentioning that in an algal culture irradiated with light of intensity *F* (Fig. 1) there is a layer at a certain distance below the surface where the light intensity will be *E*. Assuming that the culture is sufficiently dense to absorb practically all the incident light, we can say that all the light penetrating below the layer '*E*' will be utilized with maximum efficiency; the cells in these layers, however, do not show their maximum photosynthetic capacity. Above the layer '*E*' the cells will constantly show their maximum photosynthetic capacity, but they will waste light energy. Only in the layer '*E*' will the cells photosynthesize at maximum capacity with the maximum yield of energy conversion, and these cells will produce the maximum amount of dry weight in the most efficient way. In a suspension of a given density, the position of the layer '*E*' will vary with the intensity of the incident light, and will thus often change when cells are exposed to natural light. Increase in light intensity will shift the layer '*E*' downwards, thus increasing the number of cells that photosynthesize at maximum capacity. This is the reason why increasing an incident intensity which already is above light saturation still improves the yield, though this may be quite an inefficient method as far as light energy is concerned (see also Burlew, 1953, p. 17). Obviously, the best result would be obtained if, on the average, all cells would behave as if constantly in layer '*E*'. Probably the best empirical approach to this situation is obtained by violent turbulence in a suspension in which '*E*' is about half-way between the surface and the bottom.

VIII. YIELD OF ENERGY CONVERSION IN OUTDOOR AND INDOOR MASS CULTURES OF *CHLORELLA VULGARIS* (300 l. TANKS)

Outdoor mass cultures of *Chlorella* were grown in order to obtain preliminary evidence as to the efficiency with which light energy is utilized under these conditions. They were started in our group late in the summer of 1950. Most of the material on which the following statements are based was collected in 1951 by van Oorschot. Additional observations were made in 1952.

Rectangular concrete tanks, of 1 m.^{2} in area and 300 l. in capacity, were provided with Warburg's (1919) solution or a modification of the medium of Benson *et al.* (1949) (Table 4).

Completely aseptic conditions were not maintained; the tanks were cleaned thoroughly and covered with glass to keep them clean during incubation. They were inoculated with 10 l. of a pure culture. During growth the cultures received carbon dioxide regularly from a cylinder, continuous stirring being provided by a motor-driven stirrer (Pl. 2). The

Table 4. *Nutrient solutions used in tank experiments*

	Warburg solution	Benson *et al.* solution
Ca(NO$_3$)$_2$	1·00 g.	—
KNO$_3$	0·25 g.	0·50 g.
NaCl	0·15 g.	—
KH$_2$PO$_4$	0·25 g.	0·136 g.
MgSO$_4$.7H$_2$O	0·50 g.	0·50 g.
FeSO$_4$.7H$_2$O	0·02 g.	0·02 g.
Na citrate	0·04 g.	0·04 g.
Tap water	1 l.	1 l.

Table 5. *Data concerning environmental conditions, yield, chemical composition, and efficiency of light-energy conversion in outdoor mass cultures of* Chlorella (Wassink *et al.* 1953)

	Series A*	Series B†	Series C†
No. experiments‡	13	4	4
Incident energy (kcal./cm.^{2}/day)§	1757	1222	269
Yield, ash-free dry weight (g./m.^{2}/day)	7·15	4·92	2·70
Efficiency (%)‖	2·6	2·7	6·3
R-value	44·0	44·8	44·8

* 23 May to 8 October 1951; experiments made in full daylight.
† 23 August to 8 October 1951; B and C parallel experiments with reduced light intensity in C.
‡ Items below are averaged over reported number of experiments.
§ Exclusive of infrared radiation. Computed from radiation measurements by the Physics Department of the Agricultural University, Wageningen.
‖ Computed from ash-free dry weight, taking into account the R-value.

tanks were coated with black mastic inside, and in later experiments the inner surface was painted white. The average temperature was *c.* 16° C. at 9 a.m. and *c.* 25° C. at 5 p.m.

Table 5 is a summary of various series of experiments in which the algae were grown for 4–7 days and then harvested, washed, dried, weighed and sampled for the determination of their chemical composition (for further details, see Wassink *et al.* 1953). The results make it very evident that in full sunlight the average efficiency is 2·6–2·7%; in the series C in which the light intensity was about 22 % of full sunlight it was 6·3 %.

Experiments in which full daylight was weakened in several steps were carried out in 1952. By using appropriate gauze screens, the incident light intensity was decreased to 0·84, 0·50, 0·40 and 0·25 of full daylight. In most experiments the efficiency of light conversion increased regularly with decreasing light intensity. Fig. 8 shows an experiment in which the efficiency for light of relative intensity 0·84 is *c.* 4 %, that for 0·50 and 0·40 is 6–7 %, and that for 0·25 is 8–9 % of the incident light. The results of most experiments are in the same order of efficiency values.

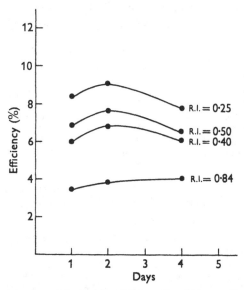

Fig. 8. Effect of weakening of the incident light on the efficiency of energy conversion in outdoor mass cultures of *Chlorella* (van Oorschot, J. L. P., 1952, unpublished). R.I. = Relative intensity compared with full daylight.

The influence of temperature on the efficiency was also studied. Since the incident light energy is above the saturation value, an effect might be expected. It turned out that no clear-cut effect of temperature was observed in densely inoculated cultures (5–10 g. cells/300 l.). Growth and efficiency in such cultures were much the same at the natural temperature (mostly 15–20° C.), and at about 30° C. (obtained with a heating coil). If, however, the cultures were inoculated thinly a marked divergence in growth and also in yield was observed, in favour of the higher temperature. The observed results can be understood by considering the location of the level with intensity E (§ VII, Fig. 1). Only cells above this level will be affected by temperature, those below will not be affected. The situation is complicated by the fact that the value of E is also temperature-dependent. But in general in a dense culture this level will be quite near the surface, so that, for example, 90 % of the cells may be in such a position that they are not affected by temperature. In dilute cultures, the E-level is considerably lower, hence an increased portion of the cells will react to temperature.

Not many other data on determinations of the efficiency of energy conversion in mass cultures are available in the literature. We may mention recent communications by Gummert *et al.* (1953), who observed an efficiency of *c.* 0·5 % of the total radiation (i.e. *c.* 1 % of the radiation capable of being utilized in photosynthesis) in large, non-sterile, greenhouse and open-air cultures. Geoghegan (1953) reports that in tubes of *c.* 7 cm. diameter, illuminated on one side at 1300 ft.c., the efficiency of light utilization was about 20 %. This appears very high in view of the fact that the incident intensity is fairly high and that the cultures must obviously be fairly dense so that respiration losses may not

Table 6. *Data concerning environmental conditions, yield, chemical composition, and efficiency of light-energy conversion in indoor mass cultures of* Chlorella (Wassink *et al.* 1953)

	Series A*	Series B*	Series C*
No. experiments†	9	3	3
Incident energy (kcal./m.²/24 hr.)‡	368	300	2650
Yield, ash-free dry weight (g./m.²/24 hr.)	7·70	5·20	19·3
Temperature (° C.)	24	32	32
Efficiency (%)§	13·3	10·9	4·7
R-value	46·0	44·2	43·6

  * Series A and B irradiated continuously with eight daylight fluorescent tubes; series C irradiated continuously with four 1000 W. incandescent lamps; series B and C consist of parallel cultures; duration of each experiment 3–6 days.
  † Items below are averaged over reported number of experiments.
  ‡ Exclusive of infrared radiation. Measured with calibrated photocell (see text).
  § Computed from ash-free dry weight, elementary composition and R-value.

be entirely negligible. Many more observations (see, for example, Davis *et al.* 1953) are available of yields in dry weight/m.²/day which could be converted into figures for efficiency of energy conversion, provided the irradiation data were available.

Indoor mass cultures were incubated in tanks similar to those used for the outdoor cultures. The irradiation was provided by either a set of eight daylight fluorescent tubes, or, for high light intensities, by four incandescent lamps of 1000 W. each (Pl. 3). Carbon dioxide was again administered from a cylinder, and the whole procedure was similar to that used outdoors except that the incident light intensity was measured at the surface by various interspaced photocell readings which were averaged and converted into absolute units by checking them with the reading of a standardized thermopile. The thermopile reading was taken with and without the filter RG 8 (Schott and Gen.), which allows an estimation of the infrared present in the incident light to be made. The cultures were illuminated continuously.

The results of these experiments are shown in Table 6. The efficiency was found to average 13·3 % for series A, the maximum being 19·7 %; series B showed an average of 10·9 % and the parallel high light intensity series C yielded 4·7 %. Temperatures were more regular, and especially in the high light intensity series they were higher than in the outdoor cultures. This may have contributed to the higher yield found in the indoor as compared with the outdoor cultures. It should not be forgotten, however, that the outdoor mass cultures, referred to above, have also occasionally yielded an efficiency of the order of 4 %.

IX. THE STUDY OF THE MASS CULTIVATION OF ALGAE

It has been remarked above (§I) that mass cultivation of algae received considerable attention only after the last war. Attempts to study the physiological implications of the subject have been made in various countries, e.g. Great Britain, U.S.A., Germany, Japan, Israel and the Netherlands. It is not known to the present author whether any work on the subject has been carried out in eastern Europe, U.S.S.R. or China. Various aspects of the problem have been tackled. Besides the basic problem of the efficiency of energy conversion, treated in some detail in the preceding sections, studies have already been made on the pattern of the daily growth of cultures, on inorganic nutrition, on the chemical composition of the algae, and on techniques for mass cultivation. An account of the information now available is given in a Carnegie report (Wassink *et al.* 1953; Kok, 1953; Gummert *et al.* 1953; Geoghegan, 1953; Davis *et al.* 1953). This report arrived too late to receive a detailed consideration in the present paper, and only a few data pertaining to the subject of energy conversion have been mentioned.

X. SUMMARY

1. General considerations about energy and food supply have directed attention to the possibility of setting up mass cultures of autotrophic micro-organisms as a technically available source of food and energy (§I).

2. Various groups of investigators have studied a number of basic problems related to this subject (§IX).

3. In the present paper, attention is concentrated upon the efficiency of energy conversion. The following points seem of importance.

4. Good field crops rarely exceed 2 % efficiency in the utilization of

radiation capable of being used for photosynthesis (or 1 % of total radiation) (§II).

5. The photosynthetic process as such, both in algae and leaf disks, is capable of converting at least 20 % of the incident light (when this is the limiting factor) into chemical matter (§III).

6. Small-scale growth experiments with algal cultures in the laboratory under conditions of light limitation showed that under these conditions the efficiency of growth is of the same order as that of photosynthesis (15–24 %) (§§IV and V). Sugar beets, cultivated under similar conditions, likewise display efficiencies of the order of 15 % (§IV).

7. Outdoor mass cultures of *Chlorella vulgaris* in natural daylight at Wageningen showed an average efficiency of 2·7 %. Weakening of the light improved the yield considerably. Improvement by increasing the temperature depended upon the density of the suspension (§VIII).

8. Indoor mass cultures of *C. vulgaris*, illuminated continuously with artificial light of low intensity, displayed efficiencies of about 13 %; on illumination with light of high intensity, the efficiency was about 5 % (§VIII).

9. The effect of certain cultural conditions, especially the nitrogen content of the medium, on the efficiency and composition of the cell material is discussed (§VI).

10. The effect of intermittency of light is discussed with reference to the stirring of mass cultures of algae (§VII).

Thanks are due to Dr van der Kerk, Utrecht, and to Prof. Coops, Amsterdam, for their co-operation in the determination of elementary analyses and heats of combustion.

REFERENCES

BENSON, A. A. *et al.* (1949). ¹⁴C in photosynthesis. In *Photosynthesis in Plants* (ed. J. Franck and W. E. Loomis), p. 381. Ames: Iowa State College Press.

BOEREMA, J. (1920). Intensiteit der zonnestraling. *Hand. I. Ned. Ind. Natuurw. Congr.* p. 99.

BURLEW, J. S. (1953). Current status of the large-scale culture of algae. In *Algal Culture from Laboratory to Pilot Plant*, p. 3. *Publ. Carneg. Instn.*

DANIELS, F. (1949). Solar energy. *Science*, **109**, 51.

DAVIS, E. A. *et al.* (1953). Laboratory experiments on *Chlorella* culture at the Carnegie Institution of Washington, Department of Plant Biology. In *Algal Culture from Laboratory to Pilot Plant*, p. 105. *Publ. Carneg. Instn.*

DEE, R. W. R. & REESINCK, J. M. (1951). Stralingsmetingen to Bandung. In *Meded. LandbHoogesch., Wageningen*, **51**, 167.

EMERSON, R. & ARNOLD, W. (1932). A separation of reactions in photosynthesis by means of intermittent light. *J. gen. Physiol.* **15**, 391.

GEOGHEGAN, M. J. (1953). Experiments with *Chlorella* at Jealott's Hill. In *Algal Culture from Laboratory to Pilot Plant*, p. 182. *Publ. Carneg. Instn.*

GUMMERT, F. *et al.* (1953). Nonsterile large-scale culture of *Chlorella* in greenhouse and open air. In *Algal Culture from Laboratory to Pilot Plant*, p. 166. *Publ. Carneg. Instn.*

KOK, B. (1948). A critical consideration of the quantum yield of *Chlorella* photosynthesis. *Enzymologia*, **13**, 1.

KOK, B. (1952). On the efficiency of *Chlorella* growth. *Acta bot. neerl.* **1**, 445.

KOK, B. (1953). Experiments on photosynthesis by *Chlorella* in flashing light. In *Algal Culture from Laboratory to Pilot Plant*, p. 63. *Publ. Carneg. Instn.*

PARDEE, A. B. (1949). Measurement of oxygen uptake under controlled pressures of carbon dioxide. *J. biol. Chem.* **179**, 1085.

RABINOWITCH, E. I. (1945). *Photosynthesis and Related Processes*, **1**. New York: Interscience.

REESINCK, J. M. (1947). Daglicht. *Electrotechniek*, **25**, no. 17.

SPOEHR, H. A. & MILNER, H. W. (1949). The chemical composition of *Chlorella*; effect of environmental conditions. *Plant Physiol.* **24**, 120.

TESCH, J. J. (1920). *Het Leven der Zee*. Amsterdam.

TILSTRA, J. F. (1952). Theoretische beschouwingen over het verband tussen rendement en wijze van belichting bij winning van zonneënergie uit een suspensie van algen. *Netherl. Organ. Appl. Sci. Res. (T.N.O.)*, The Hague.

VERMEULEN, D., WASSINK, E. C. & REMAN, G. H. (1937). On the fluorescence of photosynthesizing cells. *Enzymologia*, **4**, 254.

DE VRIES, E. (1948). *De Aarde Betaalt*. The Hague.

WARBURG, O. (1919). Über die Geschwindigkeit der photochemischen Kohlensaurezersetzung in lebenden Zellen. *Biochem. Z.* **100**, 23.

WASSINK, E. C. (1946). Experiments on photosynthesis of horticultural plants, with the aid of the Warburg method. *Enzymologia*, **12**, 3.

WASSINK, E. C. (1948). De lichtfactor in de photosynthese en zijn relatie tot andere milieufactoren. *Meded. Dir. Tuinb.* **11**, 503.

WASSINK, E. C., KOK, B. & VAN OORSCHOT, J. L. P. (1953). The efficiency of light-energy conversion in *Chlorella* cultures as compared with higher plants. In *Algal Culture from Laboratory to Pilot Plant*, p. 55. *Publ. Carneg. Instn.*

EXPLANATION OF PLATES 1–3

PLATE 1

Subirrigated beet root cultures in gravel, under daylight fluorescent tubes, used for the determination of the yield of light energy conversion during growth. White screens, surrounding the cultures, removed at the front side for taking the photograph (Glas, D. J. & Gaastra, P., 1949, unpublished).

PLATE 2

Outdoor 300 l. tanks (surface area 1 m.²) for mass cultivation of *Chlorella* in sunlight. Concrete tanks with glass covers, and stirring motor.

PLATE 3

Indoor 300 l. tanks (surface area 1 m.²) for mass cultivation of *Chlorella*. Fluorescent lights can be lowered to just above the glass cover of the tank. For high light intensities 1000 W. incandescent lamps have been used.

Plate 1.

Plate 2.

Plate 3.

THE ECONOMIC IMPORTANCE OF AUTO-TROPHIC MICRO-ORGANISMS

K. R. BUTLIN AND J. R. POSTGATE

Chemical Research Laboratory, Teddington, Middlesex

It is axiomatic that the green plants support all other forms of life. 'They accumulate chemical energy, while other organisms dissipate it' (Rabinowitch, 1945, p. 2). On land, by far the most important accumulators of chemical energy are the higher green plants; in the oceans, the phytoplankton (mainly photo-autotrophic micro-organisms) carry out this function, and on a much larger scale. If it is true that unicellular autotrophs, and probably chemo-autotrophs, lay on the line of evolution from the first 'live' molecules to the higher green plants, then, in an evolutionary sense, autotrophic micro-organisms are fundamental to the existence of macroscopic life on this planet. This, clearly, is the primary, even transcendental, factor in the economic importance of autotrophic micro-organisms. Indeed, it could be formally maintained that the title of this paper has no logical meaning, for how can one assess the economic importance of a group of organisms when, without them, there would be no economics? As pragmatists, however, we shall continue with our inquiry, for, whatever their primordial function may have been, autotrophic micro-organisms are clearly of importance both in the economy of nature and in their more immediate economic applications.

The distinction between autotrophy and heterotrophy is one of convenience rather than of biological truth; this subject is fully discussed by Woods & Lascelles (p. 1, this volume). It created difficulties in deciding on the scope of our subject. For example, our own work on sulphate-reducing bacteria has shown that most of these organisms can grow autotrophically by virtue of the enzyme hydrogenase which enables them to reduce both sulphate and carbon dioxide. Certain strains contain no hydrogenase activity and are consequently obligate heterotrophs. These strains can, however, develop hydrogenase after prolonged incubation in hydrogen and thus acquire, by adaptation or mutational change, the ability to exist autotrophically (*Chemistry Research 1951*, 1952, p. 100). Thus the distinction between autotrophy and heterotrophy can rest on a comparatively small difference in the enzymic constitution of closely related strains. For reasons of this kind, some facultative autotrophs of economic importance are more closely related to their strictly heterotrophic neighbours than to obligate autotrophs such as thiobacilli. Moreover, with facultative autotrophs it is often difficult to assess the extent to which their autotrophic faculties are involved in their economic function. For example, in the anaerobic corrosion of iron pipes in clay soils the sulphate-reducing bacteria grow heterotrophically but probably perform their 'corrosive' function mainly autotrophically; there are some photosynthetic algae which require traces of organic growth factors and therefore are not wholly autotrophic. We have been encouraged to interpret our theme broadly and will therefore include all organisms where it appears likely that their autotrophic functions are concerned in economic phenomena.

To avoid a catalogue of unrelated biochemical events and to achieve some sort of logical order, we shall, in general, discuss the economic importance of a given micro-organism under the heading of the trans-formation of the particular element—carbon, nitrogen, sulphur, iron—with which the metabolism of the organism is most commonly associated. Some organisms, notably the algae, are economically concerned in the transformation of several elements and will be discussed under several headings.

CARBON DIOXIDE FIXATION BY AUTOTROPHIC MICRO-ORGANISMS

The natural carbon cycle

The complete carbon cycle in nature (see Brian, 1953) has been simplified in Fig. 1 to emphasize the part played by autotrophic organisms. All the organic food of heterotrophs (microbes and animals) comes primarily from the activities of autotrophs. Autotrophs synthesize organic matter from carbon dioxide; heterotrophs inter-convert this organic matter for their several requirements giving off some carbon dioxide in the process; the remainder is eventually degraded to carbon dioxide by saprophytic micro-organisms, thus completing the cycle.

Photo-autotrophs are responsible for much the greater part of carbon dioxide fixation in nature. On land the chief agents are the higher green plants, but in the oceans, covering two-thirds of the earth's surface, carbon dioxide assimilation is carried out mainly by the microscopic *phytoplankton*, amongst which diatoms and dinoflagellates predominate. Rabinowitch (1945, p. 7 *et seq.*) calculated that carbon dioxide fixation in the oceans (15.5×10^{10} tons carbon yearly) is eight times greater than that by all land plants, and the 'efficiency' (fixation per unit area) is almost three times as great. He pointed out that this reduction of carbon dioxide, carried out mainly by autotrophic micro-organisms, is the largest single terrestrial chemical process; he compared it with the 10^9 tons annual output of the chemical, metallurgical and mining industries, 90 % of which is 'fossilized carbon' (coal and oil) produced by earlier photosynthesis. The economic significance of these figures needs no emphasis. On land the amount of organic synthesis from carbon dioxide by autotrophic micro-organisms is insignificant compared with that by the higher green plants, but it should not be forgotten that the chemo-autotrophs, and in particular the nitrifiers, play a vital part in providing suitable nutrients for these plants.

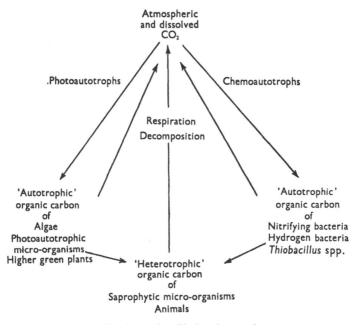

Fig. 1. A simplified carbon cycle.

Carbon dioxide-fixing autotrophic micro-organisms

The chemo-autotrophs—nitrifiers and denitrifiers, sulphur and iron bacteria—will be discussed under the various sections devoted to the transformations of the relevant elements. The photo-autotrophic green and purple sulphur bacteria (*Chlorobium* and *Chromatium*) will be described in the section on inorganic sulphur transformations; the purple non-sulphur bacteria, though of great fundamental interest, have little known economic significance and will not be further considered. In this section we shall deal in some detail with the green and blue-green algae, which are quantitatively the most influential microbiological agents of carbon dioxide fixation.

Green algae. The economically important members of the green algae include the diatoms and dinoflagellates, other constituents of the phyto-plankton, and certain genera such as *Chlorella* and *Scenedesmus* which have potential value as supplements to our food supply.

It is clear from Myers's (1951) review of the physiology of algae that the problem of autotrophy and heterotrophy in these organisms is very complicated. Some doubt must exist about the true autotrophy of most green algae, if only because few exacting tests on this property appear to have been made. We have had considerable difficulty, after consulting the literature and eminent algologists, in finding any definite answer on this point. From what evidence there is, it seems probable that some diatoms require traces of organic matter for growth, and that others are able to live heterotrophically (see Lewin, 1953), but with the proviso made above, most chlorophyllous diatoms appear to be capable of autotrophic growth in the true sense. There seems to be a little more doubt concerning the autotrophy of the Dinophyceae. With *Chlorella* and *Scenedesmus*, which have been more fully investigated, the picture is clearer; they possess both autotrophic and heterotrophic metabolisms in light and grow heterotrophically in the dark.

Blue-green algae. Morphologically and biochemically, the blue-green algae are more closely related to bacteria than to the green algae. Their chlorophyll, for example, appears to be distributed throughout the cell and not, as with green algae, segregated in chloroplasts. They are most widely distributed in soils and in fresh water but are also to be found in temperate and subtemperate seas, salt marshes, hot springs and in Antarctic lakes. Despite their general title, some species are red or violet: the Red Sea, for example, owes its colour to the presence of *Trichodesmium erythraem.* In addition to their photosynthetic abilities, certain species are capable of fixing nitrogen: their autotrophy could not very well be more complete. Fritsch (1945, p. 768) described the blue-green algae as 'an archaic group that has probably persisted during long epochs of the earth's history'. .

Natural economic activities

Figures have already been given illustrating the enormous extent of photosynthesis of organic material by the phytoplankton, beside which the other activities of photosynthetic organisms are relatively small, but certainly not unimportant.

Mineral deposits. The cell walls of diatoms contain up to 50 % silica (dry weight). When diatoms die and decay, their siliceous skeletons fall to the bottom of the water in which they have grown and, if conditions are favourable and growth continues over long periods, vast accumulations of fossilized diatoms, *diatomaceous earth*, are formed. Several such deposits are to be found in various parts of the world. Some originated in fresh waters, others in the ocean. The most extensive deposits, of marine species, are those at Lompoc, California, where the beds are 1000 ft. thick and extend over several miles. The thickest deposits so far discovered, about 3000 ft., are in the Santa Maria oil fields, California. There are large quantities in British Columbia and smaller deposits in Nova Scotia, New Brunswick, Quebec and Ontario. These diatomite products are used widely in industry, chiefly for filtration and insulating purposes (see Hull, Keel, Kenny & Gamson, 1953). They are also used as a mild abrasive in tooth pastes; with the addition of 'chlorophyll', the importance of photosynthesis in tooth hygiene is now well established, at least in the world of publicity.

Many fresh-water blue-green algae precipitate appreciable quantities of lime, sometimes producing extensive deposits. Howe (1932) gave a summary of the literature on such deposits. *Girvanella* is said to have played a part in the formation of oolites. The blue-green algae also play an important part in the disintegration of all kinds of calcareous substrates (Fritsch, 1945, p. 867).

Colonizing activity of blue-green algae. Some members of the blue-green algae (Chroococcales, *Nostoc*) grow prolifically on extensive tracts of ground, particularly in certain parts of China and Japan (Molisch, 1926, p. 104; Elenkin, 1934) and in these areas are used as food. According to Booth (1941) a community of blue-green algae forms the initial stage in plant succession in some eroded soils. Their activity in colonizing arid and rocky soils is described later (see p. 282).

Sewage disposal. Shallow lagoons are being increasingly used as an economical method for the treatment of sewage in certain parts of the United States. The purifying action is attributed to profuse algal growth, which produces and maintains a high concentration of dissolved oxygen. Bacteriological tests of some lagoons show that the coliform bacteria are rapidly destroyed; this destruction is believed to be due to production of chlorellin (Caldwell, 1946) or similar antibiotics of algal origin. The method is ideally adapted to the needs of small communities in mild climates (Myers, 1951); it was described in detail by Caldwell (1946) and Myers (1948).

Miscellaneous. Several interesting minor activities of various algae deserve mention. Blue-green algae are capable of living in strong brines (3 M-NaCl) and are frequently to be found in salt marshes. In solar salt works they serve a useful purpose by forming a carpet-like growth between salt and soil, thus preventing contamination of the salt by the black sulphide soils on which the salt is heaped (Fritsch, 1945, p. 862).

Certain dinoflagellates, notably *Gonyaulax* and *Gymnodinium*, are known to be responsible for mass mortality of fish in the Gulf of Mexico and for human poisoning by mussels along the Pacific Coast (Connell & Cross, 1950). Some marine forms of Dinophyceae, e.g. *Ceratium tripos*, are phosphorescent. Others are responsible for red snow (Suchlandt, 1916). Blue oysters owe their colour to the diatom *Navicula*.

Economic exploitation of algae

'*Fish farming.*' The phytoplankton is the basic producer of organic carbon in the sea; it supplies food for the zooplankton, and together they feed the fish, whales, etc. The productivity in terms of fish and whale oil, the two main commercial products from the sea, depends ultimately on the amount of phytoplankton produced, and this in turn is limited by the supply of nutrients, mainly by nitrate and phosphate.

There is an enormous disparity between the numbers of phytoplanktonic organisms observed in the sea and those obtained in laboratory cultures. For example, during the season of maximum productivity the number of diatoms in the sea may amount to 25 millions per litre (Harvey, Cooper, Lebour & Russell, 1935), yet in sea water enriched with nitrate, phosphate and soil extract, populations of 500 millions per litre have been obtained (Gross, 1941). Considerations such as these led to experiments in Loch Craiglin, an enclosed arm of Loch Sween, Scotland (Gross, Raymont, Marshall & Orr, 1944). The addition of sodium nitrate and superphosphate increased both the production of phytoplankton and fish. For example, in two years flounders completed growth which normally took 5–6 years. Utilization of the added fertilizer was extremely rapid; only traces remained a week after addition. These results suggested that investigations need not be confined to enclosed parts of the sea, and further experiments were carried out in Kyle Scotnish, an unenclosed arm of Loch Sween (Gross, Raymont, Nutman & Gauld, 1946). Encouraging results were obtained, but the authors suggest that experiments on a larger scale and in more suitable conditions would be necessary before the economics of 'fish-farming' could be assessed. There is obvious scope for the development of schemes for fertilizing the oceans in view of the rapidly increasing world population and the difficulties of feeding it.

Direct harvesting of marine phytoplankton. Although 15.5×10^{10} tons of organic carbon are synthesized by marine phytoplankton every year, the total annual commercial exploitation in the form of fish and whale oil amounts only to 14.5×10^6 tons (Wimpenny, 1943). The conversion of phytoplankton into fish and whales is very inefficient; there are also great losses caused by sedimentation of the phytoplankton produced. It is obvious that a greater yield of organic material could be obtained by direct harvesting of the phytoplankton at peak periods of production, and the process could be combined with addition of fertilizer. The crop might be used as food for animals; its use as human food is unlikely, for diatoms, the main constituent, contain a considerable amount of ash and silica, and their flavour is not attractive to most civilized palates. It is interesting to note, however, that members of the Kon-Tiki expedition collected plankton (phyto- and zooplankton). 'Two men on board thought plankton tasted delicious, two thought they were quite good, and for two the sight of them was enough' (Heyerdahl, 1950). The whole problem of exploitation of marine phytoplankton was discussed by Lucas (1953), who suggested that, in addition or as an alternative to its possible use as food, valuable by-products, e.g. vitamins and antibiotics, might be extracted.

Mass culture of algae. Photosynthesis in nature rarely operates in optimum conditions: the production of organic carbon, considered merely as the synthesis of so much protein, fat or carbohydrate, is grossly inefficient, handicapped as it is by the vagaries of climate, lack of nutrients at the right time and place, and the consumption of much of the photosynthetic products in other activities, e.g. the respiration of non-photosynthetic parts of green plants. In recent years, considerable labour has gone into efforts to utilize solar energy more efficiently for the production of food and other valuable products by mass growth of unicellular green algae. Most of the work has been carried out with *Chlorella*, but there are other genera, e.g. *Scenedesmus*, which might be suitable for industrial exploitation. Recent work on the problem was described by Harder & Witsch (1942), Spoehr & Milner (1948), Cook (1950), Pearsall & Fogg (1951), Geoghegan (1951), Myers, Phillips & Graham (1951) and Gaffron (1953). The present state of the research has been reviewed by Macfarlane (1951), Baldwin (1953) and Burlew (1953).

The chief points from this work are:

(i) *Chlorella* cells attain their highest photosynthetic capacity in the actively dividing state. For maximum production of organic carbon a continuous-flow method which maintains cultures in the exponential phase of growth is likely to be more useful than is static growth.

(ii) The main product is protein, carbohydrate or fat according to the stage of growth and species of alga used. For example, using *Chlorella vulgaris*, the synthesis of protein during active growth (in presence of plentiful combined nitrogen) is followed by accumulation of carbohydrate. *Chlorella pyrenoidosa*, however, accumulates fats in the later stages of growth. The relative amounts of protein, carbohydrate and fat is controlled within certain limits by the composition of the medium (Spoehr & Milner, 1949).

(iii) Mineral requirements are not exacting except for nitrate, and many natural waters would be suitable with slight supplements. For good growth air enriched with 5 % CO_2 is necessary (Spoehr & Milner, 1949): a cheap and plentiful supply of carbon dioxide would be essential for industrial exploitation.

(iv) Contamination by possibly undesirable organisms is not serious in rapidly growing mass cultures (Cook, 1950; Myers *et al.* 1951).

(v) Certain constituents of algae may be important as by-products. Witsch (1946) suggested that algae might be a source of thiamine. *Chlorella* species contain appreciable quantities of carotenoid pigments which may act as precursors of vitamin A (Pearsall & Fogg, 1951). Pratt *et al.* (1944) described the isolation of chlorellin, an antibiotic produced in cultures of *C. vulgaris*. *Chlorella* contains 0·2 % of sterols (Baldwin, 1953). Chondrillasterol, which has a potential use in cortisone synthesis, is present in *Scenedesmus obliquus* to the extent of about 0·2 % (Bergmann & Feeney, 1950). See also Burlew (1953, chs. 20–22).

(vi) *Chlorella pyrenoidosa* has been produced on a pilot plant scale using continuous-flow techniques. As a result of data obtained from these experiments a large-scale plant covering $1\frac{1}{2}$ million sq.ft. has been designed. Such a plant, it was estimated, could produce 30 tons of *Chlorella* per acre, compared with 10 tons per acre of crops grown conventionally in California (Cook, 1951). The cost of the product, assuming its use only as a supplementary food to balance animal and poultry rations, would be prohibitive in comparison with equivalent supplements such as soya-bean flour.

MICROBIOLOGICAL INORGANIC NITROGEN TRANSFORMATIONS

The flow of nitrogen compounds in nature is as essential to life as the flow of carbon compounds: the fertility of both soil and sea depend on the availability of various forms of nitrogen to different living organisms. Nitrogen transformations are very largely the result of microbiological activity; micro-organisms quickly make available the protein nitrogen of animals and plants for new generations, 'and if it were not for nitrogen fixation by bacteria and blue-green algae, all soils would long since have become completely barren' (Brian, 1953). The nitrogen cycle both in soil and sea is a very complicated process and cannot be described in detail here. In general it can be stated that the *rate* at which the various transformations take place, i.e. the rate at which the various compounds become available for their specific purposes, is a direct measure of productivity or fertility, and this largely depends on the character and efficiency of the specific organisms.

A simplified microbial nitrogen cycle is shown in Fig. 2, which forms part of the full nitrogen cycle described by many workers (e.g. Brian, 1953). Autotrophic micro-organisms participate in all the stages shown

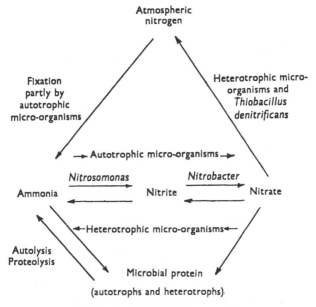

Fig. 2. A microbial nitrogen cycle.

except the denitrification of nitrate to ammonia, but their most important functions are nitrification and nitrogen fixation.

Nitrification. Perhaps the most important soil microbiological process after nitrogen fixation is the conversion of ammonia to nitrate, the form of nitrogen most available to plants and therefore of the greatest importance to animal life. Nitrification proceeds in two stages: oxidation of ammonia to nitrite by *Nitrosomonas, Nitrosococcus, Nitrosospira* and *Nitrosocystis*, and oxidation of nitrite to nitrate by *Nitrobacter, Nitrocystis* and *Bactoderma* (see Meiklejohn, 1953). Only the two genera *Nitrosomonas* and *Nitrobacter* have been studied in detail. Both types of bacteria are generally regarded as obligate chemo-autotrophs;

Dr Meiklejohn discusses the validity of this description elsewhere in this Symposium (p. 78). In the laboratory, using pure cultures, organic matter may be harmful, inert or stimulatory; in natural conditions in the soil it is possible that heterotrophs keep soluble organic material at a concentration low enough not to affect nitrification. The alleged inhibition of nitrification by organic matter was recently reviewed by Meiklejohn (1953) and discussed by Lees (1953).

Studies on the biochemistry of nitrification were reviewed by Quastel & Scholefield (1951). The soil perfusion technique of Lees & Quastel (1944) has been of great value in studying the influence of organic matter, fertilizers, etc., on soil samples in conditions approximating to those found in nature (see Chase, 1948; Lochhead, 1952). Many workers have studied the nitrifying bacteria in pure culture, while numerous investigations of nitrification in soils have been carried out in field conditions. Lees (1953) has argued 'that the results obtained from soil studies are in entire agreement with those obtained from culture studies and that the weight of the evidence suggests that under field conditions nitrification is largely, if not entirely, accomplished by the activities of *Nitrosomonas* and *Nitrobacter*'.

The view that nitrification in soil, especially in tropical countries, may be photochemical rather than microbiological, is now not valid. Joshi & Biswas (1948) found no evidence that sunlight promoted nitrification and concluded that it had no practical significance in Indian soils. In the sea, photochemical oxidation of ammonia may occur to a limited extent, but it is generally accepted that bacteria are chiefly responsible, though the evidence is confusing (ZoBell, 1946, p. 151). In view of the fact that ammonia, nitrite and nitrate are commonly found in surface waters (see Sverdrup, Johnson & Fleming, 1942, p. 243), it is surprising that nitrifying organisms have rarely been found in the surface layers of ocean waters, though often in bottom deposits. This failure may be due to faulty isolation techniques. It is often assumed that nitrate is formed near the sea bed and carried up to the surface. There is no conclusive evidence on this point, which is of considerable economic importance to the fertility of the photosynthetic (phytoplanktonic) zone, on which depends the yield of fish, whales, etc. There is a striking difference between the nitrogen cycles in soil and sea. In soils there are usually appreciable nitrogen reserves even at the periods of greatest crop production, but in the sea the production of living matter is limited by the supply of nitrogen (and phosphorus). For example, it has been found that in the English Channel at the height of summer, when growth of diatoms is at its maximum, *all* free nitrate- and ammonia-nitrogen disappears (Brian, 1953, p. 209). The economic significance of this phenomenon is that the productivity of the sea could be increased at will by the addition of nitrate (and phosphate). Experiments to increase productivity by the addition of fertilizers have been described in some detail above.

Nitrifying bacteria have been (unwittingly) used for centuries in the production of saltpetre. Marshall (1915) gave an interesting historical account of the various methods employed for its recovery from Arabian, Egyptian, Chinese and Indian soils. The most favourable conditions for natural production appear to be a hot climate with long periods without rain (to prevent loss by leaching) and primitive disposal of sewage and other nitrogenous waste materials. To-day, about 10,000 tons of refined saltpetre are extracted annually from saline efflorescences on the surface of soil in Bihar, Central India, the deposits being formed by nitrification processes in a mixture of soil, urine, cow-dung and wood ashes. The discovery of gunpowder gave an impetus to the recovery of saltpetre from suitable sites in Western Europe, chiefly stables, sheep pens, cattle sheds, disused graveyards and burial mounds. When France was unable to import nitrate during the Napoleonic wars, artificial nitre-beds of earth mixed with waste animal and vegetable matter, ashes and building materials were built inside wooden sheds. The heaps were liberally treated with urine and waste water, and aerated either by means of pipes or by turning over from time to time. After about two years the crude saltpetre was extracted with hot water. Civilization (both Western and Eastern) has now produced a more potent explosive than gunpowder, and saltpetre will no doubt be used more for its peaceful purposes, in tobacco and in fireworks, and for curing meats.

An interesting minor activity of nitrifying organisms should be mentioned. Kauffmann (1952) attributed corrosion of stone monuments partly to their activities: the organisms oxidize atmospheric ammonia to nitrous and nitric acids, and use carbonate in the stone. Corrosion results from the leaching out of calcium nitrite and nitrate.

Denitrification. The only known autotrophic denitrifying microorganism is *Thiobacillus denitrificans*, which was first isolated by Beijerinck and is widely distributed in water, mud and soil. It is a facultative anaerobe which obtains energy for the reduction of nitrate to nitrogen by oxidizing sulphur and thiosulphate. Its curious metabolism is of great fundamental interest (see Baalsrud, p. 54, this volume). There is a lack of published exact knowledge of this organism which makes impossible any reliable assessment of its economic importance, especially of the part it plays in natural denitrifying processes. Similar considerations apply to the heterotrophic denitrifying organism *Micrococcus denitrificans*. Kluyver (1953) has shown that in certain conditions this organism produces a hydrogenase which enables it to use molecular hydrogen in nitrate reduction. Certain organic substances are necessary for growth, but the cells possess an autotrophic mechanism for nitrate reduction in the presence of hydrogen.

Nitrogen fixation. In nature, the greater part of nitrogen fixation takes place through an association of bacteria with the root nodules of leguminous plants (symbiotic fixation) and by means of the heterotrophic *Azotobacter*. In recent years a third group of organisms, the photo-autotrophic blue-green algae, have been shown to play an important part in certain specialized environments. Their activities have been reviewed by Fogg (1947), Virtanen (1948) and Lochhead (1952). The difficulty of isolating the organisms in pure culture at first hindered conclusive proof of their nitrogen-fixing activity. For this reason, most of the known nitrogen-fixers belong to the family Nostocaceae, which are easier to culture than the others. With improved methods of isolation other families are being drawn into the picture (Fogg, 1951; Watanabe, Nishigaki & Konishi, 1951; Williams & Burris, 1952).

Nitrogen fixation by blue-green algae only takes place during active growth and is inhibited by ammonium salts or nitrates. Consequently, though blue-green algae are commonly found in temperate fertile soils, it is doubtful whether they contribute to the fertility of such soils. They can, however, be of great importance in poor waterlogged soils, where heterotrophic nitrogen-fixing bacteria are unlikely to flourish. Thus, in Indian rice-fields, growth of blue-green algae is very prolific and contributes notably to the fertility of the soil (De, 1939). Rice can, in fact, be grown year after year in the same soil without any addition of manure.

Another property of blue-green algae is their capacity to colonize arid soils and rocks, perhaps because of their ability to fix nitrogen and carbon dioxide and their resistance to adverse conditions of temperature, desiccation and high salt concentrations. In 1883 a volcanic eruption destroyed all visible plant life on the island of Krakatoa. The first plants to appear on the volcanic debris were blue-green algae in the form of a dark green gelatinous layer (Ernst, 1908). Fogg (1947) quoted several other examples of this colonizing ability. Singh (1950) described the reclamation of alkaline 'Usar' lands in northern India. These infertile soils support a rich growth of blue-green algae during the rainy season. Areas of about one acre were enclosed by low earth embankments, forming shallow ponds which encouraged such growths. This resulted in an increase of organic matter by 36·5–59·7 %, and of nitrogen by 30–38·4 %. In one trial the pH of the soil decreased from 9·5 to 7·6. Crops were successfully grown on the reclaimed land. This method promises to be of immense practical value to Indian agriculture and might well be applied to other barren soils.

Recent work has shown that nitrogen-fixation can be carried out by several other autotrophic or facultatively autotrophic organisms: the photosynthetic purple non-sulphur bacterium *Rhodospirillum rubrum* (Gest & Kamen, 1949; Lindstrom, Burris & Wilson, 1949); the green and purple sulphur bacteria *Chlorobium* and *Chromatium* (Lindstrom, Tove & Wilson, 1950); and autotrophic strains of sulphate-reducing bacteria (Sisler & ZoBell, 1951a). There appears to be some correlation between the ability to fix nitrogen and the presence of hydrogenase, though several organisms are known which contain hydrogenase but are unable to fix nitrogen. It is unlikely that any of the organisms mentioned have any significant economic importance by virtue of their nitrogen-fixing ability, but it is possible that this property may be important in localized special conditions and may have been very important in early geological eras.

MICROBIOLOGICAL INORGANIC SULPHUR TRANSFORMATIONS

The autotrophic bacterial sulphur cycle

The natural sulphur cycle and that part of it carried out by bacteria have been described several times (Ellis, 1932; Bunker, 1936; ZoBell, 1946; Butlin, 1953; Butlin & Postgate, 1953a); we are concerned here only with transformations by obligate or facultative autotrophs. The organisms involved and their principal reactions are summarized in the simplified microbial sulphur cycle shown in Fig. 3. Other inorganic sulphur compounds such as thiosulphate, tetrathionate and polythionates also participate in the natural cycle, but their transformations are comparatively unimportant.

The organisms involved are often associated with algae and protozoa in an ecological community which Baas-Becking (1925) described as a *sulphuretum*. The basic material on which this community subsists is hydrogen sulphide, usually formed heterotrophically by sulphate-reducing bacteria, though they may form part of it autotrophically if hydrogen is available through the activities of commensal hydrogen-producing bacteria. Changes in the environment can displace the equilibria within the *sulphuretum* and lead to the accumulation and exhaustion of various components of the sulphur cycle: e.g. sulphate, which is readily assimilated by the higher plants, may accumulate; sulphur may be precipitated; toxic concentrations of hydrogen sulphide may accumulate and destroy most forms of life; one or more of these components may become exhausted or fixed in a form in which it is not biologically utilizable. Such transformations probably occurred on a gigantic scale in early geological times and affected the present character and distribution of sulphur-containing minerals. In the oceans they can be observed on a large scale to-day (ZoBell, 1946, ch. 12), but on land

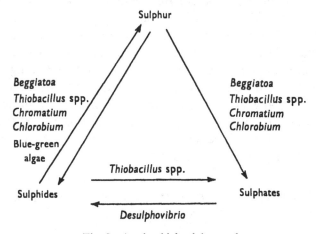

Fig. 3. A microbial sulphur cycle.

they are not so obvious because environments suitable for the formation of a *sulphuretum* are chiefly restricted to clay or waterlogged soils, muds and stagnant waters; aerated waters and fertile soils are unsuitable. Some examples of economic importance in which autotrophic sulphur bacteria are concerned are given below.

Sulphate-reducing bacteria

Sulphate-reducing bacteria are strict anaerobes. They are usually grown in a medium containing sulphate, other minerals, and an energy source such as lactate, pyruvate, malate, etc.; for good growth, the medium is enriched with materials such as yeast extract or peptone. Most strains are, however, facultative autotrophs. The nature of their autotrophic metabolism has not been studied in detail, but the established facts can be summarized briefly:

(1) Pure strains can grow autotrophically in hydrogen provided sulphate and bicarbonate are present; hydrogen can be supplied electrolytically from metallic iron in the culture (Starkey & Wight, 1945;

Butlin & Adams, 1947). Strains not containing hydrogenase are obligate heterotrophs (Adams, Butlin, Hollands & Postgate, 1951); adaptive (or mutational) formation of hydrogenase can take place in such strains, which can then grow autotrophically. Sisler & ZoBell (1951*b*) have described strains able to grow autotrophically with carbonate and hydrogen in the absence of sulphate.

(2) Autotrophic growth is slight compared with heterotrophic growth, but 'autotrophic' sulphate-reduction with hydrogen is more rapid than with any organic reducing agent (Butlin & Postgate, 1953*a*).

In a heterotrophic environment containing hydrogen, both the hetero-trophic and autotrophic metabolisms of sulphate-reducers will come into action. The growth will be predominantly heterotrophic, but, particularly if much hydrogen is present, the 'autotrophic' reduction of sulphate might well yield more sulphide than heterotrophic reduction. This mixed type of metabolism is probably of frequent occurrence in nature, and it is all but impossible to disentangle the two effects. There are, for example, many instances of mass evolution of hydrogen sulphide from polluted waters, causing not only considerable nuisance but severe damage to paint on ships and houses, and heavy mortality to fish (Butlin, 1949). The primary cause of these outbreaks is undoubtedly the presence of organic material and sulphate. There is, however, strong evidence that appreciable quantities of hydrogen and other gases are evolved during the decomposition of organic matter at the bottom of lakes and in marine sediments (Kusnetzov, 1935; Sugawara, Shintani & Oyama, 1937; Waksman, 1941; ZoBell, 1946). It is therefore highly probable that part of the malodorous activities of sulphate-reducing bacteria are the result of their autotrophic metabolism following heterotrophic growth. The crude culture of sulphate-reducers inhabiting the lower layers of the Black Sea, where none but bacterial life can exist, probably leads this mixed heterotrophic-autotrophic existence.

There are other large-scale natural activities of sulphate-reducing bacteria to which similar considerations apply. Bavendamm (1932) attributed the deposition of chalk in tropical seas to the bacterial reduction of calcium sulphate followed by reaction of the sulphide with carbon dioxide. The presence of sodium carbonate in salt lakes and deposits is probably the result of similar reactions (Gubin & Tzechom-skaya, 1930; Verner & Orlovskii, 1948). Miller (1950) suggested that sulphate-reducers may have been concerned in the formation of sulphide ores (see also Bunker, 1936, p. 36). One important indirect consequence of their activities is the presence of sulphur compounds, probably derived from hydrogen sulphide, in petroleum. Hydrogen sulphide produced by these organisms also contaminates coal gas stored in water-sealed gas holders, and the prevention of such contamination either by removal of sulphate or by inhibition of sulphate-reduction is expensive because of the huge volumes of water involved. No satisfactory solution to this problem has been found.

Corrosion of buried iron pipes. Serious losses are caused by the corrosion of buried iron pipes conveying water, gas and petroleum, chiefly in waterlogged clay soils. A total annual cost to Great Britain of £5 million for maintenance and replacements is probably an underestimate; in the U.S.A. the annual cost is variously estimated at $200 million (Logan, 1948, p. 446) and $600 million (Uhlig, 1949). The proportion of these costs attributable to anaerobic corrosion cannot be estimated, but in Great Britain it must be large, since three-quarters of the soils are clay.

Most anaerobic corrosion is characterized by the presence of sulphide in the corrosion products, and it is now widely accepted that sulphate-reducing bacteria are largely responsible for this type of corrosion. The theory of cathodic depolarization by these bacteria was first suggested by von Wolzogen Kühr & van der Vlugt (1934), and its general validity has been examined by several workers (e.g. Bunker, 1939; Starkey & Wight, 1945); the evidence in favour of it was summarized by Starkey (1947), who later (1953) presented evidence against it. The theory can be briefly stated as follows: the surface of wet iron contains numerous electrochemical systems, the cathodic elements of which (in neutral conditions and in the absence of oxygen) are polarized by hydrogen. The enzyme hydrogenase enables sulphate-reducing bacteria to use the hydrogen for reducing sulphate and in so doing enables the corrosion mechanism to operate, iron going into solution at the anodic elements. Thus, the combination of bacteria and sulphate performs a function similar to that of oxygen in aerobic corrosion.

The removal of hydrogen in this type of corrosion is carried out by the autotrophic metabolism of the sulphate-reducers, though the bacteria may have grown heterotrophically. That significant corrosion can also occur in completely autotrophic cultures was shown by Adams & Farrer (1953) in experiments extending over two years; the failure of Spruit & Wanklyn (1951) to detect significant corrosion in their autotrophic cultures is perhaps explained by the shorter duration (8–56 days) of their experiments. Conditions are rarely autotrophic in the field, though Farrer (personal communication) has observed severe corrosion, typical of *Desulphovibrio desulphuricans*, of a ferrous pipe buried 18 ft. in clay containing no detectable organic matter.

Corrosion is more rapid in heterotrophic cultures, doubtless owing to the presence of greater numbers of bacteria; even in such conditions the bacteria will utilize hydrogen preferentially for sulphate-reduction (Postgate, 1951). In clay soils conditions are almost invariably heterotrophic, and the availability of organic matter will greatly increase the number of bacteria on the metal surface and hence the rate of corrosion. Other factors affecting the corrosion rate are the availability of sulphate and variations in the pH and redox potential of the soil; corrosion in field conditions is usually more severe in alternating anaerobic and aerobic conditions than in wholly anaerobic conditions. The direct action of hydrogen sulphide on the corrosion rate of mild steel appears to be slight (*Chemistry Research 1952*, 1953, p. 12), but the presence of elementary sulphur, formed by the aerobic oxidation of sulphide during dry periods (von Wolzogen Kühr, 1938), may cause exceptionally rapid corrosion (Farrer & Wormwell, 1953). The iron sulphide formed round the corroding metal, being cathodic to metallic iron, itself encourages corrosion (Stumper, 1923), and may well play an important part in the corrosion process. It is indeed probable that in some, perhaps most, natural conditions the secondary effects of sulphate reduction are more important than cathodic depolarization by the bacteria. Starkey (1953), as a result of field experiments with steel specimens in a marine environment, considered it 'unlikely that the principal role of sulphate-reducing bacteria in corrosion is that proposed by von Wolzogen Kühr'.

Methods of preventing underground corrosion were reviewed by Butlin & Vernon (1949). Perhaps the best method is to make the pipeline the cathode of an electric system, either by an applied current or by connecting it to 'sacrificial' anodes of magnesium, aluminium or zinc. One effect of applying this 'cathodic protection' is to charge the outside of the pipe with hydrogen. Though the bacteria can and do utilize the hydrogen, depolarization leads to (a) a high pH round the pipe, preventing further bacterial action, and (b) corrosion of the anodes but not of the pipe.

Anaerobic microbiological corrosion can also occur beneath marine growths on the hulls of ships and on ferrous installations such as jetties, piers, etc. Starkey (1953) cultivated sulphate-reducing bacteria in a medium with barnacle tissue as the only organic material. Severe corrosion of the interior surface of water pipes is frequently found and is associated with nodular incrustations (tubercles) formed by electrochemical action which is sometimes initiated by iron bacteria (see section on iron transformations); the metal beneath the tubercles is invariably pitted or graphitized. Bunker (1940) detected sulphide and sulphate-reducers in tubercles and Butlin, Adams & Thomas (1949) found in several freshly obtained samples that the layers of tubercle close to the metal contained up to 2·5 % of sulphide, which was rapidly oxidized on exposure to air. They readily obtained pure cultures of sulphate-reducing bacteria from enrichment cultures in an inorganic medium incubated in hydrogen. In these instances it was possible that both growth and corrosion were mainly autotrophic. It seems probable therefore that autotrophic sulphate-reducers can be responsible for some internal corrosion of ferrous water pipes supplementary to that caused electrochemically during the formation of the tubercles.

Formation of oil deposits. The formation of paraffinic hydrocarbons in heterotrophic cultures of sulphate-reducing bacteria was reported by Jankowski & ZoBell (1944) and in autotrophic cultures by Sisler & ZoBell (1951*b*). Both reports suggest that the bacteria may play a part in the formation of petroleum, but their precise importance is still an open question (Stone & ZoBell, 1952). On the other hand, there is evidence that they are concerned in the liberation of adsorbed oil from certain substrates and may thus help the accumulation of oil in reservoirs (ZoBell, 1947).

Industrial production of sulphur. The possibility of developing an industrial process for the microbiological production of sulphide and hence (by chemical oxidation) of sulphur, has been discussed recently (Butlin, 1953; Butlin & Postgate, 1953*a*). The most promising method, now being examined in our laboratory, involves the anaerobic digestion of sewage sludge supplemented with sulphate. The small quantity of hydrogen evolved during digestion would undoubtedly be used for autotrophic sulphate reduction, but the process must be almost wholly heterotrophic.

Another technically possible process is the bacterial reduction of sulphate with hydrogen as the reducing agent, thus using the autotrophic metabolism of sulphate-reducing bacteria. Senez (1953) examined the rate and amount of sulphide produced by sulphate-reducing cultures in inorganic medium enriched with yeast extract to promote growth and using hydrogen as reducing agent. He noted that the addition of glass wool greatly improved both the rate and amount of sulphate-reduction and that growth was mainly restricted to the glass surfaces (see also ZoBell, 1943). Hydrogen gas is an eminently suitable material for microbiological sulphate-reduction; the rate of reduction is more rapid than with organic hydrogen-donors and it should lend itself readily to continuous fermentation techniques. However, the present cost of hydrogen in Great Britain is prohibitive and supplies are inadequate, but if plentiful and cheap supplies become available, as may possibly happen within the next few years, its use should be considered.

Sulphide-oxidizing bacteria

Several types of autotrophic or facultatively autotrophic micro-organisms are able to oxidize sulphide: the aerobic autotrophs, e.g. *Beggiatoa* and *Thiothrix*; the coloured sulphur bacteria *Chromatium* and *Chlorobium*, which are anaerobic photosynthetic facultative autotrophs; certain thiobacilli; and certain diatoms (*Pinnularia*) and blue-green algae (*Oscillatoria*), which, like the purple sulphur bacteria, store sulphur granules within the cell (Nakamura, 1938). The relationship between all these organisms and the Myxophyceae is discussed by Pringsheim (1949 *a*). They play a useful part in the economy of nature by removing hydrogen sulphide. In this connexion the most important appear to be the coloured sulphur bacteria. There are numerous species but we shall refer only to the two genera *Chromatium* (purple) and *Chlorobium* (green).

There are many reports of the mass growth of *Chromatium* spp. giving rise to 'red waters' and 'bloody seas' (Gietzen, 1931; Bavendamm, 1932; ZoBell, 1946). In 1952 we encountered a striking example of their mass growth in a heavily polluted, stinking, stagnant stream at Kempton Park, Middlesex. In an attempt to clean it, the owners hosed in almost half its volume of tap water. Over the next few days the water became pink and after ten days a spectacular milky crimson. The water was rich in cells of *Chromatium* bearing sulphur granules and was much less offensive than before treatment.

There are fewer reports of 'green waters' perhaps because *Chromatium* and *Chlorobium* usually grow together and the purple colour pre-dominates. Bicknell (1949) reported on the distribution of *Chlorobium limicola* in Sodon Lake, Michigan, where the organisms had appeared in great numbers. Microscopic examination indicated that an almost pure culture existed at a depth of 25 ± 5 ft., below which the water was rich in hydrogen sulphide, but within and above this depth hydrogen sulphide was undetectable. Apparently hydrogen sulphide was continuously formed in the lake mud and was removed when it reached the *Chlorobium* zone.

These examples of the removal of hydrogen sulphide by the coloured sulphur organisms suggest that a deliberate control of the nuisance caused by polluted waters might be feasible. At present insufficient is known of the conditions necessary for the growth of these organisms in natural environments. The required conditions are apparently simple:

the presence of hydrogen sulphide, inorganic salts and sunlight. It is likely that research would yield promising results in this field.

Sulphur formation in Cyrenaican lakes. The commonest source of sulphide for the development of sulphide-oxidizing organisms is the bacterial reduction of sulphate. At least one natural process is known in which a combination of sulphate-reducing and sulphide-oxidizing bacteria is responsible for the formation of appreciable quantities of elementary sulphur. In 1950 we investigated certain lakes in Cyrenaica. These were characterized by mass growth of *Chromatium* and *Chlorobium* in the shallow waters, the deeper lake waters being crude cultures of sulphate-reducing bacteria, and a sulphur mud 6–9 in. deep lay on the bottom. One small lake containing half a million gallons produced about 100 tons of sulphur annually, which was harvested by local Senussi. The expedition has been reported in detail elsewhere (*Chemistry Research 1950*, 1951; Butlin & Postgate, 1953*b*). Laboratory experiments showed that a combination of pure cultures of sulphate-reducing bacteria with *Chlorobium* or *Chromatium* (all isolated from the lakes) formed sulphur from sulphate.

The sulphate-reducers obtained from the lakes did not possess any hydrogenase activity and were therefore obligate heterotrophs, yet only traces of organic matter were present in the lake waters. Pure culture experiments showed, however, that these sulphate-reducing strains could grow in autotrophic media in combination with *Chlorobium* or *Chromatium*. From this it appeared likely that the two latter synthesized organic matter both for the growth of the sulphate-reducers and for sulphate-reduction. Thus, although the continuous supply of a small quantity of organic matter by the springs feeding the lakes was presumably utilized, it was possible that, in addition, sulphate was converted to sulphur by autotrophic mechanisms, using energy derived ultimately from sunlight. The continuous formation of sulphur in Cyrenaican lakes suggests the possibility of exploiting natural waters to augment the rapidly diminishing supplies of natural sulphur. The composition of the lake waters is comparatively simple; the only other requirement is sunlight. This problem has been discussed elsewhere (Murzaev, 1950; Butlin, 1953; Butlin & Postgate, 1953*b*), but it is clear that much more detailed knowledge of the symbiosis of sulphate-reducing and sulphide-oxidizing organisms is required before any exploitation is possible.

Origin of sulphur deposits. Hunt (1915) attributed the formation of the Sicilian sulphur deposits to bacterial sulphate-reduction in marine conditions resembling those now prevailing in the Black Sea. It is generally believed that most of the non-volcanic deposits were formed by microbiological action, though the evidence is necessarily speculative. Our observations on the Cyrenaican lakes suggest a possible autotrophic mechanism for sulphur deposition in which sulphate-reducing and sulphide-oxidizing bacteria are concerned. It is important to note, however, that several other mechanisms are possible in principle. Sulphide formed from sulphate or decomposing proteins could be oxidized by *Thiobacillus* (Senez, 1951) or *Beggiatoa*, by nitrite produced by nitrifying organisms (Iya & Sreenivasaya, 1944), or by atmospheric oxidation catalysed by iron salts (Subba Rao, Iya & Sreenivasaya, 1947). None of these appeared to be important in the Cyrenaican lakes.

Isotope fractionation. Microbiologically produced sulphur from various sources (including the Cyrenaican lakes) has a different isotopic constitution from the sulphur of the sulphate from which it is produced, and from volcanic and meteoric sulphur (Macnamara & Thode, 1951). This is due to preferential utilization of ^{32}S by micro-organisms, and consequently it is possible to deduce the origin of sulphur deposits from their isotopic constitution. An interesting aspect of this work quoted by Pirie (1953) deserves mention. Sulphur, sulphate and sulphide in very old deposits have the 'meteoric' isotope ratio, but in deposits laid down less than 800 million years ago the sulphur and sulphide have the 'biological' isotope ratio. This indicates that biological sulphur metabolism took place to a significant extent 300 million years before the earliest abundant fossil records of the Cambrian age (*c.* 500 million years ago).

Sulphur-oxidizing bacteria

Beggiatoa and *Chromatium* autotrophically oxidize sulphide to sulphur, which appears as granules inside the cell; in sulphide-deficient conditions the sulphur is further oxidized to sulphate. Some *Chlorobium* spp. are capable of oxidizing sulphide, through sulphur, to sulphate (Larsen, 1952). Certain *Thiobacillus* spp. can oxidize sulphide to sulphate, though preliminary chemical oxidation to sulphur may be necessary (Parker & Prisk, 1953), but economically the most important characteristic of *Thiobacillus* spp. is the ability to oxidize elementary sulphur to sulphate. With bicarbonate as carbon source, *T. thio-oxidans* can produce, from sulphur and mineral salts, as much as 10 % sulphuric acid, and can survive at a pH of 0·6. This remarkable property has been put to beneficial use in agriculture; it can also be very destructive.

Agricultural uses. Three uses of *T. thio-oxidans* were described by Waksman (1927, p. 614) and Starkey (1950): (i) Most phosphate fertilizers are relatively inefficient because a large proportion of the phosphate is not readily available to plants. The availability can be increased by making a compost of soil, phosphate and sulphur; the sulphuric acid formed from sulphur by *T. thio-oxidans* produces soluble phosphates. The reaction occurs readily but the method is not widely practised. (ii) Sulphur dressings are used to neutralize infertile 'black alkali' soils. The acid produced also improves the physical consistency of the soils by flocculating colloids. (iii) The control of potato scab by the application of heavy dressings of sulphur (300–1000 lb./acre) is widely used in America; the lowering of pH inhibits the infecting organism *Actinomyces scabies*. In all these examples inoculation with *Thiobacillus thio-oxidans* is not essential, though it may yield improved results (see Martin, 1921).

Corrosion of concrete. Internal corrosion of concrete sewers is widespread. It occurs only above sewage level and is always associated with the evolution of hydrogen sulphide from sewage; it frequently results in complete disintegration of the concrete. According to Parker (1945a, b, 1947) the corrosion is the result of a combination of chemical and microbiological reactions. While the primary corrosive agent is hydrogen sulphide, formed mainly by sulphate-reducing bacteria, the corrosion process proceeds chiefly by the action of various *Thiobacillus* spp. on thiosulphates, polythionates and sulphur produced chemically (Parker, 1947). Three types of *Thiobacillus* spp. appear to be involved: the obligate autotroph *T. concretivorus*, which is similar to *T. thio-oxidans* and is responsible for the final disintegration by sulphuric acid; *Thiobacillus* '*X*', which resembles *T. thioparus*; and a miscellaneous collection of facultatively autotrophic 'M' strains. Their properties have been described in detail by Parker & Prisk (1953). Several examples of corroding concrete manholes have been examined in our laboratory; corrosion was always associated with production of hydrogen sulphide, and crude cultures of the *T. thio-oxidans* type invariably developed in enrichment cultures from the corrosion products. A more picturesque example was reported by Taylor & Hutchinson (1947): two concrete Mouchel water-cooling towers were put out of action for several months by a corrosion mechanism closely resembling that described by Parker.

Corrosion of stone. Much of the decay of stonework in cities is undoubtedly due to chemical action by the acid-polluted atmosphere. Some examples can be attributed to microbiological action, but caution should be observed in accepting claims based only on the isolation of 'aggressive' organisms; their presence in excessive numbers should be established, with contributory evidence of the effect of their metabolism, e.g. formation of acid, sulphate, etc. Some well-authenticated examples of microbiological decay of stone monuments were given by Pochon, Coppier & Tchan (1951). They presented evidence that sulphide produced by sulphate-reducing bacteria in the soil beneath the monuments seeped up through the stone to its surface, where a combination of atmospheric oxidation and attack by *Thiobacillus* spp. led to acid formation and severe corrosion.

Corrosion of rubber. Serious deterioration of fire hoses was experienced during the Second World War. The cause of the trouble was traced to acid formation by the action of *T. thio-oxidans* on sulphur present in the vulcanized rubber lining. The remedy suggested was efficient drying of the hoses after use or the employment of rubber containing less than 0·1 % free sulphur (Thaysen, Bunker & Adams, 1945).

Acid mine waters. The iron-laden acid waters from coal mines present a difficult problem of disposal. They corrode pumps and pollute streams and rivers. It is estimated that the equivalent of 3 million tons of concentrated sulphuric acid from this source pours into the Ohio River and its tributaries annually (Starkey, 1950). Colmer, Temple & Hinkle (1950) isolated an organism from the acid waters which not only, like *T. thio-oxidans*, produced acid in thiosulphate broth, but also oxidized ferrous to ferric iron autotrophically (Temple & Colmer, 1951). Microscopically it was indistinguishable from *T. thio-oxidans* and bore no resemblance to the well-known iron bacteria. Temple & Colmer suggested the specific name *Thiobacillus ferro-oxidans* n.sp. for the organism. Leathen, Braley & McIntyre (1953) isolated a similar, perhaps identical, organism but could not confirm acid formation in thiosulphate broth. According to Murdock (1953), workers at the Mellon Institute have shown that it is quite practicable to reduce the effect in Pennsylvanian coal mines to small proportions by minimizing the amount of water entering the pits and particularly by diverting the flow of water from sulphur-bearing materials. The problem is not so serious in British coal mines, but a recent report (Allen, W. T., personal communication) described a Scottish mine effluent containing ferrous sulphate, sulphuric acid and ferric hydroxide, with a pH of about 3, coming from a coal seam lying alongside an iron sulphide seam. The corrosion damage to pipes and pumps was severe, and the polluting effect on nearby rivers was described as excessive.

Acid waters are also a considerable problem in South African gold mines. In order to reduce corrosion these waters are treated underground with lime, but this treatment is not only expensive but creates the further problem of how to dispose of millions of gallons of water saturated with calcium sulphate (Sutherland, S., personal communication). In South Africa, a country short of water, it would be most desirable to use this treated water again. Chemical methods for the removal of the sulphate ion are expensive, and a microbiological method, using sulphate-reducing bacteria, might be developed.

Desulphurization of oil shales and oil. The high sulphur content of some oil shales makes the extracted oil unusable. Bunker (1936, p. 27) described unsuccessful attempts to remove organic sulphur compounds from Kimmeridge oil shales (5 % sulphur) with *T. thio-oxidans*. He suggested that both oil shales and oil might be successfully desulphurized with a more suitable strain of *Thiobacillus*. The desulphurization of oil from oil wells is becoming increasingly expensive as the low-sulphur wells are exhausted, and an efficient microbiological method would be welcomed.

MICROBIOLOGICAL TRANSFORMATIONS OF INORGANIC IRON

All living organisms require iron, usually in trace amounts. Its availability is therefore of great importance, and its removal or release by micro-organisms can have important biological effects. We shall deal only with the more obvious economic effects caused by the formation of masses of ferric hydroxide (iron deposits, fouling of pipes, etc.), but perhaps the indirect, less obvious biological effects due to iron deficiencies are the more important.

Bacteria involved in iron transformations

Microbiological transformation of iron compounds in nature are brought about in two ways. (i) By specific organisms which use the oxidation of ferrous to ferric iron as a source of energy for growth: these organisms include the 'true iron bacteria', some blue-green algae and *T. ferro-oxidans*, and are described in more detail below. (ii) By a group of non-specific organisms, which act by altering the pH, oxidation-reduction potentials and carbon dioxide content of soils and waters and by decomposing organic iron compounds, leading to the solution or precipitation of iron compounds. Very many types of bacteria are included in this group. They are probably more important in iron transformations than the first group, but as the vast majority are heterotrophs, they will not be considered in detail.

'*True iron bacteria.*' The iron bacteria are a group of micro-organisms whose taxonomic and biochemical status has been the subject of much critical discussion (Winogradsky, 1888, 1922; Molisch, 1910; Cholodny, 1926; Cataldi, 1937, 1939; Pringsheim, 1949*b*, *c*). In reading through the literature it is almost impossible 'to separate established facts from

an almost overwhelming mass of rash and controversial statements'
(Pringsheim, 1949c). Fortunately, Pringsheim has done a great deal to
create some order out of chaos. A résumé of the present position is
given below.

(1) The iron bacteria derive their energy from an oxidation which may
be represented as

$$4FeCO_3 + O_2 + 6H_2O = 4Fe(OH)_3 + 4CO_2 - \Delta F_{298} = 40 \text{ kcal.}$$

The energy yield and the efficiency of carbon assimilation are low,
consequently cell growth necessitates a very large iron turnover.
Starkey (1945a) calculated that the ratio of precipitated ferric hydroxide
to weight of cell material formed is 500:1. The ferric hydroxide accumulates round the cells. Many iron bacteria can use manganese salts; some
require both iron and manganese.

(2) The autotrophy of these bacteria is not proved except perhaps for
Gallionella (Winogradsky, 1888; Lieske, 1911; Teichmann, 1935;
Sartory & Meyer, 1948). Most of them grow better with supplements
such as beef extract, peptones or yeast extract. *Gallionella*, however, is
inhibited by organic matter in laboratory media, though in nature it
grows in water contaminated with organic matter. According to
Pringsheim (1949b) considerable growth of the iron bacteria can be
obtained with only minute quantities of organic material, and it is
probable that most of their energy is in fact obtained by the oxidation
of inorganic iron (or manganese).

(3) The taxonomy of the iron bacteria is confused. Pringsheim (1949b)
demonstrated convincingly that *Sphaerotilus natans*, *Leptothrix ochracea*
and *Cladothrix dichotoma* were pleomorphic forms of the same organism.
He listed a large number of dubious species and genera (Pringsheim,
1949c).

Thiobacillus ferro-oxidans. This organism is undoubtedly autotrophic
and oxidizes ferrous to ferric iron. It differs not only morphologically
from the true iron bacteria but also in its facility to grow at pH 2·0–2·5.
Reference has already been made in a previous section to this *Thiobacillus* as the organism responsible for the acid nature of bituminous
mine waters.

Blue-green algae. Some Mycophyceae deposit iron hydroxide in or
upon their sheaths. This property is common among the filamentous
Lyngbya.

Economic activities

Iron-ore deposits. As early as 1836, Ehrenberg (1836) suggested that
micro-organisms play an important part in the formation of bog iron

ore. A detailed account of the various theories advanced to explain the deposition of iron ores was given by Harder (1919). Both chemical and biological agencies were concerned. He described three types of micro-organisms probably in part responsible: the autotrophic iron bacteria (for doubts of their autotrophy, see above), heterotrophs which cause the deposition of ferric hydroxide when either inorganic or organic iron salts are present, and heterotrophs which use organic iron compounds and precipitate ferric hydroxide. Direct evidence for participation of micro-organisms is meagre, because bacterial remains, and especially those of the iron bacteria, are necessarily short-lived. However, there is some slight evidence of this kind. Casts of the organisms are still visible in some recent bog iron deposits. Molisch (1910, p. 59) found abundant remains of iron thread bacteria in four out of sixty-one samples of bog iron ore. Harder considered it probable that much of the granular ferric hydroxide in bog iron ore was laid down by the iron bacteria but that a great deal was precipitated chemically.

According to Pringsheim (1949c) the *Leptothrix* modification of *Sphaerotilus natans* occurs in abundance where acid bog waters containing large quantities of ferrous iron run into alkaline soils. This organism does not precipitate manganese and produces a very pure type of ferric hydroxide in the form of yellowish red masses. Starkey & Halvorson (1927) and Thiel (1928) considered that heterotrophic organisms of the non-specific type are more important as agents depositing iron ore than the 'autotrophic' iron bacteria (see also Starkey, 1945b). According to Naumann (1924) and Lönnerblad (1933), the blue-green alga *Paracapsa*, a member of the unicellular Chroococcales, produces considerable iron deposits.

ZoBell (1946, p. 103) has reviewed iron transformations in marine conditions.

Fouling and corrosive activities. The growth of the iron bacteria is favoured by the presence of ferrous salts and of some organic material, and by slightly acid conditions, especially in the presence of dissolved carbon dioxide. In these conditions, large quantities of ferric hydroxide are precipitated and can cause considerable trouble by fouling streams, blocking water systems, filters and so on. In severe cases, such phenomena are given the evocative name of 'water calamity'; the process has been less happily named 'biofouling'. O'Connell (1941), Hastings (1937) and Starkey (1945b) gave many examples.

Outbreaks in water are sometimes accompanied by a reddish turbidity, an unpleasant flavour in the water, foul odours and considerable mortality to fish (Exodus vii, 20–21). Many other water organisms contribute to these disasters, but the iron bacteria are probably the only ones which use chemo-autotrophic mechanisms.

'Tuberculation' in iron pipes—the formation of nodular incrustations on the interior surface which slows up and sometimes prevents the flow of water—is sometimes caused by iron bacteria. In most such cases the source of the iron for bacterial growth is the water flowing through the pipe; the pipe does not act as a source of iron and no corrosion due to the iron bacteria occurs, though sulphate-reducing bacteria may develop inside the tubercle and cause corrosion (see p. 287). In certain conditions, however, it is possible that through their growth on limited regions of the pipe surface the iron bacteria form differential aeration cells and thus initiate electrochemical corrosion; as the size of the tubercle increases, conditions inside the tubercle become more anaerobic, and the corrosion process becomes self-augmenting (Olsen & Szybalski, 1949). Sometimes tuberculation is wrongly attributed to iron bacteria. Butlin, K. R. & Adams, M. E. (unpublished observations) have examined a large number of tubercles from corroded iron pipes and found iron bacteria in one or two cases only; in the negative cases the corrosion can be attributed to electrochemical mechanisms, sometimes augmented by subsequent development of sulphate-reducing bacteria.

The growth of iron bacteria may be prevented by chlorination; in water pipes growth is sometimes, but not invariably, prevented by lining the pipe with cement or bitumen (Starkey, 1945*b*).

MICROBIOLOGICAL TRANSFORMATIONS
OF OTHER ELEMENTS

Most elements of biological importance (oxygen, hydrogen, phosphorus, calcium, sodium, magnesium, etc.) undergo transformations brought about partly or wholly by biological systems. The continual replenishment of atmospheric or dissolved oxygen during photosynthesis is an example of the importance of photo-autotrophs in the oxygen cycle. Micro-organisms contribute to the formation of siliceous deposits and deposits of calcium and sodium carbonates; sulphide formation by autotrophic sulphate-reducing bacteria can contribute to the phosphorus cycle by releasing soluble phosphate from insoluble ferric phosphate (Hayes & Coffin, 1951). In general, however, the part played by micro-organisms in the transformations of elements other than those already dealt with has been little studied and virtually nothing is known of the specific function of autotrophs in such transformations.

DISCUSSION

The function of the autotrophic micro-organisms in the transformation of carbon, nitrogen, sulphur and so on is a legacy of their part in the evolution of higher organisms. It is not necessary to assume that the earliest forms of life were autotrophs, but it is reasonable to suppose that autotrophs were among the early living things and that their ability to produce a net synthesis of organic matter provided conditions suitable for the evolution of the majority of heterotrophs. This is not to say that their metabolism was more simple than the heterotrophs. In fact, such evidence as there is suggests that evolution to heterotrophy involved a simplification of metabolic function owing to loss of synthetic abilities (Lwoff, 1943). The autotrophs of to-day represent various mechanisms by which living organisms were able to conduct this net synthesis of organic matter; one such mechanism, photo-autotrophy, became dominant on this planet and permitted the evolution of aerobic life as commonly encountered.

Thus the chemo-autotrophs and certain branches of the photo-autotrophs represent evolutionary culs-de-sac: paths of physiological evolution which led nowhere in comparison with the photo-autotrophy of green plants and with heterotrophy. (In other parts of the universe conditions may have favoured the evolution of complex organisms dependent on different autotrophic mechanisms, but speculation on the nature of the sulphate-reducing equivalent of *Homo sapiens* would not be appropriate here.) The autotrophs still perform their primaeval role of synthesizing organic compounds of carbon, nitrogen, sulphur and so on, and in assessing their present importance in nature it is necessary to consider the extent to which these functions are supplemented by the heterotrophic micro-organisms. In the fixation of carbon dioxide, of course, the importance of the photo-autotrophs is incalculable and, though on land the higher plants provide nearly all the fixed carbon dioxide, in the sea the photosynthetic autotrophic micro-organisms of the phytoplankton are all-important. The transcendental importance of the photo-autotrophs in carbon dioxide fixation should not, however, obscure the importance of other autotrophs in the transformations of other elements. Rabinowitch (1945, p. 4) wrote of the chemo-autotrophs and photo-autotrophs: 'Conceivably these peculiar modes of auto-trophic life may have played a greater role in earlier geological ages, when chemical activity on the surface of the earth was more widespread and violent. They are consequently of considerable interest in specula-tions as to the origin and development of life on this planet. *In the contemporary cycle of the living matter on earth, these processes are of no consequence'* (our italics). In our view this is a superficial conclusion. The contemporary role of chemo-autotrophs in the sulphur cycle, for example, is too obvious to need emphasis here. Nitrification, normally a function of chemo-autotrophs, is essential for the growth of higher green plants. Similarly, the functions of autotrophs in the transformations of iron, phosphorus and other elements have been supplemented but not supplanted by heterotrophs. It is an exaggeration to say 'photosynthesis by green plants alone prevents the rapid disappearance of all life from the face of the earth' (Rabinowitch, 1945).

In the day-to-day economy of industrial and agricultural civilization the autotrophic micro-organisms are only of immediate importance in so far as they transform elements of economic interest to man. Sometimes their activities are beneficial, as in nitrification, or in nitrogen-fixation by the blue-green algae in tropical rice fields; sometimes they are disastrous, as in the corrosion of iron installations by *Desulphovibrio desulphuricans*. No useful generalization can be formulated concerning this sense of their economic activities, but one point of long-term interest is worth making. The evolution of industrial civilization on this planet has led to a squandering of the more accessible forms of chemical energy (represented by fuels, sulphur, food and other materials of high chemical potential), most of which have accumulated autotrophically over geological eras of time. Already this phenonemon is being reflected in world shortages of various classes of food and raw materials. As these concentrated reserves of accumulated chemical energy become exhausted it may well become increasingly necessary to utilize deliberately the chemo- and photo-autotrophic micro-organisms for their replenishment.

We gratefully acknowledge the help and advice of many authorities during the preparation of this paper. Particularly are we indebted to Dr H. G. Thornton, F.R.S., for information on the nitrifying bacteria, to Dr G. E. Fogg for information on algae, and to Dr C. E. Lucas for information on marine phytoplankton. Much of the work surveyed in this paper was unfamiliar to us and we should like to emphasize that any errors of fact or judgement were made by ourselves. This paper is published by permission of the Director, Chemical Research Laboratory.

REFERENCES

ADAMS, M. E., BUTLIN, K. R., HOLLANDS, S. J. & POSTGATE, J. R. (1951). The role of hydrogenase in the autotrophy of *Desulphovibrio*. *Research, Lond.* 4, 295.

ADAMS, M. E. & FARRER, T. W. (1953). The influence of ferrous iron on bacterial corrosion. *J. appl. Chem.* 3, 117.

BAAS-BECKING, L. M. G. (1925). Studies on the sulphur bacteria. *Ann. Bot., Lond.* 39, 613.

BALDWIN, J. M. (1953). The production and utilization of microbiological protein. A review of the literature. *Rep. Coun. sci. industr. Res. Aust.* no. T11.

BAVENDAMM, W. (1932). Die mikrobiologische Kalkfällung in der Tropischen See. *Arch. Mikrobiol.* 3, 205.

BERGMANN, W. & FEENEY, R. J. (1950). Sterols of algae. I. The occurrence of chondrillasterol in *Scenedesmus obliquus*. *J. org. Chem.* 15, 812.

BICKNELL, A. K. (1949). The occurrence of a green sulphur bacterium in Sodon Lake. *Lloydia*, 12, 183.

BOOTH, W. E. (1941). Algae as pioneers in plant succession and their importance in erosion control. *Ecology*, 22, 38.

BRIAN, P. W. (1953). Microbiology, Cantor Lectures. *J. Roy. Soc. Arts*, 51, 194.

BUNKER, H. J. (1936). *A review of the Physiology and Biochemistry of the Sulphur Bacteria*. Dep. sci. industr. Res., Chem. Res. spec. Rep. no. 3. London: H.M. Stationery Office.

BUNKER, H. J. (1939). Microbiological experiments in anaerobic corrosion. *J. Soc. chem. Ind., Lond.* 58, 93.

BUNKER, H. J. (1940). Microbiological anaerobic corrosion. *Chem. & Ind.* 18, 412.

BURLEW, J. S. (1953). Editor of *Algal Culture, from Laboratory to Pilot Plant. Publ. Carneg. Instn.*

BUTLIN, K. R. (1949). Some malodorous activities of sulphate-reducing bacteria. *Proc. Soc. appl. Bact.* no. 2, 39.

BUTLIN, K. R. (1953). The bacterial sulphur cycle. *Research, Lond.* 6, 184.

BUTLIN, K. R. & ADAMS, M. E. (1947). Autotrophic growth of sulphate-reducing bacteria. *Nature, Lond.* 160, 154.

BUTLIN, K. R., ADAMS, M. E. & THOMAS, M. (1949). Sulphate-reducing bacteria and internal corrosion of ferrous pipes conveying water. *Nature, Lond.* 163, 26.

BUTLIN, K. R. & POSTGATE, J. R. (1953a). Microbiological formation of sulphide and sulphur. Symp. '*Microbial Metabolism*', VIth Int. Congr. Microbiol. p. 126.

BUTLIN, K. R. & POSTGATE, J. R. (1953b). The microbiological formation of sulphur in Cyrenaican lakes. Symp. *Inst. Biol. Lond.* (in the Press).

BUTLIN, K. R. & VERNON, W. H. J. (1949). Underground corrosion of metals: causes and prevention. *J. Instn Wat. Engrs*, 3, 627.

CALDWELL, D. H. (1946). Sewage oxidation ponds: performance, operation and design. *Sewage Wks J.* 18, 433.

CATALDI, M. S. (1937). Aislamento de *Leptothrix ochracea* en medios solidos a partir de cultivos liquidos. *Folia biol., B. Aires*, p. 337.

CATALDI, M. S. (1939). *Estudio fisiológico y sistemático de algunas Chlamydobacteriales*. Thesis, Buenos Aires.

CHASE, F. E. (1948). A preliminary report on the use of the Lees and Quastel soil perfusion technique in determining the nitrifying capacity of field soils. *Sci. Agric.* 28, 315.

Chemistry Research 1950 (1951). London: H.M. Stationery Office.

Chemistry Research 1951 (1952). London: H.M. Stationery Office.

Chemistry Research 1952 (1953). London: H.M. Stationery Office.

CHOLODNY, N. (1926). *Die Eisenbakterien. Beitrage zu einer Monographie*. Jena: Fischer.

COLMER, A. R., TEMPLE, K. L. & HINKLE, M. E. (1950). An iron-oxidizing bacterium from the acid drainage of some bituminous coal mines. *J. Bact.* 59, 317.

CONNELL, C. H. & CROSS, J. B. (1950). Mass mortality of fish associated with the protozoan *Gonyaulax*. *Science*, 112, 359.

COOK, P. M. (1950). Symp. '*The Culturing of Algae*', p. 53. Yellow Springs: Kettering Foundation.

COOK, P. M. (1951). Chemical engineering problems in large-scale culture of algae. *Industr. Engng Chem. (Industr.)*, 43, 2385.

DE, P. K. (1939). The role of blue-green algae in nitrogen fixation in rice fields. *Proc. Roy. Soc.* B, **127**, 121.

EHRENBERG, C. G. (1836). Vorläufige Mittheilungen über das wirklige Vorkommen fossiler Infusorien und ihre grosse Verbreitung. *Ann. Phys., Lpz.* **38**, 213.

ELENKIN, A. A. (1934). Bau und geographische Verbreitung der essbaren Alge *Nematonostic flagelliforme. Sovetsk. Bot.* **4**, 89.

ELLIS, D. (1932). *Sulphur Bacteria.* London: Longmans Green.

ERNST, A. (1908). *The New Flora of the Volcanic Island of Krakatoa.* Cambridge University Press.

FARRER, T. W. & WORMWELL, F. (1953). Corrosion of iron and steel by aqueous suspensions of sulphur. *Chem. & Ind.* **31**, 106.

FOGG, G. E. (1947). Nitrogen fixation by blue-green algae. *Endeavour*, **6**, 172.

FOGG, G. E. (1951). Studies on nitrogen fixation by blue-green algae. II. Nitrogen fixation by *Mastigocladus laminosus. J. exp. Bot.* **2**, 117.

FRITSCH, F. E. (1945). *Structure and Reproduction of the Algae*, **2**. Cambridge University Press.

GAFFRON, H. (1953). Food from algae. *Research, Lond.* **6**, 222.

GEOGHEGAN, M. J. (1951). Unicellular algae as a source of food. *Nature, Lond.* **168**, 426.

GEST, H. & KAMEN, M. D. (1949). Studies on the metabolism of photosynthetic bacteria. IV. Photochemical production of molecular hydrogen by growing cultures of photosynthetic bacteria. *J. Bact.* **58**, 239.

GIETZEN, J. (1931). Untersuchungen über marine Thiorhodaceen. *Zbl. Bakt.* (2. Abt.), **83**, 183.

GROSS, F. (1941). Food production by fish and oyster farming. *Nature, Lond.* **148**, 71.

GROSS, F., RAYMONT, J. E. G., MARSHALL, S. M. & ORR, A. P. (1944). A fish-farming experiment in a sea loch. *Nature, Lond.* **153**, 483.

GROSS, F., RAYMONT, J. E. G., NUTMAN, S. R. & GAULD, D. T. (1946). Application of fertilizers to an open sea loch. *Nature, Lond.* **158**, 187.

GUBIN, W. & TZECHOMSKAYA, W. (1930). Über die biochemishe Sodabildung in den Sodaseen. *Zbl. Bakt.* (2. Abt.), **81**, 396.

HARDER, E. C. (1919). Iron depositing bacteria and their geologic relations. *Prof. Pap. U.S. geol. Surv.* no. 113.

HARDER, R. & WITSCH, H. VON (1942). Bericht über Versuche zur Fettsynthese mittels autotropher Mikroorganismen. *Forschungsdienst*, **16**, 270.

HARVEY, W. H., COOPER, L. H. N., LEBOUR, M. V. & RUSSELL, F. S. (1935). Plankton production and its control. *J. Mar. biol. Ass. U.K.* **20**, 407.

HASTINGS, A. B. (1937). *Biology of Water Supply.* Econ. Ser. Brit. Mus. no. 7A.

HAYES, F. R. & COFFIN, C. C. (1951). Radioactive phosphorus and exchange of lake nutrients. *Discovery*, **10**, 78.

HEYERDAHL, T. (1950). *The Kon-Tiki Expedition.* London: Allen and Unwin.

HOWE, M. A. (1932). The geologic importance of the lime-secreting algae. *U.S. Dep. Interior: Shorter Contr. Gen. Geol. Prof. Paper 170*, E. 57.

HULL, W. Q., KEEL, H., KENNY, J. & GAMSON, B. W. (1953). Diatomaceous earth. *Industr. Engng Chem. (Industr.)*, **45**, 256.

HUNT, W. F. (1915). The origin of the sulphur deposits of Sicily. *Econ. Geol.* **10**, 543.

IYA, K. K. & SREENIVASAYA, M. (1944). Preliminary study of the bacterial flora associated with sulphur deposits on the east coast (Masulipatam). *Curr. Sci.* **13**, 316.

JANKOWSKI, G. T. & ZOBELL, C. E. (1944). Hydrocarbon production by sulphate-reducing bacteria. *J. Bact.* **47**, 447.

302   K. R. BUTLIN AND J. R. POSTGATE

JOSHI, N. V. & BISWAS, S. C. (1948). Does photo-nitrification occur in the soil? *Indian J. agric. Sci.* **18**, 115.

KAUFFMANN, J. (1952). Rôle des bactéries nitrifiantes dans l'altération des pierres calcaires des monuments. *C.R. Acad. Sci., Paris,* **234**, 2395.

KLUYVER, A. J. (1953). Some aspects of nitrate reduction. Symp. '*Microbial Metabolism*'. *VIth Int. Congr. Microbiol.* p. 71.

KUSNETZOV, S. I. (1935). Microbiological researches in the study of the oxygenous regimen of lakes. *Verh. int. Ver, Limnol.* **7**, 562.

LARSEN, H. (1952). On the culture and general physiology of the green sulphur bacteria. *J. Bact.* **64**, 187.

LEATHEN, W. W., BRALEY, S. A. & McINTYRE, L. D. (1953). The role of bacteria in the formation of acid from certain sulphuritic constituents associated with bituminous coal. II. Ferrous iron-oxidizing bacteria. *Appl. Microbiol.* **1**, 65.

LEES, H. (1953). Nitrification in soil and culture. *Abstr. Comm. VIth Int. Congr. Microbiol.* **3**, 124.

LEES, H. & QUASTEL, J. H. (1944). A new technique for the study of soil sterilization. *Chem. & Ind.* **22**, 238.

LEWIN, J. C. (1953). Heterotrophy in diatoms. *J. gen. Microbiol.* **9**, 305.

LIESKE, R. (1911). Beiträge zur Kenntnis der Physiologie von *Spirophyllum ferruginium*, Ellis, einem typischen Eisenbakterium. *Jb. wiss. Bot.* **49**, 91.

LINDSTROM, E. S., BURRIS, R. H. & WILSON, P. W. (1949). Nitrogen fixation by photosynthetic bacteria. *J. Bact.* **58**, 313.

LINDSTROM, E. S., TOVE, S. R. & WILSON, P. W. (1950). Nitrogen fixation by the green and purple sulphur bacteria. *Science,* **112**, 197.

LOCHHEAD, A. G. (1952). The nitrifying bacteria. *Annu. Rev. Microbiol.* **6**, 185.

LOGAN, K. H. (1948). Corrosion by soils. In *The Corrosion Handbook* (ed. H. H. Uhlig). London: Chapman and Hall.

LÖNNERBLAD, G. (1933). Zur Kenntnis der Eisenausfällung der Pflanzen. *Bot. Notiser,* p. 402.

LUCAS, C. E. (1953). Marine phytoplankton and their exploitation. *Symp. Biol. Prod. of the Sea. Inst. biol.* (in the Press).

LWOFF, A. (1943). *L'Évolution Physiologique.* Paris: Hermann.

MACFARLANE, W. A. (1951). *Summary of Recent U.S. Work on Chlorella as a Source of Food,* etc. U.K. Scientific Mission, Memo. No. 28/51; B.C.S.O. (N.A.), Report no. 769.

MACNAMARA, J. & THODE, H. G. (1951). The distribution of ^{34}S in nature and the origin of native sulphur deposits. *Research, Lond.* **4**, 582.

MARSHALL, A. (1915). *Explosives, their Manufacture, Properties, Tests and History.* London: Churchill.

MARTIN, W. H. (1921). A comparison of inoculated and uninoculated sulphur for the control of potato scab. *Soil Sci.* **11**, 75.

MEIKLEJOHN, J. (1953). The nitrifying bacteria: a review. *J. Soil Sci.* **4**, 59.

MILLER, L. P. (1950). Formation of metal sulphides through the activities of sulphate-reducing bacteria. *Contr. Boyce Thompson Inst.* **16**, 85.

MOLISCH, H. (1910). *Die Eisenbakterien.* Jena: Fischer.

MOLISCH, H. (1926). *Pflanzenbiologie in Japan auf Grund eigener Beobachtungen.* Jena: Fischer.

MURDOCK, H. R. (1953). Industrial wastes. *Industr. Engng Chem. (Industr.),* **45** (2), 101 A.

MURZAEV, P. M. (1950). On possible methods of hastening the natural process of formation and accumulation of native sulphur. (In Russian.) *C.R. Acad. Sci., U.R.S.S.* **72**, 343.

MYERS, J. (1948). Studies of sewage lagoons. *Publ. Wks, N.Y.* **79**, 25.

MYERS, J. (1951). Physiology of the algae. *Annu. Rev. Microbiol.* 5, 157.

MYERS, J., PHILLIPS, J. N. & GRAHAM, J. R. (1951). On the mass culture of algae. *Plant Physiol.* 26, 539.

NAKAMURA, H. (1938). Über die Kohlensaureassimilation bei nideren Algen in Anwesenheit des Schwefelwasserstoffes. *Acta phytochim., Tokyo*, 10, 271.

NAUMANN, E. (1924). Über *Paracapsa siderophila* n.g., n.sp. als Ursache einer auffälligen limnischen Eiseninkrustration. *Ark. Bot.* 18, no. 21.

O'CONNELL, W. J. (1941). Characteristics of microbiological deposits in water circuits. *Proc. Amer. Petrol. Inst., Annu. Mtg*, 22, 66.

OLSEN, E. & SZYBALSKI, W. (1949). Aerobic microbiological corrosion of water pipes. *Acta chem. scand.* 3, 1094, 1106.

PARKER, C. D. (1945 a). The corrosion of concrete. 1. The isolation of a species of bacterium associated with the corrosion of concrete exposed to atmospheres containing hydrogen sulphide. *Aust. J. exp. Biol. med. Sci.* 23, 81.

PARKER, C. D. (1945 b). The corrosion of concrete. 2. The function of *Thiobacillus concretivorus* (nov.sp.) in the corrosion of concrete exposed to atmospheres containing hydrogen sulphide. *Aust. J. exp. Biol. med. Sci.* 23, 91.

PARKER, C. D. (1947). Species of sulphur bacteria associated with the corrosion of concrete. *Nature, Lond.* 159, 439.

PARKER, C. D. & PRISK, J. (1953). The oxidation of inorganic compounds of sulphur by various sulphur bacteria. *J. gen. Microbiol.* 8, 344.

PEARSALL, W. H. & FOGG, G. E. (1951). The utilization of algae for industrial photosynthesis. *Food. Sci. Abstr.* 23, 1.

PIRIE, N. W. (1953). Ideas and assumptions about the origin of life. *Discovery*, 14, 238.

POCHON, J., COPPIER, O. & TCHAN, Y. T. (1951). Rôle des bactéries dans certaines altérations des pierres des monuments. *Chim. et Industr.* 65, 496.

POSTGATE, J. R. (1951). The reduction of sulphur compounds by *Desulphovibrio desulphuricans*. *J. gen. Microbiol.* 5, 725.

PRATT, R., DANIELS, T. C., EILER, J. J., GUNNISON, J. B., KUMLER, W. D., ONETO, J. J., STRAIT, L. A., SPOEHR, H. A., HARDIN, G. J., MILNER, H. W., SMITH, J. H. C. & STRAIN, H. H. (1944). Chlorellin, an antibacterial substance from *Chlorella*. *Science*, 99, 351.

PRINGSHEIM, E. G. (1949 a). The relationship between bacteria and Myxophyceae. *Bact. Rev.* 13, 47.

PRINGSHEIM, E. G. (1949 b). The filamentous bacteria *Sphaerotilus, Leptothrix, Cladothrix* and their relation to iron and manganese. *Phil. Trans.* B, 233, 453.

PRINGSHEIM, E. G. (1949 c). Iron bacteria. *Biol. Rev.* 24, 200.

QUASTEL, J. H. & SCHOLEFIELD, P. G. (1951). Biochemistry of nitrification in soils. *Bact. Rev.* 15, 1.

RABINOWITCH, E. I. (1945). *Photosynthesis and Related Processes*, 1. New York: Interscience.

SARTORY, A. & MEYER, J. (1948). Contribution à l'étude de l'évolution physiologique de deux bactéries ferrugineuses. Leurs facteurs d'énergie et de synthèse. *C.R. Acad. Sci., Paris*, 226, 443.

SENEZ, J. (1951). Problèmes écologiques concernant les bactéries des sédiments marins. *Vie et Milieu*, 2, 5.

SENEZ, J. (1953). Sur l'activité et la croissance des bactéries anaérobies sulfato-réductrices en cultures semi-autotrophes. *Ann. Inst. Pasteur*, 84, 595.

SINGH, R. N. (1950). Reclamation of 'Usar' lands in India through blue-green algae. *Nature, Lond.* 165, 325.

SISLER, F. D. & ZOBELL, C. E. (1951 a). Nitrogen fixation by sulphate-reducing bacteria indicated by nitrogen/argon ratios. *Science*, 113, 511.

SISLER, F. D. & ZoBELL, C. E. (1951b). Hydrogen utilization by some marine sulphate-reducing bacteria. *J. Bact.* **62**, 117.

SPOEHR, H. A. & MILNER, H. W. (1948). *Chlorella* as a source of food. *Yearb. Carneg. Instn*, **47**, 100.

SPOEHR, H. A. & MILNER, H. W. (1949). The chemical composition of *Chlorella*. Effect of environmental conditions. *Plant Physiol.* **24**, 120.

SPRUIT, C. J. P. & WANKLYN, J. N. (1951). Iron/Sulphide ratios in corrosion by sulphate-reducing bacteria. *Nature, Lond.* **168**, 951.

STARKEY, R. L. (1945a). Precipitation of ferric hydrate by iron bacteria. *Science*, **102**, 532.

STARKEY, R. L. (1945b). Transformations of iron by bacteria in water. *J. Amer. Wat. Wks Ass.* **24**, 381.

STARKEY, R. L. (1947). Sulphate reduction and the anaerobic corrosion of iron. *Leeuwenhoek ned. Tijdschr.* **12**, 193.

STARKEY, R. L. (1950). Relations of microorganisms to transformation of sulphur in soil. *Soil Sci.* **70**, 55.

STARKEY, R. L. (1953). The relationship of sulphate-reducing bacteria to iron corrosion in the marine environment. *Abstr. Comm. VIth Int. Congr. Microbiol.* **3**, 1007.

STARKEY, R. L. & HALVORSON, H. O. (1927). Studies on the transformations of iron in nature. 2. Concerning the importance of microorganisms in the solution and precipitation of iron. *Soil Sci.* **24**, 381.

STARKEY, R. L. & WIGHT, K. M. (1945). *Anaerobic Corrosion of Iron and Steel*. New York: American Gas Association.

STONE, R. W. & ZoBELL, C. E. (1952). Bacterial aspects of the origin of petroleum. *Industr. Engng Chem. (Industr.)*, **44**, 2564.

STUMPER, R. (1923). La corrosion du fer en présence du sulfure de fer. *C.R. Acad. Sci., Paris*, **176**, 1316.

SUBBA RAO, M. S., IYA, K. K. & SREENIVASAYA, M. (1947). Microbiological formation of elemental sulphur in coastal areas. *Abstr. Comm. IVth Int. Congr. Microbiol.* p. 157.

SUCHLANDT, O. (1916). Dinoflagellaten als Erreger von rotem Schnee. *Ber. dtsch. bot. Ges.* **34**, 242.

SUGAWARA, K., SHINTANI, S. & OYAMA, T. (1937). Dissolved gases in some Japanese lakes and their significance in lake metabolism. *J. chem. Soc. Japan*, **58**, 890.

SVERDRUP, H. U., JOHNSON, M. W. & FLEMING, R. H. (1942). *The Oceans*. New York: Prentice-Hall.

TAYLOR, C. B. & HUTCHINSON, G. H. (1947). Corrosion of concrete caused by sulphur-oxidizing bacteria. *J. Soc. chem. Ind., Lond.* **66**, 54.

TEICHMANN, E. (1935). Vergleichende Untersuchungen über die Kultur und Morphologie einiger Eisenorganismen. Dissertation, Prag. Deutsche Universitat.

TEMPLE, K. L. & COLMER, A. R. (1951). The autotrophic oxidation of iron by a new bacterium: *Thiobacillus ferro-oxidans*. *J. Bact.* **62**, 605.

THAYSEN, A. C., BUNKER, H. J. & ADAMS, M. E. (1945). 'Rubber acid' damage in fire hoses. *Nature, Lond.* **155**, 322.

THIEL, G. A. (1928). A summary of the activities of bacterial agencies in sedimentation. *Repr. nat. Res. Coun., Wash.* ser. no. 85, 61.

UHLIG, H. H. (1949). The cost of corrosion to the U.S. *Chem. Engng News*, **27**, 2764.

VERNER, A. R. & ORLOVSKII, N. V. (1948). The role of sulphate-reducing bacteria in the salt conditions of soils in Baraba. (In Russian.) *Pochvovedenie*, p. 553.

VIRTANEN, A. I. (1948). Biological nitrogen fixation. *Annu. Rev. Microbiol.* **2**, 485.

WAKSMAN, S. A. (1927). *Principles of Soil Microbiology*. London: Baillière, Tindall and Cox.

WAKSMAN, S. A. (1941). Aquatic bacteria in relation to the cycle of organic matter in lakes. *Symposium on Hydrobiology*, p. 86. Madison.

WATANABE, A., NISHIGAKI, S. & KONISHI, C. (1951). Effect of nitrogen-fixing blue-green algae on the growth of rice plants. *Nature, Lond.* 168, 748.

WILLIAMS, A. E. & BURRIS, R. H. (1952). Nitrogen fixation by blue-green algae and their nitrogenous composition. *Amer. J. Bot.* 39, 340.

WIMPENNY, R. S. (1943). The Nation's Food. VI. Fish as food. 1. The biology of the sea fisheries. *Chem. & Ind.* 21, 230.

WINOGRADSKY, S. (1888). Über Eisenbakterien. *Bot. Ztg,* 46, 262.

WINOGRADSKY, S. (1922). Eisenbakterien als Anorgoxydanten. *Zbl. Bakt.* (2. Abt.), 57, 1.

WITSCH, H. VON (1946). Wachstum und Vitamin B_1-Gehalt von Süsserwasseralgen unter verschiedenen Aussenbedingungen. *Naturwissenschaften,* 33, 221.

VON WOLZOGEN KÜHR, C. A. H. (1938). The unity of the anaerobic and aerobic iron corrosion process in the soil. *Water, Den Haag,* 22, 33, 45.

VON WOLZOGEN KÜHR, C. A. H. & VAN DER VLUGT, I. S. (1934). The graphitization of cast iron as an electrobiochemical process in anaerobic soils. *Water, Den Haag,* 18, 147.

ZOBELL, C. E. (1943). The effect of solid surfaces upon bacterial activity. *J. Bact.* 46, 39.

ZOBELL, C. E. (1946). *Marine Microbiology*. Waltham, Mass: Chronica Botanica Company.

ZOBELL, C. E. (1947). Bacterial release of oil from oil-bearing materials. *World Oil,* 126, 36; 127, 35.

Printed in the United States
By Bookmasters